D0629453

THE MICROBIOLOGY OF TERRESTRIAL ECOSYSTEMS

B. N. RICHARDS
DEPARTMENT OF ECOSYSTEM MANAGEMENT
UNIVERSITY OF NEW ENGLAND
ARMIDALE, NSW
AUSTRALIA

Copublished in the United States with
John Wiley & Sons, Inc., New York

Longman Scientific & Technical,
Longman Group UK Limited
Longman House, Burnt Mill, Harlow
Essex CM20 2JE, England
and Associated companies throughout the world

Copublished in the United States with
John Wiley & Sons, Inc., 605 Third Avenue, New York, NY 10158

First published 1987

British Library Cataloguing in Publication Data

Richards, B. N.
The microbiology of terrestrial ecosystems.
1. Soil micro-organisms
I. Title
576'.190948 QR111

ISBN 0-582-45022-5

Library of Congress Cataloging-in-Publication Data

Richards, B. N. 1928 –
The microbiology of terrestrial ecosystems.

Includes index.
1. Soil microbiology. 2. Soil ecology. I. Title.
QR111.R49 1987 576'.15'09148 86–7138

ISBN 0-470-20706-X (USA only)

Set in 10/11pt. Linotron Times Roman.
Produced by Longman Singapore Publishers (Pte) Ltd.
Printed in Singapore.

For Elizabeth

Contents

Preface

This book enlarges on the concepts and principles outlined briefly in an earlier publication, *Introduction to the Soil Ecosystem*. Like the former work, it emphasizes the role of soil microorganisms and their interactions with plants in the functioning of terrestrial ecosystems. It is designed to give students of ecology an understanding of plant – microbe relationships in an ecosystem context, and to bring to students of microbiology a wider appreciation of the microbial contribution to ecological processes. It should prove useful for both undergraduate and graduate students in the biological sciences and in such applied fields as agriculture, forestry and natural resources management.

The book differs in several respects from standard works on soil microbiology. I have adopted what is essentially a systems approach throughout, examining the role of soil microbes both in the soil – litter subsystem and in the wider context of the ecosystem and, where applicable, the biosphere. It is for this reason that plant – microbe interactions have received proportionally more emphasis than is usual in a book of this kind. In the same vein, the idea of a 'functional, factorial' approach to the study of ecosystems is introduced and discussed, albeit briefly. This concept had its origins in the attempt by V. V. Dokuchaev to establish soil properties as functions of environmental factors, and was refined and extended to encompass the ecosystem as a whole by Hans Jenny. It seems to me the most appropriate means of dealing with the principle of interacting limiting factors at the ecosystem level. Its relevance as a conceptual framework for the study of ecosystems is likely to become more widely appreciated, following renewed interest in the ideas advanced by A. S. Watt on pattern and process in plant communities, and the reappraisal of successional theory during the past decade.

Preparation of this book began during a period of study leave spent in the Department of Microbiology at La Trobe University, Melbourne. I take this opportunity to thank the head of that department, Professor J. S. Waid, for making this possible.

Department of Ecosystem Management
University of New England
Armidale, NSW

14 August 1985.

Acknowledgements

We are grateful to the following for permission to reproduce copyright material:

Academic Press Inc. (London) Ltd for our fig. 8.1b from a plate in *Annals of Botany* N.S. 18, 385 & Table 7.2 from Table 1 p 8 (Harley & Smith 1983); Professor G. Bond for our figs 8.1a,b from Plates Ia,b (Stewart 1966); Blackwell Scientific Publications Ltd for our figs 1.4 from fig 1 p 79 (Fenton 1947), 3.4 from fig 3 p 190 (Swift 1977a), 6.4 from fig 5.1d p 95 (Nye & Tinker 1977), & 7.5 from fig 1 p 263 (Nicolson 1967); Cambridge University Press for our Table 2.1 from Table 2 p 266 (Lewis 1973); The Ecological Society of America for our fig 4.4 from fig 1 p 323 (Olson 1963); The Institute of Foresters of Australia for our Table 7.4 from Table 2 p 4 (Lamb & Richards 1971); Iowa State University Press for our Table 1.2 from Table 6.1 (Buol et al 1980); IUFRO Congress Secretariat for our fig 7.11 from (Bowen & Theodorou 1967); Macmillan Publishing Co Inc. for our figs 1.5 from fig 12.4 p 313 (Brady 1974) & 3.6 from fig 2.9 p 26 (Whittaker 1975); Martinus Nijhoff for our figs 3.1 from fig 2 p 332 (Richards & Bevege 1969) & 4.1 from fig 1 p 237 (Jones & Griffiths 1964); The National Research Council of Canada for our fig 7.8a,b & c from figs 1,2 & 3 p 1832 (Kinden & Brown 1975); The Editor, *New Phytologist* for our figs 6.2 from fig 6 p 738 (Foster 1982), 7.3 from fig 10 p 692 (Chilvers & Gust 1982) & 7.4b from Plate 1, no. 4 (Cox & Tinker 1976); Pergamon Press Ltd for our figs 3.2 based on data from (Jones & Richards 1977) & 6.1 from fig 4 pp 446–7 (Greaves & Darbyshire 1972) & Tables 5.1 from portions of Tables 4 & 5 pp 161–8 (Jones & Richards 1977) & 6.3 from Table 3 p 122 (Bowen & Theodorou 1979); North Holland Publishing for our fig 4.2 derived from pp 76–84 (Edwards & Heath 1963) & Table 4.3 from Table 1 p 8 (Macfayden 1963); The Society of American

Foresters for our Table 7.1 modified from Table 2 p 409 (Richards & Wilson 1963); Springer Verlag for our figs 2.2 & 5.1 from figs 1.2 p 16 & 1.3 p 19 (Charley & Richards 1983) & our Tables 6.1 from Table 1 p 283 (Foster & Rovira 1978) & 8.2 from Table 8.3 p 334 (Gibson & Jordan 1983); Professor W.D.P. Stewart for our fig 8.5a from Plate 3 (Stewart 1966); The Swedish Natural Science Research Council for our fig 5.8 from fig 8 p 520 (Reiners 1981); UNESCO for our figs 4.5a and 4.6 from diagrams in (Weigert 1970); University of California Press for our figs 1.12 from fig 1 p 39 (Kevan 1965) & 7.12 from fig 4 p 222 (Harley 1965) & for our Table 7.6 from table 1 p 222 (Harley 1965); John Wiley & Sons Ltd for our figs 4.7 from fig 1.1 (Bolin 1981), 5.10 from fig 9.18 p 546 (Jenkinson 1981) & 5.12a, b from figs 7.2 p 406 & 7.3 p 407 (Greenland & Hayes 1981); The Williams & Wilkins Company Baltimore for our Table 6.2 from Table 1 p 65 (Wallace & Lochhead 1949).

We have attempted unsuccessfully to trace the copyright holders for fig 7.10 and would appreciate any information that would enable us to do so.

1

Introduction to the soil ecosystem

In most natural ecosystems a large part of the sun's energy fixed each year by photosynthesis is returned as shed plant tissues to the soil, where decomposition processes release organically bound nutrients for reuse by green plants. In a managed ecosystem, much of the primary production is diverted into materials harvested by man. This is usually achieved by maintaining the ecosystem in a successional stage, which alters the fluxes of energy and nutrients from those that pertain in the undisturbed system. The results of such deflections from the 'metabolic norm' should become apparent during the mineralization of organic residues in the soil, since it is here that the major functional processes of ecosystems, namely energy flow and nutrient cycling, interact most strongly. A knowledge of energy and nutrient dynamics in the soil ought therefore to be of assistance in predicting the ecological consequences of disturbing natural ecosystems, and the environmental impact of utilization and management practices. It is one purpose of this book to provide a background to such enquiries, by outlining the role of soil organisms in ecosystem function.

Except in special circumstances, the solar energy captured by photosynthesis (gross primary production) is the total amount of energy available for all the energy-consuming processes of ecosystems. These include biosynthesis, growth, nutrient uptake and transport, the energy needed for these purposes being provided by the oxidation of photosynthate during respiration. The soil is deficient in photosynthetic organisms and so does not have the capacity to capture significant quantities of solar energy itself. Instead, it depends on energy rich substances produced elsewhere, that is, on the energy contained in plant and animal residues. The decomposition of these residues, or litter, is brought about by a sequence of biological processes involving both microbes and the soil fauna. The soil – or to be more precise, the soil–litter subsystem – may thus be regarded as a control valve, or gate, through which must pass most of the nutrients taken up by the primary producers and much of the energy fixed by them as well. Not only this, but the rate at which energy and nutrients flow through this gate frequently governs the productivity of the whole ecosystem. This regulatory role is particularly important in systems which operate on a very low nutrient capital, and in those where decomposition processes are limited by environmental extremes. In such situations the level of community productivity is likely to be determined by organic matter turnover in the soil–litter subsystem.

In this context, the operation of the soil–litter subsystem is the key to understanding total ecosystem function. As indicated above, it is the site of mineralization processes that release nutrients previously bound in the plant biomass, so that they are once again available for recycling through above-ground portions of the ecosystem. Mineralization is brought about by soil animals and microorganisms which depend upon a regular accession of plant and animal residues to provide them with food and energy. A detailed examination of soil biology and biochemistry is therefore an essential feature of any comprehensive program of ecosystem analysis.

1.1 The functional significance of soil organisms

Most of the microbes in terrestrial ecosystems are found in the soil, and indeed the microbiology of such ecosystems can be largely equated with soil microbiology. This is not to say that microorganisms do not occur elsewhere, such as on the surfaces of leaves of green plants, but their numbers and biomass are small compared to the corresponding figures for soil. Except in extreme habitats restricted to lichens, algae and mosses, the primary producers in terrestrial ecosystems are vascular plants (seed plants, ferns and allied forms). In most mature ecosystems of this kind, much of the organic matter fixed each year by these plants is returned directly to the soil. This is true irrespective of whether the system is a natural grassland, a grazed pasture, a virgin forest, or a forest managed for timber production. Only under intensive grazing is a significant amount of the primary production diverted from the producer–decomposer pathway.

Energy flow and organic matter decomposition

The flow of energy through the system is closely related to the processes of accumulation and decomposition of organic matter. The amount of organic matter fixed by an ecosystem per unit of time is a measure of its productivity. Primary productivity is that fixed by photosynthesis, and it has two components: net productivity which is the rate at which organic matter accumulates within the system,[1] and gross productivity which is the rate at which it is fixed by the system. Gross productivity is therefore equivalent to net productivity plus the amount of phytosynthate used per unit of time through respiration by the primary producers. These definitions ignore any part of the gross production which might be excreted into the environment. In terms of mass this may be relatively insignificant, yet it can be of great importance functionally in terrestrial ecosystems, because microorganisms in the vicinity of plant roots (i.e. in the rhizosphere) use plant exudates as a source of energy and nutrients. This topic will be discussed more fully in Chapter 6, while the contribution of soil organisms to the energy budget of ecosystems is dealt with in Chapter 4.

The process of organic matter decomposition in soils is greatly influenced by environmental factors. An increase in soil temperature can stimulate the metabolic activity of the microflora and hasten mineralization (i.e. conversion to carbon dioxide) of the organic matter; concomitantly, there is an increase in the rate of energy flow through the system. Subjecting a soil to alternate cycles of wetting and drying can also promote the breakdown of organic matter, and such a pattern of decomposition may be of particular significance in regions which experience a monsoon climate. The question of organic matter breakdown in soil is treated in some detail in Chapter 4; for the present it will suffice to say that the decay of plant and animal residues is a complex process, involving both microbes and the soil fauna (Fig. 1.1). Animals occupy the dual roles of consumers and decomposers, some preying on other animals or grazing on plants and microorganisms, others being detritus feeders.

Two major groups of microbes are involved in organic matter decomposition, the fungi and the bacteria. Both use the same basic mechanisms to decompose insoluble substrates, namely the hydrolysis of complex compounds by exoenzymes (see Ch. 2). As S. D. Garrett pointed out in 1963 however, the physical organization of fungi gives them an advantage over bacteria in the breakdown of cellulosic plant remains, which constitute a major fraction of the organic matter component of terrestrial ecosystems. Bacteria have no intrinsic mechanism for penetrating plant tissues, and their progress as cellulose decomposers is limited to surface erosion, because the rate at which they break down their substrate is proportional to the rate at which exoenzymes are produced and diffuse out from the bacterial colonies. Fungi, on the other hand, supple-

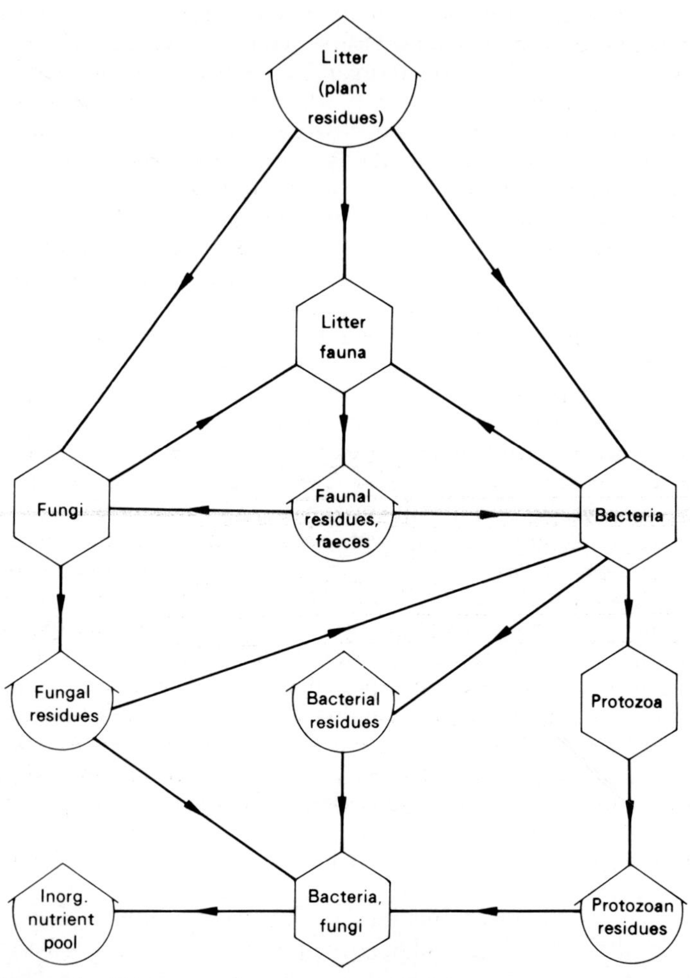

Fig. 1.1 Diagrammatic representation of the detritus food web in soils, greatly simplified.

ment the action of exoenzymes with mechanical pressure from their elongating hyphae, and furthermore their filamentous habit permits them to ramify throughout dead plant tissues with relative ease. Only in anaerobic habitats, such as waterlogged peats and in the rumen of cattle, do cellulolytic bacteria predominate over fungi.

It would be a gross oversimplification to assert that organic matter decomposition is the only contribution of fungi and bacteria to ecosystem function. As will be seen in Chapter 7, there are some fungi, the so-called mycorrhizal fungi, which are generally incapable of decomposing organic matter, yet which greatly influence the process of nutrient cycling in many ecosystems. These fungi satisfy their energy requirements by entering into a close anatomical and physiological association with the roots of many plants, so that a small fraction of the photosynthate produced by the plants is passed directly to the fungi. This 'sacrifice' on the part of plants is recompensed by the fungi providing them with mineral nutrients at rates greatly in excess of those pertaining in the absence of the mycorrhizal fungi. In turn, enhanced plant growth results in greater deposition of leaf litter on the soil, and greater amounts of root detritus. The end result is that the soil system receives an increased input of energy due to the presence of mycorrhizal fungi, even though these fungi may not themselves be able to release the energy contained in the plant residues.

There are many other processes in soils which involve microorganisms. Bacteria in particular contribute in numerous ways. The oxidation of organic compounds is only one of the many means at the disposal of bacteria for satisfying their energy needs. A study of macroscopic organisms, that is, plants and animals, could leave one with the impression that only two kinds of energy-yielding metabolic processes existed in the living world – photosynthesis and respiration – but various groups of bacteria have evolved other mechanisms for capturing and utilizing energy for biosynthesis and growth. These will be described in Chapter 2, and their functional significance in ecosystems will be considered in subsequent chapters, but reference to Table 1.1 will indicated something of the range of energy-yielding processes which exists in the microbial world.

Nutrient cycles

The exchanges of chemical elements between the living and non-living parts of the ecosystem constitute what are known as nutrient or mineral cycles. On a global scale they are referred to as **biogeochemical cycles**. Although microbial biomass may be a relatively small fraction of the total system biomass, microbial activity is of paramount importance in the circulation of the elements. This will be discussed again in Chapter 5, but for the present it can be inferred, from the great variety of microbial oxidations shown in Table 1.1, that bacteria, in company with fungi, not only make a major contribution to nutrient cycling (and energy flow) in particular ecosystems, but also have great significance for the geochemical cycles of the elements in the biosphere as a whole. There are several reasons, apart from their metabolic diversity, why bacteria and fungi are such potent agents of geochemical change. Because of

Table 1.1 Energy-yielding oxidations in microbial metabolism[†]

Reductant	*Oxidant*
Sugars	O_2
Ethyl alcohol	O_2
H_2	O_2
NH_4^+	O_2
NO_2^-	O_2
H_2S	O_2
S, $S_2O_3^{2-}$	O_2
Fe^{2+}	O_2
Sugars and other organic substrates	NO_3^-
Sugars and other organic substrates	NO_2^-
Sugars and other organic substrates	SO_4^{2-}, SO_3^{2-}
Sugars and other organic substrates	$S_2O_3^{2-}$
H_2, CO, organic acids, alcohols	CO_2
Sugars and related compounds[‡]	
Sugars	
Sugars	
Sugars	
Sugars, organic acids	
Sugars, starch, pectin	
Amino acids	

[†] Only those organisms using organic compounds as reductants (H donors) are decomposers.

[‡] In this and subsequent oxidations, the oxidant is an organic compound, but for the sake of simplicity the individual compounds are not listed. In any event, because of the complexity of the reactions and the variety of end products, it is not always possible to quote specific H acceptors (oxidants).

Products	*Organism*
CO_2, H_2O	Protozoa, fungi, many bacteria
Acetic acid, H_2O	Acetic acid bacteria
H_2O	Hydrogen bacteria
NO_2^-, H_2O	Nitrifying bacteria
NO_3^-	Nitrifying bacteria
S, H_2O	Thiobacilli
SO_4^{2-}	Thiobacilli
Fe^{3+}, H_2O	Iron bacteria
NO_2^-, H_2O	Dentrifying bacteria
N_2, N_2O, H_2O	Dentrifying bacteria
S^{2-}, H_2O	*Desulphovibrio*
SH^-, H_2O	*Desulphovibrio*
CH_4, H_2O	Methane bacteria
Lactic acid, ethyl alcohol, CO_2	Lactic acid bacteria
Ethyl alcohol, CO_2	Yeasts
Acetic, succinic and lactic acids, formic acid or H_2 and CO_2, ethyl alcohol	*Escherichia*
Butanediol, lactic acid, formic acid or H_2 and CO_2, ethyl alcohol	*Enterobacter*
Propionic, succinic and acetic acids, CO_2	*Propionibacterium, Veillonella*
Butyric and acetic acids, CO_2, H_2	*Clostridium*
Acetic acid, NH_3, CO_2	*Clostridium*

their small size, they have a very large surface-to-volume ratio that permits rapid interchange of materials between their cells and the environment. Equally important is their extremely rapid rate of reproduction, generation times in bacteria being measured in minutes and hours.[2] In addition, they are ubiquitous in distribution, being found in every conceivable habitat on the surface of the Earth.

1.2 Regulation in ecosystems

A mature or climax ecosystem may be regarded as an open system in steady state, that is a condition independent of time in which production and consumption of each component are equally balanced, the concentration of all components within the system remaining constant even though there is continual change. It is true that over short periods of time ecosystems appear to show considerable change, but if cognizance is taken of the time factor the steady-state condition is seen to apply. Thus in a deciduous forest where leaf shed occurs in autumn, there are large fluctuations in the amount of leaf litter on the forest floor during the course of a single season. However, since there is no net accumulation over a period of years, production (leaf fall) being equal to consumption (leaf decomposition), the litter layer may be regarded as a steady state system.

Ecosystems, being open systems, do not strictly speaking obey the second law of thermodynamics, which states that systems in isolation spontaneously tend towards states of greater disorder, that is, their entropy increases. In contrast to closed (isolated) systems, ecosystems increase in order as they are mature towards a climax or steady state, that is, their entropy decreases. A system in steady state is a stable system and its stability is maintained by regulatory processes which act in such a way as to minimize entropy. This tendency for a mature ecosystem to resist change, and to return to a steady state if disturbed, is termed **homeostasis**. Homeostatic controls sometimes involve outside factors (organism × environment interactions), but in many instances self-regulating processes (organism × organism interactions) are responsible. A major purpose of this book is to demonstrate the role of plant-microbe interactions in the functioning of terrestrial ecosystems.

Ecosystem stability

Implicit in the concept of homeostasis is the idea that stability is the norm and that instability is atypical. However, there is a growing body of ecological literature which points to the importance of periodic natural disturbance in determining structural patterns and modifying functional processes in terrestrial ecosystems. Indeed, J. H. Connell and R. O. Slatyer (1977) contend that disturbance is so widespread that no community has yet been shown to have reached a steady-state equilibrium. In 1947, A. S. Watt proposed that environmental factors interact with intrinsic characteristics of vegetation to produce alternating developmental (upgrade) and destructive (downgrade) phases of plant communities, and this process creates distinctive patterns in the landscape. This perception of community dynamics implies that the climax or steady state encompasses proportional representation of all phases characteristic of a particular vegetation. Watt termed it the 'phasic equilibrium'. F. H. Bormann and G. E. Likens (1979) visualize it as a structural and floristic mosaic of varying complexity which they have designated the **shifting mosaic steady state**. The traditional view of ecological stability given above, namely that a stable system will tend to return to its steady state when disturbed, is clearly much too narrow.

Many different definitions of stability have been proposed. Most relate to structural stability and are concerned not only with the system's ability to restore its structure following disturbance but also with its capacity to resist environmental change by maintaining constancy in the kinds and numbers of organisms present. Our concern however, is primarily with functional stability, which may be defined as the system's ability to withstand changes in the rates of energy flow and nutrient cycling and to restore the *status quo* after these processes have been disrupted. This definition assumes that functional processes are maintained in steady state in undisturbed ecosystems but it must be pointed out there is little evidence to support such a premise. Indeed, if internal and external factors interact to determine pattern in vegetation, as Watt proposed, it is pointless to seek such evidence. Instead, we should attempt to ascertain the variability of functional processes in the system during all its constituent phases, that is, the frequency and magnitude of changes in energy and nutrient flux throughout the whole course of its development. This should lead to an understanding of two measures of functional stability that are critical to management, namely **persistence**, which is the capacity of the system to maintain the rates of key processes within defined limits despite fluctuating environmental factors, and **resilience**, which is the ability of the system to return to its original state after disturbance.

Ecosystem management

One purpose of this book is to provide an ecological perspective on the utilization, conservation and management of the Earth's renewable biological resources. The stability of managed ecosystems, discussed briefly above, is relevant in this context. In little more than 10 000 years, man's impact on his environment has increased enormously, and today about two-thirds of the Earth's land surface is subject to his activities, in other words is managed more or less intensively for agriculture, forestry, mining or outdoor recreation. Man has become a significant geochemical agent, modifying the biophysical environment on the scale of natural geochemical forces.

Despite efforts to halt the rate of human population growth, there will be a great many more people to feed, clothe and shelter in the twenty-first century than there are today. The long term capability of our crop, pasture and forest lands to meet these needs is critical to man's survival. On a global scale, supplying future food requirements will call for intensification of crop production systems at the expense of more extensive agronomic practices, and the sacrifice of other potentially arable lands presently carrying managed pastures or natural plant communities. However, as G. W. Cox and M. D. Atkins (1979) have emphasized, attempts to achieve a revolutionary increase in food production through expansion of mechanical agriculture, without taking cognizance of its social and ecological implications, are likely to have unforeseen repercussions. Even in highly industrialized societies, such an approach has caused some degree of environmental degradation and social unrest. The same may be said of the intensification of forest management practices to achieve higher outputs of wood and wood products.

Among the deleterious ecological consequences of intensified utilization of land-based biological resources may be listed the depletion of soil organic matter, the exhaustion of native nutrient pools, the deterioration of soil structure, the loss of soil fertility, the eutrophication of adjacent aquatic ecosystems, and the pollution of even remote ecosystems by pesticide residues. Furthermore, highly mechanized agriculture relies heavily on supplementary inputs of energy and nutrients (and often water), thereby creating or exacerbating problems of resource allocation.

This is not meant to imply that managed ecosystems are necessarily unstable. They are in fact subject to the same ecological principles and obey the same natural laws that govern systems undisturbed by man. In order to ensure their stability, therefore, we must seek an understanding of these basic laws and principles. An essential prerequisite of this is an appreciation of the regulatory role of soil microbes in the functional processes of ecological systems.

1.3 Ecosystem models

An ecosystem may be illustrated graphically as a series of compartments or storages linked by flows of energy and materials. Figure 1.2 is a generalized diagram of a hypothetical ecosystem in the

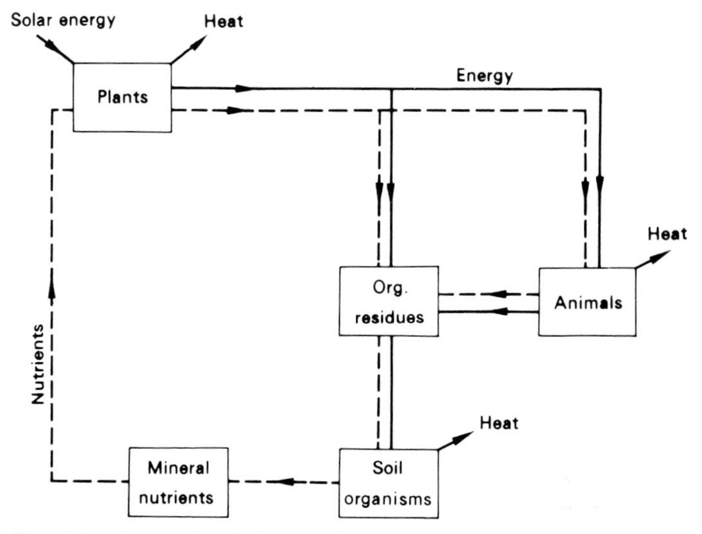

Fig. 1.2 Generalized picture of an ecosystem, showing the major pathways of energy and nutrients.

steady-state condition, with a closed nutrient cycle (i.e. no gains from weathering, rainfall, etc., nor losses by leaching, gaseous diffusion, etc.). Storages are represented by boxes, flows by arrows. The potential energy available for metabolic work is indicated by the storage of organic matter in plant and animal residues, while low-energy compounds resulting from respiration (mineral nutrients, CO_2) are shown as an inorganic pool.

A simplified version of the **energy circuit diagrams** introduced by H. T. Odum in 1967 is used throughout this book, to illustrate the flow of energy and materials in the ecosystem in a more definitive manner. Different shaped modules (Fig. 1.3) represent ecosystem components each having a specified structure and function, and linked by pathways along which regulated flows of energy are directed. The six modules shown are particularly useful in the context of microbial ecology, for they enable a clear distinction to be made between the major nutritional groups of organisms described in Chapter 2. Energy may flow either alone from a **primary source** indicated by a circle, for example as light, or from secondary sources in association with materials such as inorganic nutrients or organic matter; those **potential energy storages** are depicted by tank-like symbols. Major energy flows are shown as solid lines and pathways of low-energy compounds (i.e. inorganic nutrients) as broken lines.

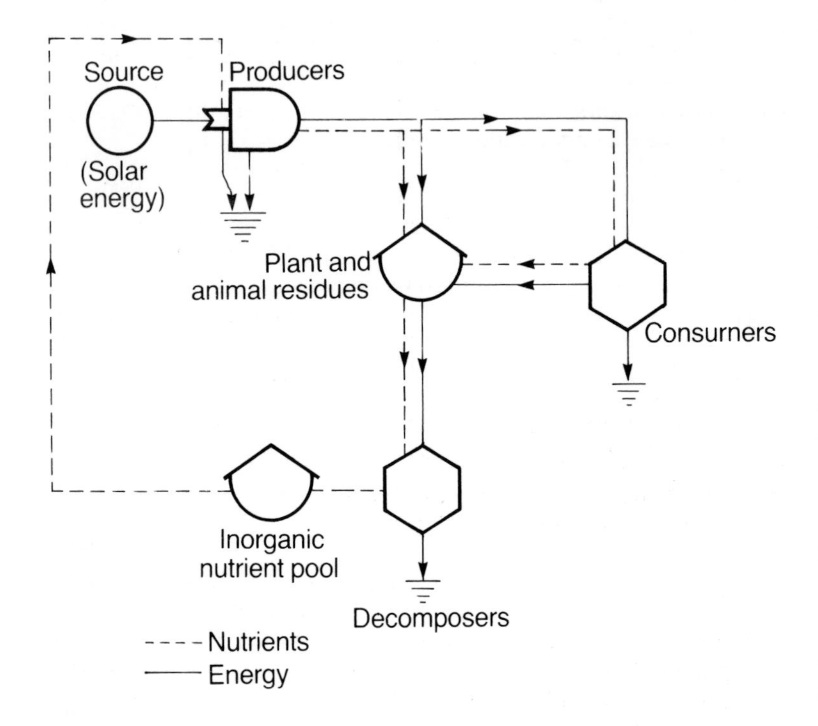

Fig. 1.3 Energy circuit diagram of a hypothetical ecosystem in the steady state condition with a closed nutrient cycle. Major energy flows are depicted as solid lines, while the pathways of low energy compounds (inorganic nutrients) are shown as broken lines. (Based on the notation of Odum (1967) – refer to text for explanation.)

The hexogonal symbol represents a **heterotrophic population** dependent on a flow of energy from an outside source such as plant or animal detritus. The bullet-shaped symbol depicts an **autotrophic population**, for example green plants which capture solar energy by the process of photosynthesis and store it in biomass. The regulatory mechanism in the system in the **work gate**, symbolized as a valve controlling the interaction of two energy flows. When the two input flows are maintained at the work gate in constant concentrations, the output is a function of their product. When the supply of one of the reactants is limited, increases in the concentration of the other will lead to progressively diminishing increments in output, producing a response curve which is characteristically a rectangular hyperbola. As illustrated in Fig. 1.3, limiting plant nutrients act in this fashion, controlling the flow of solar energy, and thus the productivity of the ecosystem, by means of a work gate. The remaining symbol, an arrow directed towards earth, is a **heat sink** and indicates the dispersion of potential energy into heat that occurs during every spontaneous process, according to the second law of thermodynamics; it is often used to represent maintenance energy.

The physiological processes basic to ecosystem function are photosynthesis, respiration and nutrient absorption. As indicated earlier, the solar energy captured by photosynthesis (gross primary production) is normally the total amount of energy available for all the energy-requiring reactions of the system, although some ecosystems have important auxiliary energy sources as well. Energy fixed by photosynthesis does biologically useful work through the process of respiration. This work includes that needed to maintain the structure and organization of the ecosystem as well as that needed for the more obvious processes of biosynthesis, growth and nutrient uptake.

Figure 1.3 is greatly simplified, and shows only the major energy pathways. To understand how the system functions, we need to identify and measure as many of the individual storages and flows as possible. Lack of knowledge of all the individual processes involved may prevent us from doing this, but the energy network diagram is still useful because it enables us to direct attention to the energy flows which are of particular interest, even if others remain unidentified as miscellaneous respiration. Throughout this book, much use will be made of such graphic models. Illustrating structure and function in this way has a unifying influence on ecological thought, since the most diverse ecosystems can be depicted in the same fashion.

1.4 Soil as an ecosystem

The mineral and organic matter fractions of the soil are part of the abiotic environment of most terrestrial ecosystems. In other words, soil constitutes a subsystem of a larger system. We can, however, regard the soil as an ecosystem in its own right, and study the relationships between its structure and function as we would the larger system. One point of difference immediately emerges, however: the producer component is relatively insignificant in the soil, algae being the only photosynthetic organisms present. The soil ecosystem does not therefore have the capacity to capture a substantial amount of solar energy, and so depends on energy-rich substances brought in from outside. As pointed out previously, these materials are in the form of plant and animal residues, especially the former. Such a situation fits the concept of ecosystem even though the soil system is not self-contained. All ecosystems derive energy from beyond their boundaries: some are directly dependent on solar energy, while others depend on energy derived from another, usually larger, system. Furthermore, as will be seen later, there are subsystems in the soil which obtain energy from neither the sun nor organic detritus, but from inorganic compounds. While such subsystems contribute only insignificant amounts to the energy budget of the soil ecosystem, they nevertheless play a vital role in the geochemical cycles of the elements; their activities in this regard form the subject matter of much of Chapter 5.

Environmental components of the soil ecosystem

The abiotic or environmental portion of the soil ecosystem has several recognizable components: the mineral fraction, organic matter, soil water and the soil atmosphere.

The mineral fraction

Examination of the physical nature of soil reveals that it consists of particles, which are named according to their size, as follows: > 2 mm diameter, gravel; 2–0.2 mm, coarse sand; 0.2–0.02 mm, fine sand; 0.02–0.002 mm, silt; < 0.002 mm, clay. The relative proportions of the various sized particles in a soil determine its **texture**. Thus we speak of a sand (> 90% sand), a loam (approx. 50% sand, 20% clay, 30% silt), a clay (> 40% clay), or various combinations of these terms, such as sandy clay, silty clay loam, and so on. Gravel consists of relatively unweathered fragments of the parent rock, whereas sand nearly always comprises grains of single minerals, mainly quartz, derived from rock weathering. Clay, however, is made up of minerals which are different from those found in unweathered rocks. Silicates of calcium, magnesium and potassium make up the majority of primary rock minerals, and when these break down, releasing the metal ions, the silicate fraction remains to be converted into secondary clay minerals.

It is the chemical composition of the mineral fraction of soil which largely determines its fertility. In a strongly leached soil, developed on highly weathered acidic rocks, some 90–95 per cent of the sand may be quartz, while in a soil arising from a moderately weathered basic rock, quartz may comprise only 60 per cent of the same fraction, and minerals such as felspars, augite and olivine make up the rest.

The **cation exchange capacity** of the soil is determined primarily by the density of the negative charges on the surfaces of clay minerals and amorphous organic matter, and the cations adsorbed by these colloidal particles constitute the exchange phase of the soil. Most soil reactions take place in this thin layer of ionized solution which is tightly held on the solid soil particles. The adsorbed ions can be exchanged for others brought near the colloid surface by moving water, or produced there by biological activity. The carbon dioxide evolved during respiration of soil organisms dissolves in water to form carbonic acid, which dissociates to produce hydrogen ions that may exchange with calcium, potassium or other ions of the exchange phase. The ions so released into the soil solution are available for uptake by plants and microorganisms.

Soil organic matter

Rock weathering will not produce a soil in the absence of organisms. The residues of dead plants and animals sooner or later find their way into the soil. Some of these residues decay on the surface, and only the end-products enter the soil, but some are incorporated by the action of earthworms and other animals before decomposition begins. The final stage of organic matter breakdown is the more or less amorphous material known as **humus**. Between 50 and 80 per cent of soil organic matter may be adsorbed on clay surfaces, giving rise to what is known as the clay–organic complex; the remainder is relatively undecomposed plant material.

Soils vary greatly in the amount of organic matter they contain, from less than 1 per cent in some sands to over 80 per cent in some peats. Just over 100 years ago, a Danish worker, P. E. Müller, recognized that there were two broad classes of organic matter, mull and mor. In a **mull** soil, such as a krasnozem associated with subtropical rainforest in eastern Australia, or a brown earth with deciduous broadleaved forest in Europe, plant remains are decomposed and well incorporated in the surface layers of soil. There is no sharply demarcated organic horizon, and the organic matter decreases gradually with depth. Mull soils are often rich in mesofauna, earthworms being especially abundant, and the upper part of the profile is loose textured and shows a good crumb structure (Fig. 1.4). **Mor** humus contrasts strongly with mull, and develops typically in podzol soils. The leaf litter forms a thick layer clearly disjunct from the mineral soil below, and although the uppermost leaves may be loosely scattered, the underlying ones are more or less decomposed and matted together with fungal hyphae. The

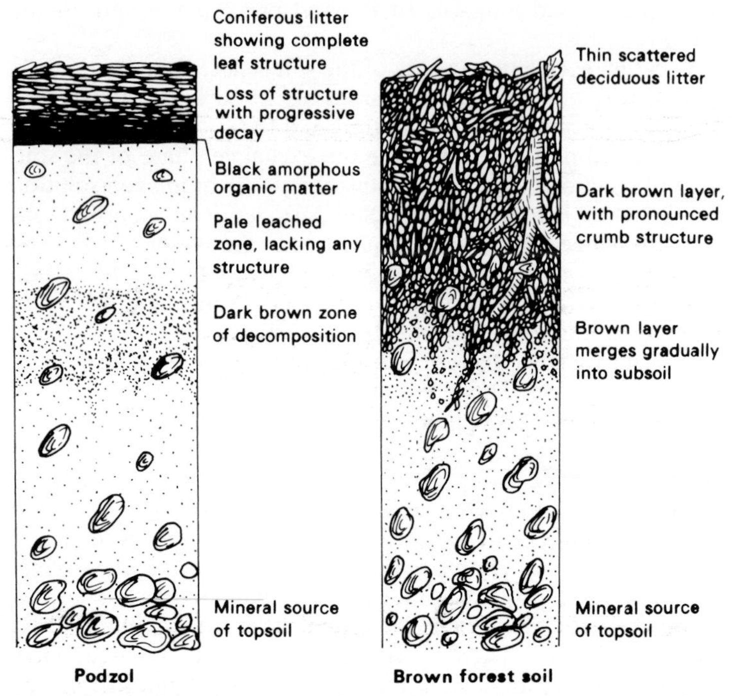

Fig. 1.4 Profiles of two forest soils, showing the contrast between mull and mor humus developing, respectively, in deciduous broadleaved and coniferous forest ecosystems respectively. (From Fenton 1947)

lower part of the organic horizon is amorphous and dark coloured, and the upper layer of mineral soil below is usually sandy and stained dark with organic matter. Animal life is not abundant (earthworms are characteristically absent) and mor soils show no signs of being aggregated into crumbs.

The organic matter on the surface of mor soils is sometimes separated into distinct layers. Three such layers may be recognized: the litter or **L-layer** which is uppermost and consists mainly of freshly fallen plant debris; the fermentation or **F-layer** where active decomposition is taking place; and the lowermost humification or **H-layer** which comprises more or less amorphous organic material. Mor humus is best seen on acid podzols in the cool temperate coniferous forests of the northern hemisphere. Although occurring on the surface of the mineral soil, it is just as much an integral part of the soil ecosystem as is the humus fraction of mull soils. Forms of forest humus intermediate between mull and mor, such as moder and duff mull (Hoover & Lunt 1952), have also been described, but for many purposes a simpler classification of surface detritus often suffices: the O_1 or A_{00} horizon (L-layer) comprising freshly fallen, relatively unaltered leaves, twigs etc.; and the O_2 or A_0 horizon (F- or F/H-layer) consisting of fragmented and partially or wholly decomposed materials (Fig. 1.5).

Because arable soils are all of the mull type, agricultural scientists do not find the classification of humus into mull and mor very useful, but ecologists and foresters have found the terms valuable and use them widely. Grassland soils, such as chernozems and black earths, are essentially mull soils in the sense that organic matter is well distributed throughout the profile. Compared to trees, however, grasses translocate a greater proportion of their photosynthate into roots, which are extensive and fine and decompose *in situ* to add to the soil organic fraction at considerable depth. This is not to say that root detritus is important only in ecosystems dominated by herbaceous plants. On the contrary, it is clear from J. S. Waid's (1974) review that root systems in general are a significant source of substrates for decomposer organisms, and even in forests the below-ground inputs to the soil sub-system may be similar in magnitude to those from above-ground (Cox *et al.* 1978). None the less, the contrast between major ecosystem types in relation to substrate sources for soil microbes should be stressed, since it is likely to influence the reactions of the systems to management practices, be they agronomic or silvicultural. According to L. E. Rodin and N. I. Basilevich (1967), the annual accession of above-ground detritus (litter) to the soil surface ranges from 25 t ha^{-1} in tropical rainforest to 1–2 t ha^{-1} in semi-shrub desert communities and shrub tundra.

Soil water

Water is held within the structural components (i.e. the soil matrix) by a variety of forces, the net effect of which is to lower the free energy of the water: in other words, the matrix forces lower the chemical potential of soil water, or the **water potential**. The presence of solutes also decreases the water potential. For an unsaturated soil, the water potential equals the sum of the matrix and solute potentials, and is always less than the energy of pure, free water at the same temperature and pressure in the same location. Relative to a free water surface, therefore, water in unsaturated soil is under negative pressure, that is, tension or suction. This pressure deficiency may be expressed in a number of different units, for example as centimetres of water (h), as atmospheres ($h/1035$), as bars (atmospheres $\times$ 0.987) or in SI units as mega-

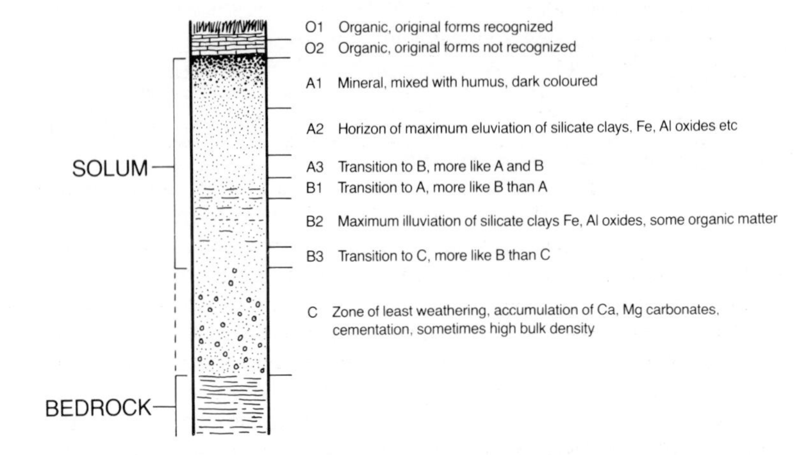

Fig. 1.5 A theoretical soil profile showing all of the horizons usually distinguished. Any particular individual profile, however, generally exhibits clearly only part of these (After Brady, 1974). Further details are as follows:

O_1 This horizon consists of loose, relatively undecomposed organic debris. It is usually absent on grasslands but present on forest soils, especially those of temperate regions, and very abundant at certain times of the year.

O_2 A partially or fully humified organic horizon. It may be matted, fibrous, or granular. On grassland it is usually absent. On forest soils two distinct layers often may be distinguished – F, the zone of fermentation and H, a markedly humified zone.

A_1 A dark coloured mineral horizon containing a relatively large amount of humified organic matter thoroughly mixed with the inorganic layer. In chernozems this layer is very thick; in podzols very thin or absent; and in other soils variable.

A_2 A light coloured mineral horizon resulting from a leaching and bleaching action (eluviation). It is well developed and very noticeable in true podzols – the gray layer or *bleicherde* of these soils. It is absent from chernozems and certain other soils especially those of arid and semi-arid regions.

A_3 A transition layer, often absent. When it is present, it is more like A than B.

B_1 Also a transitional horizon and absent from many soils, more like the B than A.

B_2 A zone of accumulation (illuviation) especially of silicate clays and iron and aluminium compounds. It is particularly prominent in true podzols as a reddish-brown horizon. The hardpan (*ortstein*) of the podzol occurs here when present as does the clay pan of certain other soils. In arid region soils the B_2 is often characterized by special structural forms, frequently columnar or prismatic.

B_3 A transitional layer. It may or may not be present.

C Unconsolidated parent material similar to that from which the solum has developed. This may come from the bedrock below or it may have been transported from elsewhere and deposited on the bedrock.

pascals (MPa = bars × 10^{-1}). A common practice in soil science is to express suction as $\log_{10}h$ and to call the derived quantity 'pF'.

As weathering progressively produces smaller and smaller particles, both the pore volume of the soil and the proportion of small pores increases. This leads not only to an increase in the total amount of water that can be held in the interstices between the soil particles, but also results in a greater fraction of the stored water being influenced by close proximity to adsorption surfaces, and therefore a reduction in the average energy status of the water molecules, that is, an increase in matrix potential. The amount of water in soil therefore depends on its texture, but in addition another property, known as **structure**, is important. Structure expresses the way in which the individual soil particles are aggregated into units of differing size, such as crumbs and clods. In the terminology of soil science, these structural units are known as peds, and their shape determines whether the basic structure of the soil is described as laminar, prismatic, blocky or spheroidal. Structure influences soil water-holding properties because it reflects the size class distribution of the pore spaces between the aggregates. Moulding and reorganization of particles of sand, silt and clay into aggregates is an essential process in soil formation. Both biological and physical agents are involved and the resulting structural properties differ markedly among major soil groups. However it is the pores between the aggregates rather than the aggregates themselves which are of greatest interest. Soil is a unique porous environment and pore size, continuity and tortuosity govern the movement of soil water and soil air, and so largely determine many of the processes that occur in the soil (Greenland & Hayes 1981). Pores range in size from above 50 μm equivalent diameter (too large to retain water against gravity), through storage pores of 50 μm–500 nm which hold water against gravity and allow it to move towards plant roots, to very fine pores less than 500 nm which retain water against both drainage and the suction exerted by roots.

During or immediately after heavy rain, all the pore spaces may be filled with water, but if the soil is allowed to drain freely under the influence of gravity, the larger pores empty of water and become filled with air. Following this initial drainage, the soil moisture content stabilizes at a level where gravitational forces are more or less balanced by matrix forces, and the soil is said to be at **field capacity**. As the soil dries out still further, organisms have increasing difficulty extracting water from it, since they must exert a suction large enough to overcome the retentive forces of the matrix before water will move into them. Finally a stage is reached when plants can no longer remove water and the soil is then said to be at the **permanent wilting point**. Soils vary a great deal in the amount of water available to organisms between field capacity and wilting point, and in the strength with which it is held. Furthermore, some microorganisms have the capacity to extract water from soils too dry to support the growth of plants.

Soil water always contains a variety of solutes including the nutrient ions required by both plants and microorganisms. 'Soil solution' is the term applied to all the water in the soil and its dissolved materials, other than that held on colloid surfaces by electrical forces. The main anions in the soil solution are nitrate and bicarbonate, and in arid soils chloride and sulphate also; phosphate and silicate are normally present at much lower concentrations.

The soil atmosphere

The composition of the soil atmosphere is the resultant of a number of processes which are proceeding concurrently. Respiration of plant roots and soil organisms utilizes oxygen and produces carbon dioxide, and the concentration gradients so formed cause oxygen to diffuse in from the soil surface and carbon dioxide to diffuse out into the air above. In dry soils, and in litter, the carbon dioxide concentration seldom exceeds 0.5 per cent (cf. the average concentration of the Earth's atmosphere, 0.03 per cent) but in wet soils where diffusion is hindered and microbial activity is stimulated, it increases several-fold and may even reach 10 per cent for short periods.

Soil aeration depends to a large extent on the arrangement of the soil particles with respect to one another, that is, on structure. Mull soils are characteristically well structured with many soil aggregates exceeding 0.25 mm in diameter, so that these soils contain many pores larger than 50 μm. It is the proportion of such large pores, which will not hold water by capillarity and which are therefore normally filled with air, that determines how well aerated is a soil. Without these relatively large pores, movement of gases through the soil would be restricted, and microbial and root respiration thereby inhibited.

1.5 The soil profile

As already indicated, the transformation of rock into soil involves not only the comminution of rocks into smaller particles as a result of physical and chemical weathering, but also requires the incorporation of organic matter derived from the activities of plants, animals and microorganisms. The continual interaction of all these processes usually leads to the differentiation in soil of more or less distinct horizontal layers, termed **horizons**, which collectively constitute the **soil profile** (Fig. 1.5). Soil-forming processes contributing to the development of the profile involve additions to and losses from the soil, and translocations and transformations within it (Buol *et al.* 1973). They are summarized in Table 1.2.

Among the more important processes involved in profile development are mineral dissolution and leaching. The more soluble silicate minerals such as felspars are hydrolysed and converted to secondary clay minerals, according to the following reaction (R. E. White 1979): (see page 23).

Table 1.2 Some processes of soil formation

Category	*Process*	*Description*
Addition	Enrichment	Acquisition of substances by the solum from an outside source
	Accumulation	Addition of mineral particles to the soil surface by wind or water
	Litter fall	Deposition of plant residues on the soil surface
Loss	Leaching	Removal of solutes from the soil body
	Erosion	Removal of material usually from the surface layer
Translocation	Eluviation	Removal of substances from part of the profile as in the A horizon of a podzol
	Illuviation	Movement of material into part of the profile as in the B horizon of a podzol
	Decalcification	Removal of $CaCO_3$ from one or more horizons

Table 1.2 (con't) Some processes of soil formation

Category	*Process*	*Description*
	Calcification	Concentration of $CaCO_3$ in a subsoil horizon
	Salination	Concentration of soluble salts such as sulphates and chlorides of Ca, Mg, Na and K in certain horizons
	Desalination	Removal of soluble salts from salty horizons
	Akalization (solodization)	Concentration of Na^+ ions on the exchange sites in soil
	Dealkalization	Leaching of Na^+ ions and salts from horizons of Na^+ accumulation
Transformation	Decomposition	Breakdown of organic materials
	Humification	Conversion of raw organic matter into humus
	Mineralization	Oxidation of organic matter with release of inorganic ions
	Immobilization	Incorporation of inorganic ions into organic substances by biosynthesis
Transformation and translocation	Podzolization (silicification)	Chemical migration of Al^{3+} and Fe^{3+} and/or organic matter resulting in the concentration of silica in the eluviated horizon
	Lateritization (desilicification)	Chemical migration of silica out of the solum resulting in the concentration of Fe- and Al-sesquioxides therein, with or without the formation of ironstone (laterite) and concretions
	Gleization	Reduction of iron under anaerobic (waterlogged) conditions to produce a bluish to greenish grey soil matrix (often mottled with yellow, brown and black) and Fe and Mn concretions

Source: After Buol *et al.* (1980)

$$M^+ \text{ (silicate)}^- + H_2O \leftrightharpoons H^+ \text{ (silicate secondary mineral)}^- + M^+OH^-$$

The reaction proceeds to the right more rapidly in an acid environment, where hydrogen ions are readily available to neutralize the hydroxyl ions of the strong base, MOH. The sources of these H-ions are carbonic acid formed by CO_2 dissolving in soil water, and the carboxyl and phenolic groups produced during the humification of organic matter.

Once dissolved, elements are subject to leaching by water percolating through the developing soil profile. The mobility of any particular element depends on its solubility in water which in turn is influenced by pH. Calcium is the most mobile of the nutrient cations while the sesquioxides of aluminium and iron are relatively immobile. The leaching of Al^{3+} and Fe^{3+} from Al_2O_3 is enhanced by the formation of soluble organo-metal complexes, that is, by chelation.

The several horizons which develop as a result of these soil-forming processes are designated by a particular notation. The simplest model recognizes a litter layer (L), an eluvial horizon (A), and illuvial horizon (B), and a parent material horizon (C). More complex notations (Fig. 1.5) are needed to encompass all aspects of profile development.

Some important soil groups

Podzols and podzolics. In cool, humid, forested regions with abundant rainfall, iron and aluminium compounds are leached from the surface soil (A horizon) and accumulate, along with clays, in the subsoil (B horizon), which grades gradually into comminuted but otherwise unaltered parent material (C horizon). Soils of this kind, which are usually acidic and grey or brown in colour, are called **podzols** and **podzolics**. In the northern hemisphere, podzols tend to form on siliceous and permeable parent materials, whereas podzolics develop on calcareous and clayey materials. In warmer climates, podzolics may occur on siliceous rocks. Podzols are found typically under coniferous forests or heath, where they are characterized by the presence of mor humus; podzolics are more likely to have humus of an intermediate (moder) type.

Brown earths and brown forest soils. While podzols develop best on acid quartz sands under conifers and heathland communities, these vegetation types do not lead to podzolization on other parent material, even though they may still produce a mor humus. The reason for this is that the mobilization of iron and aluminium is brought about not so much by humic acids released from the mor, but rather by the formation of water-soluble complexes between Fe^{2+} and Al^{3+} ions and the polyphenols washed off living plant leaves. The ferrous polyphenols are adsorbed on clay particles and thus can leach readily only through profiles low in clay. Although this retention mechanism is counteracted to some degree by the tendency for free polyphenols to disperse clays and so facilitate their leaching, this is not the chief means of clay eluviation in soils of higher clay content. In such soils, clay particles are moved down the profile mainly by the process of 'lessivage', which involves the mechanical transport of dispersed clay down cracks or channels by percolating water, and its deposition as oriented films on soil peds in the B horizon. In temperate continental climates the soils so formed, which characteristically have a well developed crumb structure throughout, are called **brown earths** or **brown forest soils**. They grade into podzolics in situations where the translocation of adsorbed iron oxides gives the subsoil a darker colour than the A horizon.

Latosols and krasnozems. In the wet tropics and subtropics, intense leaching removes all the iron and aluminium oxides from the surface horizon and results in their accumulation either as gravel or cemented masses on the underlying clay horizons. Such lateritic soils, as they are called, are red or yellow in colour. In the advanced stages of this process, on flat terrain where erosion is minimal, clay minerals decompose to sesquioxides and silica, and the structural and textural differentiation of B horizons evident earlier in the sequence is lost. The more highly weathered of the lateritic soils, which are often found forming on acidic rocks (e.g. granite and quartzite) are known as red earths or **latosols**. Less highly weathered members of this group, which retain a recognizable illuvial clay horizon, are called red loams or **kraznozems**. In the high rainfall zone of eastern Australia, kraznozems form on basic parent materials such as basalts and andesites, and typically support tropical or subtropical rainforest. Morphologically, these soils resemble the typical mull soils of the broadleaved forests of cool temperate regions of the northern hemisphere, the brown earths or brown forest soils.

Saline and alkali soils. In areas of low rainfall, associated with shrubs and sparse grasses, leaching is not a dominant process, and soluble salts accumulate in the upper part of the profile as a result of water rising to within a few feet of the surface by capillarity. Sulphates and chlorides of calcium and sodium are the most common salts, and these may on occasion accumulate in amounts sufficient to form a crust on the soil surface. These saline soils are called **solonchaks** or white alkali soils. With increasing rainfall, or a lowering of the water table, leaching causes a dilution of the soil solution. If the calcium reserves are low, sodium ions tend to replace calcium on the exchange complex and, eventually, the soil solution will contain sufficient sodium carbonate (the CO_3^{2-} ion deriving from respiratory CO_2) to cause the pH to rise above 9.0, leading to the dispersal of clay and humus. In consequence, the surface horizon becomes dark brown or black, and such soils are known as **solonetz** or black alkali soils. Further leaching of clay and organic matter leads to the formation of an illuvial clay horizon which has a characteristic prismatic structure of vertical columns with rounded tops. The overlying eluvial horizon is loose and porous, and slightly acid in reaction. At this stage the soil is called a **solod**. These saline and alkali soils frequently occur in the same area, their distribution being determined mainly by the depth of the water table which in turn depends on relief: solonchaks occur in low-lying parts, solods on higher ground, and solonetz soils in between.

Prairie soils and chernozems. In intermediate, sub-humid climates, permanent grasslands are found on a variety of soils, but in the northern hemisphere they are particularly associated with prairie soils, chernozems (black earths) and chestnut soils. **Prairie soils** develop at the wet end of the spectrum and show evidence of considerable leaching. **Black earths** are less strongly leached so that calcium carbonate tends to accumulate in the subsoil. With increasing aridity, the black earths grade into **chestnut soils** which are characterized by the accumulation of gypsum (calcium sulphate) below the carbonate horizon. These grassland soils all have a pronounced, friable crumb structure which is very water stable.

1.6 Soil classification

There is no universally accepted scheme of classifying soils, and attempts to develop hierarchical systems of classification are frustrated by the fact that soil is a continuum and boundaries consequently artificial. Certainly most of the systems proposed so far make inadequate provision for intergrades between clearly recognizable soil categories. Most countries have devised their own system, to suit their individual needs. All have something in common however, in that classification is usually based on:

(a) the recognition of distinctive horizons;

(b) whether or not the horizon boundaries are sharp or diffuse;
(c) inferred homologies between horizons in different soils.

One scheme which has been widely used for many years groups soils into three orders: zonal soils which are determined largely by climate; intrazonal soils which reflect locally dominant environmental conditions such as poor drainage, and which cross zonal soil boundaries; and azonal soils which lack differentiation into horizons. Sub-orders based usually on a combination of climate and vegetation type are also recognized (except in azonal soils), but it is the next level of subdivision which is most useful for describing the habitat of soil organisms. This is the level of the 'great soil group', such as podzol, grey brown podzolic, latosol, chernozem, krasnozem, brown earth, humic gley etc., some of which were described in the previous section. In 1975, a long series of attempts to develop a more objective system of classification culminated in the publication by American soil scientists of the USDA Soil Taxonomy, which recognizes ten soil orders based on the number and nature of soil horizons present and their chemical composition (determined in some instances by laboratory analyses).

It is not always easy to equate the two systems, although there is some degree of equivalence between them. Thus azonal soils are called entisols in the USDA system. Great soil groups of semi-humid grasslands, that is, soils with dark brown to black surface horizons such as prairie soils and chernozems, fall into the order mollisol. Well developed podzols (Fig. 1.4) are termed spodosols. Young (postglacial) soils with moderate horizon development, found in cool humid regions of the northern hemisphere, such as the brown forest soil in Fig. 1.4, are included in the order inceptisol. Soils with better development of horizons, but still relatively high in metallic cations, such as the grey-brown podzolics, are known as alfisols. Strongly leached soils of low cation status and with marked texture contrast between horizons, and usually occurring on Pleistocene or older land surfaces in warm humid climates, for example red-yellow podzolics, are placed in the order ultisol. The highly weathered red and yellow laterites and latosols of very old landscapes in the tropics and sub-tropics are classified as oxisols.

Further details on soils, soil formation and soil processes may be found in the works of H. D. Foth (1978), E. W. Russell (1973) and D. J. Greenland and M. H. B. Hayes (1981); the development of the soil ecosystem is discussed in Chapter 3. For more information on ecology, texts such as E. P. Odum (1971) and R. H. Whittaker (1975) should be consulted.

1.7 The soil biota

Soil organisms may be conveniently classified on the basis of size into the microbiota (algae, protozoa, fungi and bacteria), the mesobiota (nematodes, springtails, small arthropods and enchytraeid worms), and the macrobiota (earthworms, molluscs, and the larger enchytraeids and arthropods). The macrobiota might also be taken to include the roots of plants, burrowing rodents, reptiles and amphibia. However, plants can hardly be regarded solely as soil organisms, while vertebrates, together with many insects, are only temporary soil inhabitants. Plant roots, none the less, are important components of the soil ecosystem by virtue of their interrelationships with microorganisms. Other members of the macrobiota are not however within its province, but a brief outline of the major groups which are included is given below.

1.7.1 Soil microorganisms: the microbiota

Detailed descriptions of the different kinds of microorganisms found in the soil are beyond the scope of this book. The generalized account which follows is meant to provide some background in microbiology for those readers with little or no exposure to this field. It may be supplemented by reference to Alexopoulos and Mims (1979, Bold *et al.* (1980), Brock (1979), Stanier *et al.* (1977), and Webster (1980). Most attention is given to the bacteria and fungi, since these two groups of microbes are those most intimately concerned with energy flow and nutrient transfer in terrestrial ecosystems.

The boundary line between plants and animals becomes blurred when one considers organisms of microscopic dimensions. Microbiologists use this as a rationale for proposing a separate kingdom to include all those creatures which can be distinguished from typical plants and animals by virtue of their simpler level of biological organization. The members of this group (the Protista) are either unicellular or coenocytic[3] or, if multicellular then they lack the extensive differentiation into distinctive tissues which is a characteristic of plants and animals in the adult stage. The Protista as thus defined comprises two quite distinct groups of organisms, the bacteria and cyanobacteria (blue-green algae) on the one hand, and the fungi, protozoa and the algae proper on the other. Since the advent of the electron microscope, it has been possible to relate this division of protists into two groups to differences in cellular organization. There are two quite different kinds of cells among living organisms, namely the eukaryotic cell, which is the structural unit of all plants and animals, and of the fungi, protozoa and true algae, and the prokaryotic cell, found in all bacteria and cyanobacteria. Apart from being generally smaller, the prokaryotic cell differs from the eukaryotic in several important respects, the most outstanding differences being its lack of proper nucleus (i.e. the DNA is not enclosed within a membrane) and the absence of membrane-bounded respiratory organelles or mitochondria. For the sake of convenience, protists possessing eukaryotic cells are often called higher protists, while those having prokaryotic cells are referred to as lower protists. Genetic recombination in prokaryotes is unidirectional or virus-mediated, whereas in eukaryotes it generally involves karyogamy (nuclear fusion) followed by meiosis. The difference in cellular organization between the two groups is so fundamental and far-reaching that it has been proposed as the basis for two 'superkingdoms', the Prokaryota and the Eukaryota (Whittaker & Margulis 1978).

Classification of protists

The higher protists are differentiated on the basis of energy source, feeding habit and structure (Table 1.3). The boundaries between the various groups are not well defined and there is much overlapping between categories. For example, many unicellular protozoa are similar in morphological detail to the unicellular algae but they lack chloroplasts or photosynthetic pigments. The lower protists are classified according to their energy source and structure, and whether or not they possess organs of locomotion.

1.7.2 Higher protists: algae, protozoa and fungi

The presence or absence of chloroplasts is the primary basis of classification in the higher protists, separating the photosynthetic algae from the protozoa and fungi in terms of both cell structure and physiology. The fungi and protozoa are chemotrophic organisms, that is they depend on chemical sources of energy for their life processes (see Ch. 2), and it is not easy to make a clear separation between them. Certainly, typical members of either group are quite distinctive, but there are transitional forms, classified as slime moulds, which have some of the features of both groups. Most of the protozoa are, however, unicellular (or acellular) and motile, whereas fungi are generally filamentous and non-motile. Furthermore, while all fungi absorb their nutrients from solution, many protozoans can ingest solid particles. Some authorities (e.g. Whittaker 1959; Alexopoulos & Mims 1979) consider that the fungi are a group of taxonomic status comparable with plants and animals, and should therefore be placed in a separate kingdom.

Table 1.3 Characteristics of the major groups of protists

Group	*Energy source*	*Feeding habit*	*Structure*	*Flagella*
Higher Protists				
Algae	Photosynthetic	Absorb dissolved nutrients (except carbon)	Unicellular, or multi-cellular filaments or colonies	Mainly present
Protozoa	Chemosynthetic	Ingest solid particles, or absorb dissolved nutrients (including carbon)	Unicellular without cell walls	Mainly present
Fungi	Chemosynthetic	Absorb dissolved nutrients (including carbon)	Mainly filamentous and coenocytic	Mainly absent
Lower Protists				
Cyanobacteria†	Photosynthetic	Absorb dissolved nutrients (except carbon)	Unicellular or filamentous	Absent
Bacteria	Chemosynthetic‡	Absorb dissolved nutrients (including carbon)	Mainly unicellular, also filamentous	Present or absent

† Also known as blue-green algae.
‡ Green and purple bacteria can photosynthesize but – unlike the algae and cyanobacteria – without evolution of oxygen.

Algae

The algae are predominantly aquatic organisms, occurring in lakes, rivers and swamps, and in the oceans. Some however occur in terrestrial habitats especially if these are moist. Free swimming and free floating algae, together with protozoa and small animals and some fungi and bacteria, make up the community known as plankton. As the primary source of energy and food for fish and other aquatic animals, the algae (phytoplankton) are of great significance in marine and lacustrine ecosystems. In recent years, there has been a better appreciation of the fact that many species regularly occur in the soil, not only at the surface but also at depth. Others are found on the faces of stones and rocks, or on the bark of trees, even though these habitats may be subject to intermittent desiccation.

Seven divisions of algae are recognized on the basis of their pigmentation, the chemistry of their cell walls, the nature of their stored food reserves, and the kind of flagella they possess, if any. Only one division, the **Chlorophycophyta**, or green algae, occurs in soil to any great extent. Many soils, however, also contain diatoms (**Chrysophycophyta**), a group that is better known as the main component of marine phytoplankton. Algae are frequently unicellular, the cells occurring singly or in association as colonies, though in some species the cells are arranged in filaments.

Protozoa

The majority of protozoa are parasitic on or in animals, some causing serious diseases such as malaria. Other protozoans live in mutalistic relationships with higher organisms. Certain flagellates, for example, live in the guts of termites and play a vital role in the digestion of wood by these insects. Free-living protozoa are abundant in the soil, in fresh water and in the sea.

Four main groups of protozoa are recognized. Amoeboid forms (amoebae and rhizopods), many of which move by extending finger-like projections of cytoplasm, or pseudopodia, are placed in the **Sarcodina**. Those which have as organs of locomotion numerous, short hairlike cytoplasmic extensions known as cilia, are classified as **Ciliophora** (Infusoria). Members of the **Mastigophora** are motile by means of one or more longer, whiplike appendages, called flagella.[4] Amoebae, rhizopods, flagellates and ciliates are all common in soil, where they feed on bacteria. The fourth group is the **Sporozoa**, representatives of which are all parasitic; their mechanism of locomotion is in many cases not yet clearly understood.

Fungi

Like the protozoa, the fungi are chemotrophic higher protists. Although their most common habitat is the soil, many of the less advanced species such as water moulds are entirely aquatic. While some of the more primitive forms closely resemble the flagellates, the typical fungus possesses a very distinctive and characteristic form. Its vegetative body usually consists of microscopic branched filaments, 3–8 µm in diameter, known as **hyphae** (Fig. 1.6). Collectively this system of branching hyphae is called a **mycelium**. At certain stages in the life-cycle, the mycelium bears minute propagules called **spores**, which germinate to produce new hyphae. Not all fungi have a mycelium, however. The yeasts, for example, are a group of fungi that have globose or ellipsoidal cells which reproduce by budding.

A fungus colonizing a fresh substrate grows vegetatively for a short time, and then may produce one or more kinds of asexual spore, that is a spore which does not result from nuclear fusion.

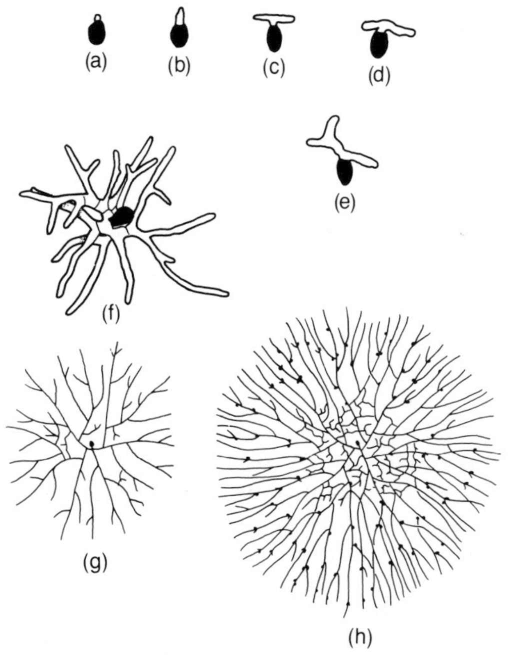

Fig. 1.6 Stages in the development of a fungal mycelium. Germination of a spore of *Coprinus sterquilinus* to form a radially expanding colony in a plate culture. (From Buller 1909–34)

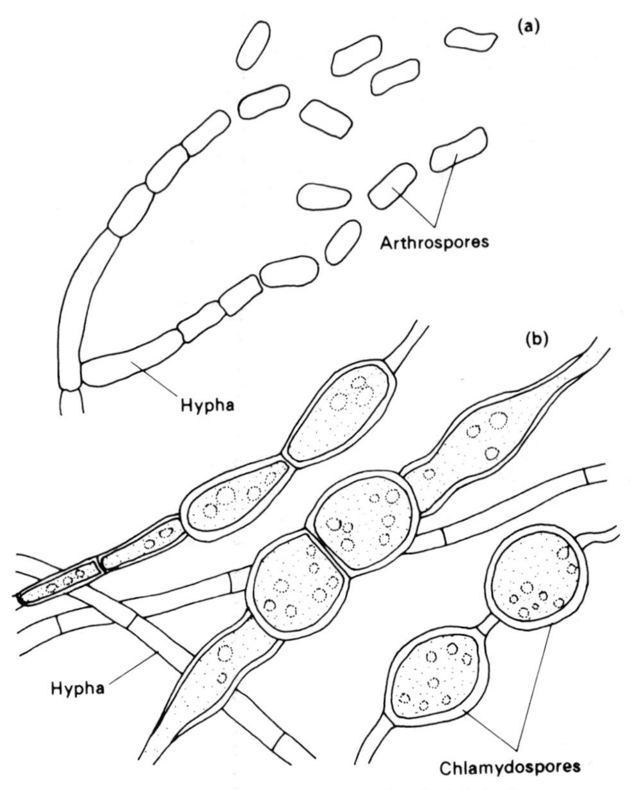

Fig. 1.7 Asexual reproduction in fungi. (a) Arthrospores or oidia, produced by hyphal fragmentation. (b) Chlamydospores, organs of survival each formed by developing a thick, protective wall around a hyphal cell.

Later, as the mycelium ages or when environmental conditions become unfavourable for further vegetative growth or when sufficient reserves have been amassed, many fungi reproduce sexually, forming distinctive spores following the fusion of two nuclei. In some species sexual reproduction is the only form of sporulation; others have no known sexual stages and produce only asexual spores.

A common method of asexual reproduction among filamentous fungi is by fragmentation of hyphae: any fragment of a vigorously growing mycelium is capable of developing into a new individual if the environment is favourable, and this is the method most commonly used for propagating fungi in the laboratory. One form of hyphal fragmentation, which occurs in some fungi in response to an unfavourable environment, is the rounding-off and separation of parts of the hyphae to form spore-like bodies known as **arthrospores** or **oidia** (Fig. 1.7).

The most important method of asexual reproduction in fungi is however, by spores. Basically, there are only two ways in which asexual spores can be formed – in sac-like structures called sporangia, or directly on the mycelium. Spores borne in sporangia are designated **sporangiospores** while those borne directly on hyphae are known as **conidia** (Fig. 1.8). Conidia are never motile. Sporangiospores of some species are however flagellated and resemble the zoospores of algae. Such motile sporangiospores are in fact called zoospores; they differ from other kinds of asexual spore in another important respect, in that they do not possess cell walls. Zoospores occur only in the most primitive aquatic families; conidia are characteristic of more advanced groups.

In addition to asexual spores, which are usually relatively short-

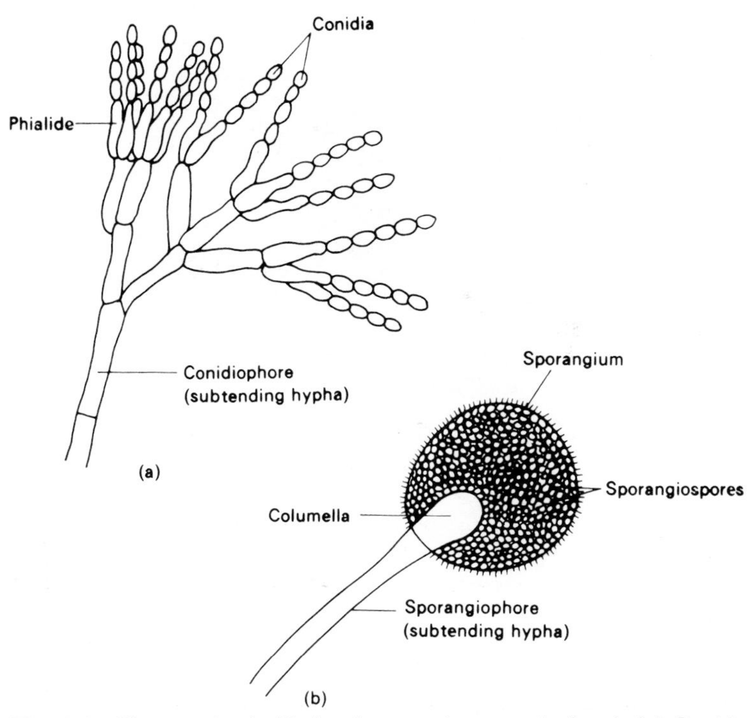

Fig. 1.8 The two basic kinds of asexual spores in fungi. (a) Conidia of *Penicillium*, spores borne directly on the hyphae. (b) Sporangiospores of *Rhizopus*, produced in sac-like structures known as sporangia.

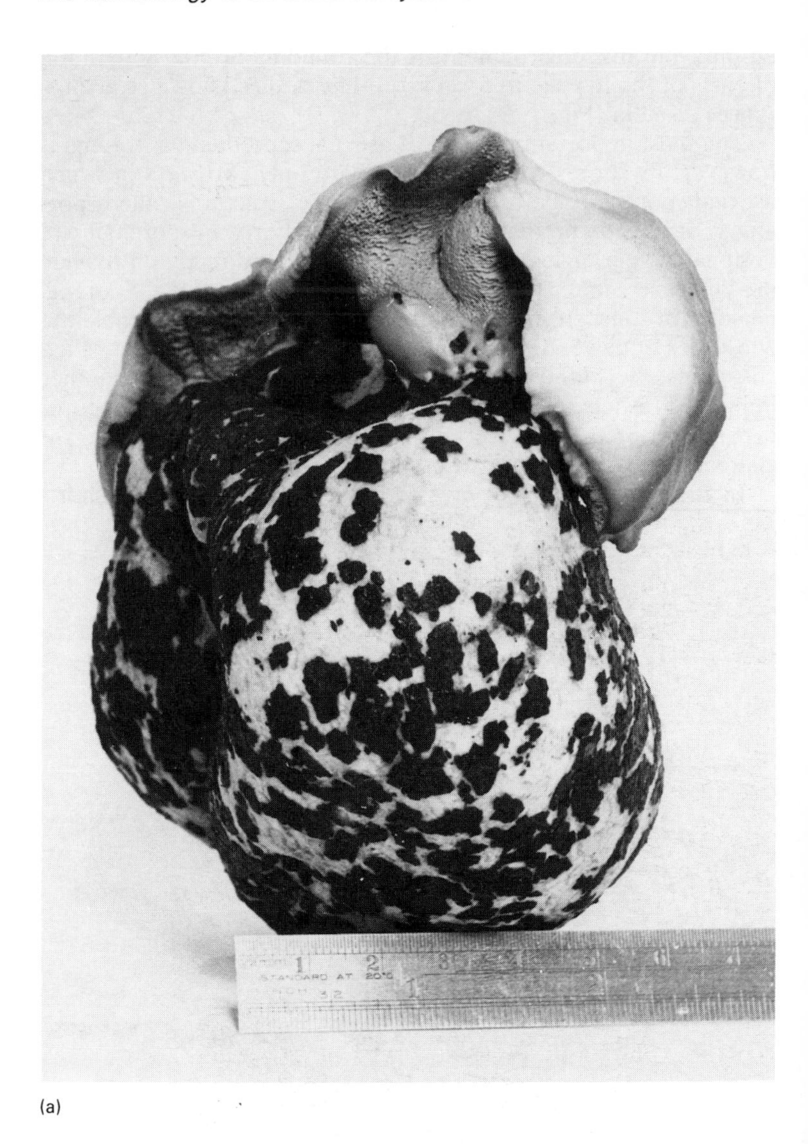

(a)

Fig. 1.9 Vegetative structures of fungi. (a) sclerotium of *Polyporus mylittae* showing fruiting bodies (basidiocarps) produced on germination. The upper edge of the scale is marked in centimetres, the lower in inches; this species produces some of the largest sclerotia known. (Photo by V. A. Raszewski (b) rhizomorphs. (Photos by kind permission of H. J. Hudson)

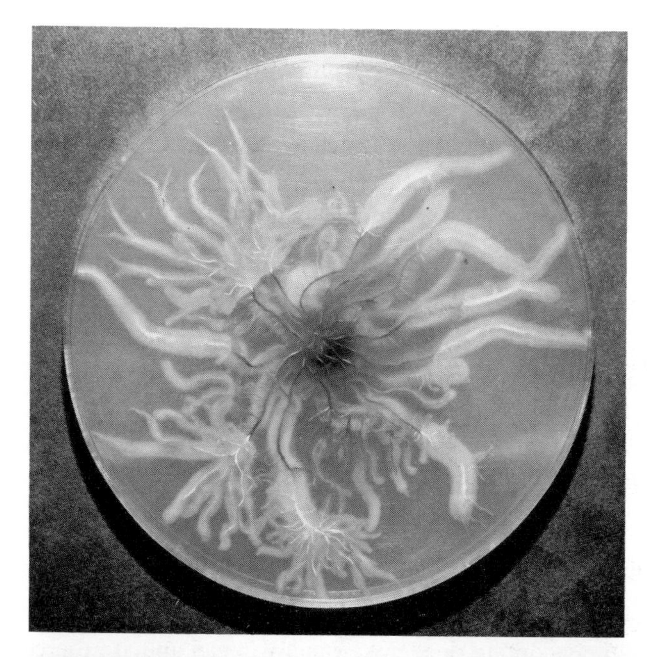

20mm

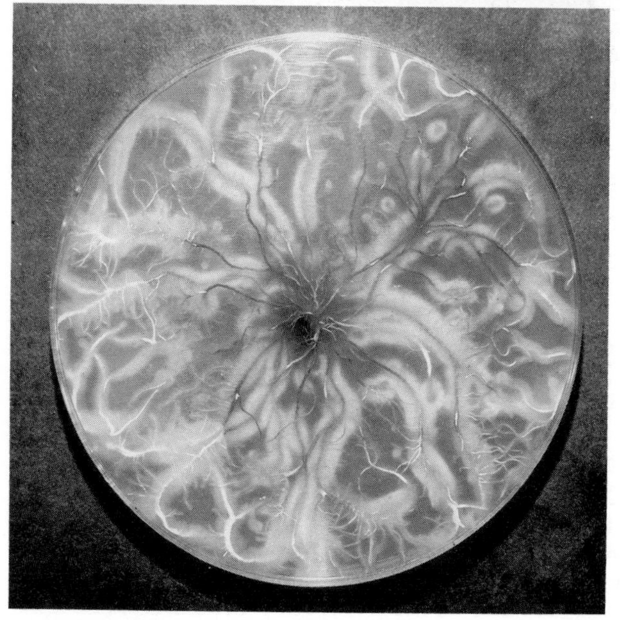

20mm

(b)

lived and not well adapted for survival, a mycelium may produce resistant structures which can survive adverse conditions for long periods of time, and which recommence growth when the environment again becomes favourable. The two most common survival organs are chlamydospores and sclerotia. **Chlamydospores** are formed by the laying down of a thick wall around hyphal cells or spores. They may develop from intercalary or terminal cells, and differ from conidia and sporangiospores not only by virtue of their thick protective wall but also because they are generally not deciduous (Fig. 1.7). A **sclerotium** is a firm, more or less globose, mass of hyphae which upon germination gives rise to a sexual fruiting body or to asexual spores or to a mycelium. Another vegetative structure produced by some fungi is the **rhizomorph**, an organ of spread which also has some survival value. Rhizomorphs are cord-like bodies formed by the aggregation of numerous hyphae, structurally resembling elongated sclerotia; both structures are illustrated in Fig. 1.9.

Fungi are formally described as osmotrophic, achlorophyllous members of the Kingdom Myceteae (*sensu* Alexopoulos & Mims 1979) whose filamentous somatic structures have cell walls typically composed of chitin and glucans other than cellulose. The 'cells' of septate fungi may contain one, two or many nuclei and are therefore, strictly speaking, coenocytic rather than cellular. The septa may be solid but more usually have a central pore large enough to permit the migration of nuclei and other organelles.

Classification of fungi

Owing to their number and wide range of form, there is no general agreement as to the best method of classifying fungi, and there is controversy among mycologists about the limits and names of taxa higher than the rank of genus. For many years three classes and one 'form-class' were recognized, namely the classes Phycomycetes, Ascomycetes and Basidiomycetes, and the form-class Deuteromycetes (Fungi Imperfecti). Most authorities now accept that the Phycomycetes does not represent a natural unit and have discarded this as a formal name. Alexopoulos and Mims (1979) classify the fungi proper (i.e. the typically filamentous forms, as distinct from the slime moulds – see below) into two divisions, the Mastigomycota which includes *inter alia* the class Oomycetes, and the Amastigomycota which contains the class Zygomycetes along with the Ascomycetes, Basidiomycetes and Deuteromycetes. The Mastigomycota contains all the true fungi that produce flagellated, motile cells; the Amastigomycota possess no motile cells. Webster (1980) recognize five subdivisions in the Eumycota or true fungi: Mastigomycotina (comprising three classes including the Oomycetes and Chytridiomycetes); Zygomycotina (two classes including the Zygomycetes); Ascomycotina (Ascomycetes); Basidiomycotina (Basidiomycetes); and Deuteromycotina (Fungi Imperfecti).

The **Oomycetes** are an economically important group. Some forms have aquatic characteristics, such as the production of zoospores, but the class as a whole shows evidence of adaptation to terrestrial habitats. The most advanced species are terrestrial obligate parasites of plants, passing almost their entire life-cycle within the tissues of their hosts. The **Zygomycetes** are terrestrial fungi which produce a characteristic non-motile spore known as a **zygospore**, as a result of sexual fusion. The group includes many common moulds, some of which are responsible for the spoilage of foodstuffs. From the point of view of soil ecology, three orders are particularly important. The Mucorales include several widespread genera of soil fungi responsible for the decomposition of plant detritus and animal dung. The Endogonales contains the single family Endogonaceae, which comprises a number of genera that that infect the roots of higher plants to form dual absorbing organs known as mycorrhizas; these are discussed in Chapter 7. The Zoopagales include the so-called predacious fungi (see sect. 3.5) which occur in soil and decaying plant material, where they destroy amoebae and small invertebrates such as rhizopods and nematodes.

The remaining members of the Amastigomycota are distinguished from the Oomycetes and Zygomycetes by the regularly septate nature of their hyphae. In addition, there is a marked tendency towards aggregation of the hyphae into large, complex structures in which the identity of the component parts is lost. Two classes are recognized, the Ascomycetes and the Basidiomycetes, and a third 'form-class', the Deuteromycetes.

The **Ascomycetes** form a large group, occurring in a wide variety of habitats. Many are saprophytic (saprotrophic) soil inhabitants while others are important agents of plant disease. The yeasts, which are unicellular ascomycetes, form the basis of the baking, brewing and wine-making industries. The primary characteristic of the class is the **ascus**, a sac-like structure containing typically eight ascospores; the ascospores (but not the ascus) are the products of sexual fusion. Many Ascomycetes produce their asci in fruiting bodies known as ascocarps. The typical asexual spores of filamentous ascomycetes are conidia. These are borne on special hyphae known as conidiophores and are sometimes contained in fruiting bodies.

The family Eurotiaceae (Aspergillaceae) contains some of the most common and widely distributed of fungi. The conidial or 'imperfect' stage dominates the life-cycle. *Aspergillus* and *Penicillium* are two genera of great economic significance as spoilage organisms and as commercial sources of antibiotics and enzymes. Their conidia are very numerous and readily dispersed by air. The protoplasm of some aspergilli and penicillia develops a low solute potential and consequently they are able to absorb water and grow at low water potentials such as are found in dry environments and in concentrated solutions of sugars or salts; this enables them to exploit habitats unavailable to most other fungi and to bacteria. A further group contains the genus *Chaetomium*, which includes some common soil fungi having an important role in the breakdown of cellulosic plant remains, especially in acid soils where bacterial activity is at a minimum. Another widespread ascomycete is *Sordaria fimicola*, a coprophilous species occurring on the dung of various animals in the soil.

There are many fungi which have septate hyphae and which, so far as is known, reproduce only by asexual means. Since these fungi apparently lack a sexual stage, they are commonly referred to as imperfect fungi and are classified as **Deuteromycetes** or Fungi Imperfecti. This class is a heterogeneous assemblage of species which bear no phylogenetic relationship to one another. Their conidial stages are generally very similar to the conidial stages of some well known Ascomycetes, and it is assumed that with relatively few exceptions the imperfect fungi represent the conidial (asexual) stages of Ascomycetes whose sexual stages either rarely occur in nature and have not yet been found, or have disappeared from the life-cycle during the course of evolution. A few species, which produce only vegetative structures, represent imperfect stages of Basidiomycetes. The conidial stages of many Ascomycetes (e.g. *Penicillium* and *Aspergillus*) are, for the sake of convenience, also classified as Deuteromycetes. The group contains many fungi of economic importance, including most of the fungal pathogens of man, many serious plant pathogens, many industrially important fungi, and many common soil saprophytes.

The large class known as **Basidiomycetes** has such common fungi as mushrooms, puffballs, bracket fungi and rusts. The distinguishing feature of the group is the **basidium**, usually a club-shaped structure which bears four exogenous sexual spores known as basidiospores; basidia are usually produced within a complex fruiting body called a basidiocarp. The Basidiomycetes are a diverse group of fungi occupying a wide range of habitats. Many species form mycorrhizas with the roots of forest trees (see Ch. 7). Some play an important part in the breakdown of leaf litter and woody debris in forests; others are characteristic of grassland soils. The large and important family Polyporaceae includes the majority of the wood destroying fungi, many of which cause heart rot of living trees, while others are saprotrophic on woody residues in the forest (see Ch. 4). The basidiocarps of most species in this family are bracket-like, hence the common name bracket fungi. The family Agaricaceae includes the majority of mushrooms, the best known being the edible field mushroom, *Agaricus campestris*.

Slime moulds

This is a group of fungus-like phagotrophic organisms some of which have close affinities with the protozoa. They resemble fungi in forming spores surrounded by a definite wall, although the manner of spore formation is often very different from that found in the true fungi. During part of their life-cycle they exist in amoeboid form, unbounded by cell walls, that is as a plasmodium or pseudoplasmodium. Alexopoulos and Mims (1979) place the slime moulds in the same kingdom as the fungi (Kingdom Myceteae), assigning them to divisional status as the Gymnomycota. Webster (1980) assigns them to the division Myxomycota. Two classes, the Acrasiomycetes or cellular slime moulds, and the Myxomycetes or true slime moulds, are abundant and widespread in moist habitats on forest floors, for example in decomposing leaf litter and beneath the bark of decaying logs. The amoeboid stages of the life-cycles of both groups, known as myxamoebae, prey on bacteria and thus may have a significant ecological role in some forest ecosystems.

1.7.3 Lower protists: bacteria and cyanobacteria

The lower protists may be subdivided into a number of distinct groups on the basis of photosynthetic ability, means of locomotion and nature of the cell wall, if present. The cyanobacteria or blue-green algae are photosynthetic organisms which, if motile, exhibit gliding motility.[5] The filamentous gliding bacteria are similar but non-photosynthetic. The myxobacteria and spirochaetes are two groups of non-photosynthetic bacteria having thin and flexible cell walls; the myxobacteria show gliding motility, the spirochaetes have their own peculiar means of movement. The mycoplasmas are very small organisms, barely visible with the light microscope, which do not have cell walls. The final group is the true bacteria or eubacteria, which have rigid cell walls and, if motile, move by means of flagella (see below). Among the eubacteria there exists great versatility in metabolism, and this gives them singular importance in the functional processes of ecosystems, such as the decomposition of organic matter and the cycling of nutrient elements. For this reason, the eubacteria will be treated in more detail than the other groups.

Cyanobacteria

The cyanobacteria, formerly known as blue-green algae, include both aquatic and terrestrial forms. Although the majority are filamentous, unicellular and colonial forms are known. They have an extremely wide distribution in soils, freshwater lakes and streams, and in the oceans. Some species can fix atmospheric nitrogen, either as free living organisms or in association with certain fungi as lichens. The 'fixation' of gaseous nitrogen, that is its reduction to ammonia which is then used for biosynthesis and growth, is the province of relatively few microorganisms, yet is the ultimate source of nitrogen for all organisms (see Ch. 8).

The classification of cyanobacteria is uncertain but Stanier and Cohen-Bazire (1977) distinguish four groups (Fig. 1.10). The **Chroococcaceans** are unicellular rods or cocci, mostly immotile, that reproduce by binary fission or budding. The **Pleurocapsaleans** consist of single cells in fibrous coats, which reproduce mainly by

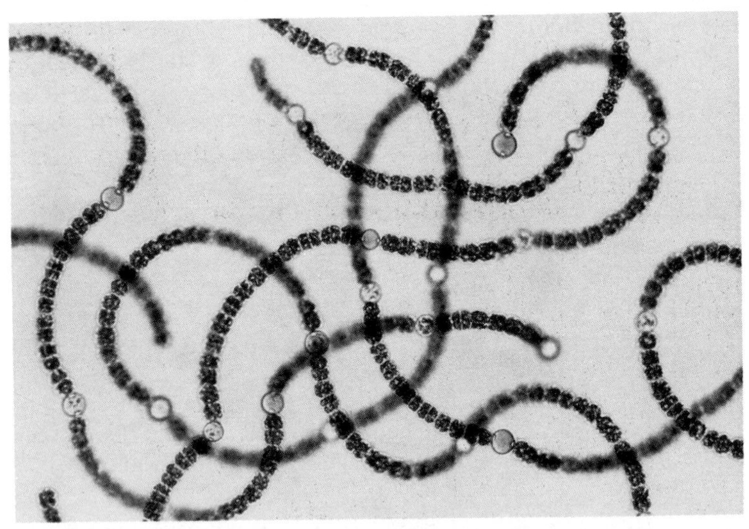

Fig. 1.10 A filamentous heterocystous cyanobacterium. The rounded thick-walled and apparently empty cells are the heterocysts, where atmospheric nitrogen is 'fixed'. (Photograph kindly supplied by Mr. R. J. Banens.)

multiple fission to form numerous daughter cells called baeocytes. Baeocytes show gliding motility.

The cells of **Oscillatorians** form long strands known as trichomes, ensheathed in hollow, tubular structures. Reproduction takes place by the fragmentation of trichomes but within trichomes the strands of vegetative cells elongate by binary fission. **Heterocystous cyanobacteria** also have trichomes but these contain occasional rounded, seemingly empty cells termed heterocysts, as well as normal vegetative cells. Gliding motility is evident and reproduction is by trichome fragmentation. Heterocysts lack an oxygen-evolving photosystem and it seems likely that the resulting anaerobic environment provides the conditions necessary for the efficient functioning of the nitrogen-fixing system (see Ch. 8), since heterocystous cyanobacteria are nitrogen fixers. *Anabaena* and *Nostoc* are typical genera, occurring free living in fresh water and also in association with ascomycetes in the lichen symbiosis (see sect. 3.5) or with certain green plants (see Ch. 8). They also occur in warm, moist soils where they produce a substance known as geosmin, which contributes to the earthy smell of damp soil. This metabolite is also formed by actinomycetes (q.v.).

Bacteria

Three basic cell shapes occur among the true bacteria – cocci, rods and spirals. **Cocci** are roughly spherical cells, and may occur singly, or in pairs, tetrads or other groupings; the particular way in which the cells are arranged in a useful taxonomic feature. **Rods** are straight or slightly curved cylinders, which are found singly or in chains of varying length. **Spirilla** are curved rods in the shape of a helix; the shorter forms, which appear comma-shaped under the microscope, are termed **vibrios**.

Bacteria which are motile are provided with one or more **flagella**, which propel the organism through a liquid medium by their rhythmic movement. Flagella may arise from the ends of the cells, in which case they are termed polar, or they may be distributed all over the cell surface, when they are known as peritrichous. Many bacterial cells are surrounded by **capsules** or slime layers, consisting of polysaccharides or, less frequently, polypeptides. A few genera of bacteria, for example *Bacillus* and *Clostridium*, have the capacity to form **endospores**, which are structures highly resistant to heat and desiccation. Certain other genera such as *Azotobacter* produce a different kind of resting cell known as a cyst, which is less distinctive than an endospore and is not as resistant to adverse environmental conditions. The majority of bacteria, however, possess no specialized resting cells, and survive by their vegetative cells persisting in a state of reduced vitality.

The Gram stain

An important differentiating characteristic of bacteria is their reaction to the Gram stain, which correlates well with certain morphological features of bacteria. Named for the Danish physician who devised it in 1884, this consists essentially of staining with crystal violet in weakly alkaline solution, adding iodine as a mordant, then washing with alcohol. Bacteria which are decolorized by the alcohol are termed Gram-negative, and those which retain the stain Gram-positive. Gram-positive bacteria have very thick walls which are dehydrated upon treatment with alcohol, causing pores in the cell walls to close and so preventing the crystal violet–iodine complex from being washed out. Most cocci, nearly all the spore-forming rods, and all actinomycetes are Gram-positive. Gram-negative forms include the spirilla, all the polarly flagellated non-spore forming rods, and most of the petritrichously flagellated non-spore forming rods.

Reproduction and colony formation in bacteria

The true bacteria normally reproduce by transverse binary fission, resulting in the formation of two equal daughter cells. On the surface of a solid medium, the progeny of a single cell remain together as a discrete aggregate known as a **colony**, the type of colony being frequently characteristic of the species (Fig. 1.11). The close proximity of a large number of cells in a bacterial colony results in a condition known as physiological crowding, in which individual cells compete with each other for available nutrients and inhibit each other by the localized accumulation of toxic waste products. The rate of breakdown of a solid medium by bacteria is determined by the rate at which enzymes diffuse out from the periphery of the colony and bring fresh substrates into solution. This contrasts markedly with the behaviour of a fungal colony, which grows by hyphal extension, the hyphae at the colony margin continually exploiting fresh substrates, and conducting nutrients to the older parts of the mycelium. The component hyphae of a fungal colony are thus less affected by physiological crowding than are the individual cells of a bacterial colony. Bacteria rapidly colonize and decompose soft animal tissues, which have a high protein content, but are less well adapted than fungi for the decomposition of plant residues which generally have a lower nitrogen status than animal matter. The development of mechanical pressure by fungal hyphae further enhances their ability to penetrate tough plant tissues, as indicated in section 1.1.

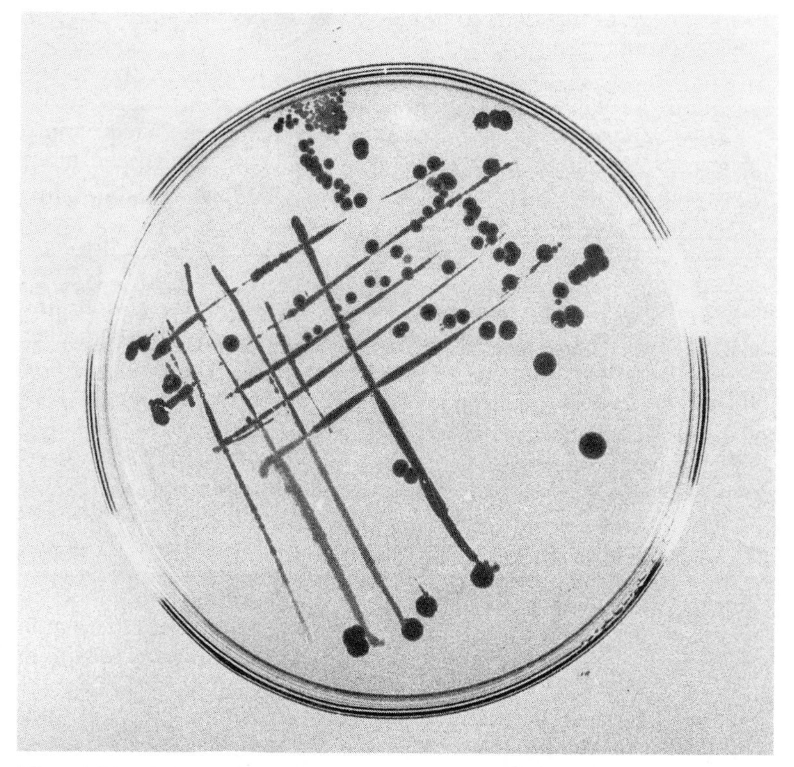

Fig. 1.11 Colony formation in bacteria. Characteristic, pigmented colonies of *Serratia marcescens* streaked on nutrient agar. The petri dish is 9 cm in diameter. (Photo by G. Wray and V. A. Raszewski.)

Classification of bacteria

The problem of classifying bacteria is made difficult because their range of form is sufficient to establish only very broad groups, and is inadequate for the definition of species, genera or even families. The basis of classification is the clone or **strain**, which is a population of genetically identical cells derived from a single cell. Structural and morphological characteristics serve as the primary means of identification but physiological and biochemical properties play a more important role than in the eukaryotes. In other words, bacteria (unlike fungi and other organisms) are classified not so much by how they look as by what they do. For this reason, various physiological groups of bacteria are discussed in later chapters, in the context of their contributions to functional processes in ecosystems. For the moment, it suffices to give a brief description of some of the groups which may be recognized on mainly morphological grounds.

Bacteria form the second of the two divisions in the superkingdom Prokaryota. The standard reference work, *Bergey's Manual of Determinative Bacteriology*, separates the bacteria into nineteen 'parts', including the mycoplasmas. Some of the more important of these, from the point of view of the soil ecologist, are described below.

The **gliding bacteria** are widespread in nature. Several categories may be distinguished. One group common in soils is typified by the genus *Cytophaga*, a rod-shaped form capable of digesting complex organic molecules including cellulose and chitin. Another group, the myxobacteria, is represented by several genera including *Myxococcus*. These organisms also occur as unicellular rods but at a certain stage in the life cycle the rods aggregate into multicellular 'fruiting bodies' containing resting cells known as myxospores. Myxobacteria derive their nutrients mainly by destroying other bacteria though a few species can utilise cellulose as a source of carbon.

The **spiral** and **curved bacteria** are a group of Gram-negative forms which includes some of the more widespread and important of all soil bacteria. While the spirilla are generally more common in aquatic habitats, one species is found in the rhizosphere (i.e. in the root zone, see Ch. 6) of tropical grasses, where it fixes atmospheric nitrogen. Originally described as *Spirillum lipoferum*, this bacterium is now classified as *Azospirillum*. Representative of the curved forms is *Bdellovibrio*, a genus of small, vibrioid organisms which are obligate parasites of other bacteria; they are widespread in soil and water and may have a regulatory role in controlling the population densities of their hosts (see sect. 3.5).

Gram-negative aerobic rods and **cocci** are represented by three widely distributed families of soil bacteria, the pseudomonads (Pseudomonadaceae), the free-living N_2-fixers (Azotobacteraceae), and the symbiotic N_2-fixers and related forms (Rhizobiaceae). The pseudomonads decompose a variety of soluble compounds released during the breakdown of plant and animal remains; they are polarly flagellated rods, aerobic or facultatively anaerobic.[6] *Pseudomonas aeruginosa* is a facultative anaerobe able to use NO_3^- as an electron acceptor as well as O_2 (sect. 2.2).

The free-living N_2-fixing bacteria (Azotobacteraceae) are obligate aerobes. *Azotobacter* is the best known genus, but *Beijerinckia* and *Derxia* are important too, especially in the tropics.

The family Rhizobiaceae contains two genera of root-infecting bacteria. One genus, *Rhizobium*, produces nodules on leguminous plants in which nitrogen is fixed symbiotically. The second, *Agrobacterium*, causes tumour-like growths known as crown galls. Root nodule symbioses are described in Chapter 8.

The better known representatives of the **Gram-negative facultatively anaerobic rods** are either constituents of the normal microflora of the mammalian intestine (e.g. *Escherichia coli*) or agents of epidemic diseases. The most important soil-inhabiting species were formerly classified as *Aerobacter* but are now placed in the genus *Enterobacter*. Some strains of *Klebsiella* can fix atmospheric nitrogen when growing anaerobically.

Perhaps the best known members of the **Gram-negative chemoautotrophs** are found in the family Nitrobacteraceae. This family contains bacteria capable of oxidizing reduced forms of inorganic nitrogen. Morphologically, they are heterogeneous, ranging from rods to cocci. Some are non-motile, some are motile with polar flagella while others have peritrichous flagella. One genus (*Nitrobacter*) reproduces by budding. Another group of related genera is typified by *Thiobacillus*, which is a polarly flagellated rod that oxidizes reduced forms of inorganic sulphur. What unites this miscellaneous collection of Gram-negative bacteria is their autotrophic mode of life, the energy needed for biosynthesis and growth being derived from oxidation–reduction reactions (see Ch. 2).

Gram-positive cocci include the micrococci, which are aerobic or facultatively anaerobic, spherical bacteria widely distributed in nature; *Micrococcus* is the most common genus found in soils.

Gram-negative cocci and coccobacilli are aerobes, represented in soil by the widespread genus *Achromobacter*.

With the exception of *Sporosarcina*, a coccoid form, all the **endospore-forming bacteria** are rods. Two well known genera are the aerobic or facultatively anaerobic *Bacillus*, and the strictly anaerobic *Clostridium*. They are widely distributed in soil, and many species can fix atmospheric nitrogen. The bacilli are a very heterogeneous collection of species, and probably should be split into two or more genera. Many can produce extracellular enzymes (sect. 2.4) which permit them to hydrolyse polysaccharides and other complex organic molecules. The clostridia are likewise very diverse in their physiology, some fermenting cellulose, some starch and pectin, others proteins and amino acids, and still others sugars.

The largest and most heterogeneous of all the subdivisions (parts) of bacteria recognized in *Bergey's Manual* embraces the **actinomycetes and related organisms**. These are all Gram-positive and are characterized by their irregular morphology. They vary from regular rods through irregular, club-shaped rods to filamentous (mycelial) forms. The coryneform bacteria are irregular rods which exhibit a peculiar mode of cell division known as 'snapping fission'; this is most pronounced in *Corynebacterium* where it results in aggregations of club-shaped cells taking on the appearance of a palisade. *Arthrobacter* is one of the most common of soil bacteria and is very resistant to desiccation and starvation; it is characteristically pleomorphic, changing in shape from rod to coccus with age. *Cellulomonas* is a bacterium with coryneform morphology that is able to digest cellulose.

The actinomycetes have a filamentous, often branching habit, and include many common soil saprophytes. In the genus *Mycobacterium* the mycelium is rudimentary, soon breaking into irregular, sometimes branched rods. The mycelial habit is more pronounced in *Nocardia* and *Actinomyces*, but after growth ceases the mycelium fragments into rods or cocci which are indistinguishable from unicellular eubacteria. In *Streptomyces* and *Micromonospora* the mycelium is permanent, and reproduction takes place by means of conidia. Streptomycetes are responsible for the characteristically earthy odour of soil, as a result of their producing volatile terpene derivatives known as geosmins (these are also produced by certain cyanobacteria, q.v.). The streptomycetes are strict aerobes, very versatile in their nutrition, being able to utilize a wide variety of organic compounds. Many produce antibiotics, including streptomycin, chloramphenicol, and the tetracyclines. The genus *Frankia* is the endophyte of the N_2-fixing nodules of non-leguminous angiosperms; this phenomenon is discussed in Chapter 8.

Finally, there are the very small almost sub-microscopic organisms known as **mycoplasmas**. Lacking cell walls, these microbes are highly pleomorphic, their form ranging from coccoid to filamentous, often much branched. Mycoplasmas are responsible for certain plant diseases, previously thought to have been caused by viruses.

1.7.4 Soil fauna: the meso- and macrobiota

A great variety of animals is found in the soil, with representatives of nearly all the animal phyla except those that are restricted to marine environments (Fig. 1.12). There are those, such as earthworms, nematodes, some mites and springtails, which spend their entire life-cycle in the soil whereas others, no less important, are present for only part of their existence. The soil fauna is usually taken to include all animals that pass one or more active stages wholly or largely in the soil or surface litter, and to exclude those species which occur there only in passive stages such as eggs, cysts

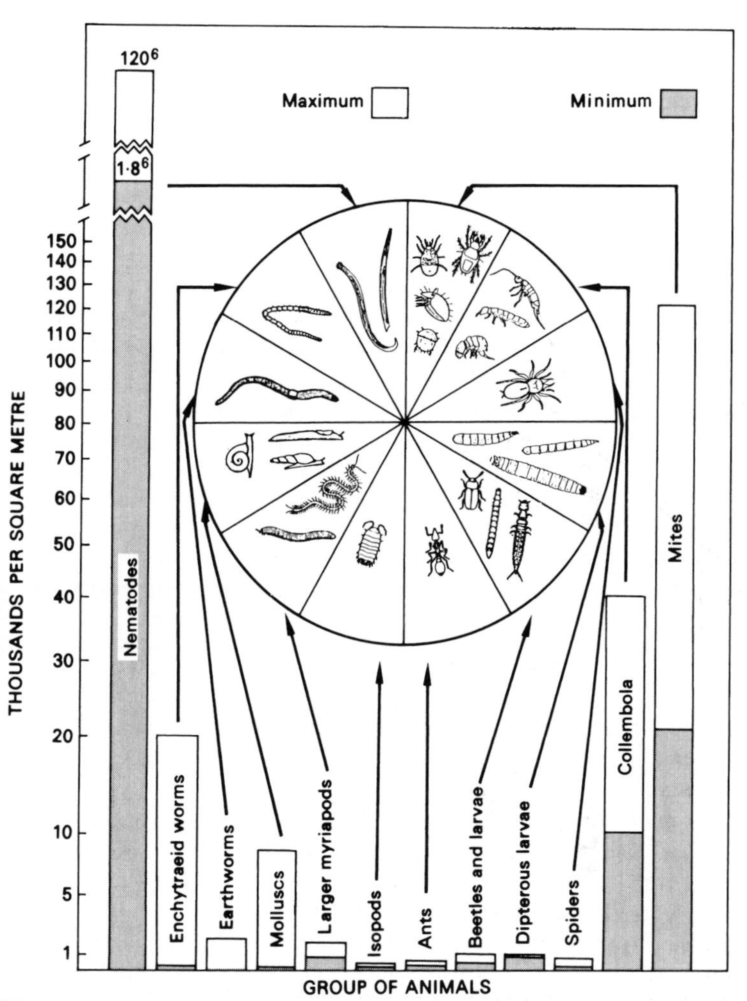

Fig. 1.12 The major kinds of animals found in the soil, showing the range of population densities likely to be encountered in a cool temperate grassland ecosystem. (From Kevan 1965)

or pupae, or which use the soil during periods of dormancy or for temporary shelter. For more detailed information, consult Wallwork (1970, 1976).

Classification of the soil fauna

The soil fauna has been classified in a number of different ways, using different criteria as the basis of subdivision. The criteria which are the most widely used are:

(a) the amount of time which the animals spend in the soil;
(b) the habitat preference of the animals;
(c) the method of feeding;
(d) the means of locomotion;
(e) the size of the animals.

None of these criteria is entirely satisfactory but together, at least, they form a basis for dealing with the very diverse assemblage of soil animals in manageable categories.

The time which different animals spend in the soil varies from permanent residency to transitory visitation. There are two main microhabitats for soil animals, aquatic (water-filled pores and the surface moisture film surrounding soil particles) and terrestrial (the soil atmosphere); animals will show a preference for one or other of these. In terms of feeding, animals may be biophagous or saprophagous. The former include carnivorous, phytophagous or microbivorous forms (see Ch. 4). Where locomotory activity is considered, a distinction is usually made between the 'burrowers', that is those animals which actively dig their own tunnels, and the 'non-burrowers' which move through existing channels.

A subdivision of the soil fauna on the basis of body size is probably the most widely used classification system, perhaps because the size classes chosen correspond quite closely with the various kinds of sampling apparatus that have been devised. Using the criterion of size the soil fauna may be divided into three categories (Fig. 1.13). Animals less than 200 μm long, that is, not visible with the naked eye, comprise the **microfauna**; all the soil protozoa are in this category. In addition to protozoa (which have already been considered under the microbiota, sect. 1.7.2) some nematodes, rotifers and mites are just small enough to be classified as microfauna. As M. J. Swift and his co-authors have pointed out (Swift *et al.* 1979), the microfauna comprises a coherent functional group since none is involved in litter comminution. The **macrofauna** consists of those animals whose length is measured in centimetres, namely vertebrates, earthworms and the larger members of the enchytraeid, mollusc and arthropod groups. Animals of intermediate size (200 μm – 1 cm) constitute the **mesofauna** (sometimes called the meiofauna). Thus defined, the mesofauna includes most of the nematodes, rotifers, springtails and mites together with various small enchytraeids, molluscs and arthropods.

Although some vertebrates may be classified as soil animals (e.g. moles, root-feeding rodents, some lizards and frogs), the

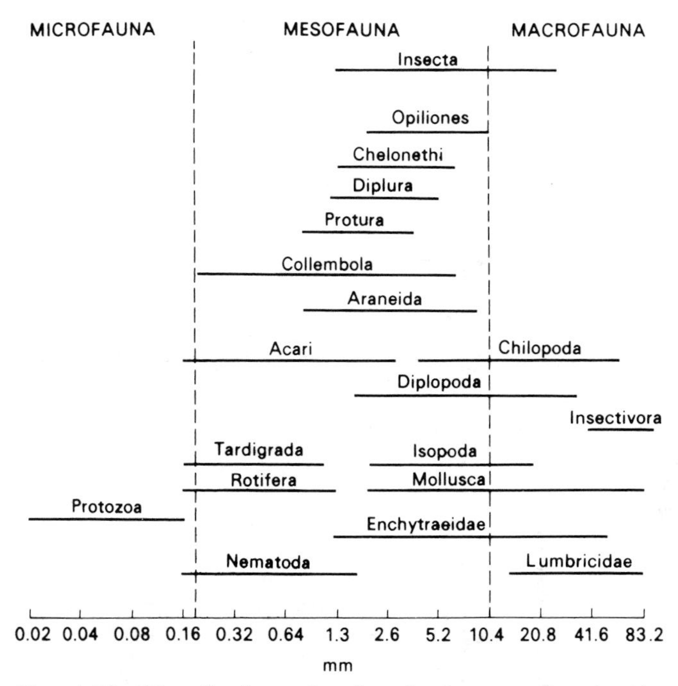

Fig. 1.13 Classification of soil animals according to size. (From Wallwork 1970) Reproduced by kind permission of the author.

greater part of the soil fauna consists of invertebrates (Table 1.4). After the protozoa, nematodes are by far the most numerous of the soil animals but the group showing the greatest diversity, that is the largest number of species, is the Arthropoda. The most commonly occurring arthropods are mites and springtails; others which are found in the soil fairly frequently are the millipedes, termites, ants, and the larvae of flies and beetles. Annelid worms, though few in species, make a significant contribution to biomass in some soils. Brief descriptions of the better known groups of soil invertebrates follow, while some typical forms are shown in Fig. 1.14.

1.7.5 Invertebrates other than arthropods

Flatworms

The phylum Platyhelminthes contains not only the well known parasitic forms, the flukes and tapeworms, but also the free-living

Table 1.4 Common groups of invertebrate soil and litter fauna

Phylum	*Class or sub-class*	*Order*	*Common name*
Platyhelminthes	Turbellaria		Flatworms
Aschelminthes	Nematoda		Nematodes (roundworms)
	Rotifera		Rotifers
Annelida	Oligochaeta		Earthworms and potworms
Mollusca	Gastropoda	Pulmonatiformes (Pulmonata)	Snails and slugs
Arthropoda	Arachnida	Scorpioidiformes (Scorpionoidea)	Scorpions
		Araneiformes (Araneida)	Spiders
		Acariniformes (Acarina)	Mites
		Pseudoscorpioidiformes (Chelonethi)	Pseudoscorpions
		Opilioniformes (Opiliones)	Harvestmen, daddy longlegs
	Crustacea	Isopodiformes (Isopoda)	Woodlice
	Myriapoda		
	Diplopoda		Millipedes
	Chilopoda		Centipedes
	Insecta		
	Apterygota	Collemboliformes (Collembola)	Springtails
	Pterygota	Isopteriformes (Isoptera)	Termites
		Lepidopteriformes (Lepidoptera)	Moths and butterflies
		Dipteriformes (Diptera)	Flies
		Coleopteriformes (Coleoptera)	Beetles
		Hymenopteriformes (Hymenoptera)	Ants, bees and wasps

Fig. 1.14 Typical members of the soil fauna, natural size. (From Russell 1973)

Key: *Chilopoda*: Centipedes 4,5,28,40. *Diplopoda*: Millipedes 19,22,41.
Arachnida: (a) Araneida, Spiders 31, 33; (b) Acarina, Mites: Gamasidae 2,14,36; Oribatidae 24,32,37,45; Tyroglyphidae 42.
Insecta: (a) Collembola, Springtails 3,6,10,13,16,20,50; (b) Lepidoptera (larva) 8; (c) Coleoptera: Staphylinoidea (adults) 7,9,17,30,35,38,44; Staphylinoidea (larvae) 15,23,27, 29,34,41; Carabidae (adult, small species) 25; Carabidae (larva, large species) 12; Elateridae (larvae), Wireworms 21,26,39,45; (d) Diptera: Bibionidae (larvae) 1,43,48,49; Cyclorrhapha (larva) 11; Anthomyiidae (adult) 18.

flatworms (**Turbellaria**). Some Turbellaria inhabit the soil: they are always found in damp situations, for example under logs and in decaying leaf litter of moist fertile soils. Perhaps the best known soil Turbellaria are the ‘land planaria’; these are not usually numerous although a fairly rich fauna may be found in the warm moist litter of subtropical and tropical rainforests, mainly in the southern hemisphere. Other turbellarians which are microscopic and which live in the water film have occasionally been recorded in large numbers in certain fertile soils. Many of the larger soil Tubellaria are carnivorous; other much smaller ones may be microbivorous while still others are carrion feeders.

Nematodes and rotifers

Both nematodes and rotifers are aquatic animals of the phylum Aschelminthes; those forms which occur in the soil are active only in water films. Most rotifers have some protective device which operates when desiccating conditions prevail: some secrete shells but many of them enter a state of 'anabiosis' in which a great deal of water is lost and their metabolism slows down considerably. It is thought that certain of the soil nematodes can survive repeated desiccations.

Rotifers are quite abundant in the litter and humus layers of forest soils, their population density reaching 10^5 m^{-2}. About a hundred soil-inhabiting species have been recorded. They rarely exceed 2 mm in length and the soil forms are usually much smaller than this. Most of the commonly occurring soil rotifers are creeping forms, many of which consume plant or animal debris although some feed on protozoa or algae.

Most of the 10 000 or so known species of **nematodes** are either free living in the sea or parasitic; of the 2000 or so fresh-water forms about 1000 are true soil inhabitants. Most soil nematodes are microscopic, transparent, threadlike animals. They are usually found in the upper 10 cm of the profile, where population densities of the order of 10^6 m^{-2} are not uncommon. They are usually more numerous in mull and grassland soils than in acid podzols; however relatively few nematode species are restricted to one particular soil type. Nematode numbers are generally higher in the vicinity of plant roots than elsewhere in the soil. Most nematodes have a high reproductive capacity: they produce large numbers of young and there are usually at least five or six generations a year.

All nematodes appear to feed on living material, the dead organic matter in the soil not forming part of their diet. Soil nematodes have varied food requirements. Some ('eelworms') extract plant sap by piercing roots, others prey on rotifers or other nematodes, while others consume bacteria, algae, fungi or protozoa. Most workers agree that nematodes cannot participate directly in the decomposition of organic matter in the soil but they may be significant ecologically as regulators of microbial populations, and as a food source for other members of the soil biota.

Earthworms and potworms

The phylum Annelida comprises the segmented worms. Several families which belong to the class Oligochaeta are soil dwellers. The soil oligochaetes of Europe belong to two families, the **Lumbricidae** or 'true' earthworms and the **Enchytraeidae** – the smaller, less well known, potworms. Lumbricids are less abundant outside the Palaearctic realm where earthworms of other families, for example the **Megascolecidae**, are more common. The Megascolecidae are chiefly southern hemisphere forms: the giant earthworms of Australia belong to this family.

Earthworms

In grassland soils and woodland mulls of the northern hemisphere earthworms number hundreds per square metre, but in acid podzols they are numbered in tens. Earthworms are not generally as widespread throughout the tropics and subtropics as they are in temperate climates. Furthermore, while northern temperate species are most abundant in neutral and alkaline soils, many tropical and southern hemisphere earthworms are well adapted to acid conditions. Earthworms can only exist in an active state under moist conditions: they require copious water for excretion, a moist skin for respiration and a fairly turgid body cavity (to act as a hydrostatic skeleton) for locomotion. Yet most earthworms have a remarkable ability to withstand desiccation and can survive several months of drought in a quiescent state. In soils where earthworms are abundant they dominate the invertebrate biomass and since they are relatively large animals their effects on the physical structure of the soil may be considerable.

The Lumbricidae fall into two main groups, those that live in surface organic horizons and ingest little mineral material, and those that live predominantly in the mineral soil. Members of the latter group ingest mineral matter when burrowing or feeding, and most of this is voided through the anus when the associated nutrients have been utilized. Some lumbricids, however, both feed on the surface and ingest mineral soil. The common European earthworm *Lumbricus terrestris* may burrow to a depth of several metres to escape unfavourable surface conditions, but most lumbricids do not tunnel more than a few centimetres below ground level, and only rarely collect food from the surface. *L. terrestris* however emerges at night, to feed on leaf fragments and other plant debris at the soil surface.

Earthworms ingest both decaying plant material and mineral soil, some being quite selective in the type of plant material they eat. Perhaps the most important effect of earthworms in the soil is the fragmentation of litter and the mixing of the small fragments of plant material with the soil. These fine particles pass out of the worm's gut bound together in crumbs which form the familiar worm casts. These casts are more water-stable than the original soil, and because of this it has been suggested that earthworms are important in promoting a more stable soil structure. It should be pointed out however, that worm casts on the surface of the soil are only stable for a short time. The mucus which worms secrete, and which binds the walls of their burrows, may also be important in promoting the structural stability of soil.

The microbial population of worm casts is generally higher than in the surrounding soil. Fresh earthworm faeces have a paste-like consistency, are poorly aerated but rich in ammonia and organic matter which is partially digested; these faeces may well act as a substrate conducive to microbial growth.

Earthworms are of special significance in the soil since both cellulases and chitinases are present in their alimentary canals: these worms thus take a direct part in the decomposition of litter in the soil. This is in addition to their indirect effects as comminutors of litter and promoters of microbial activity.

Enchytraeids

The potworms are small (1 – 5 mm in length), white oligochaete worms which are often present in the soil in very large numbers; they are not distributed uniformly but usually occur in groups. Enchytraeids are very sensitive to drought, and soil forms are most abundant in moist temperate climates. Maximum numbers are found in acid soils of high organic matter content, where populations of 10^5 m^{-2} may be found.

Enchytraeids do not possess enzymes which would enable them to digest complex plant polysaccharides and there is no evidence that any of the organic material they ingest is chemically decomposed. They probably feed mainly on bacteria and fungi. Plant fragments, fungi, bacteria and silica grains have been found in their guts and it is possible that the enchytraeids produce water-stable crumbs in the soil. As with earthworms, the faecal material produced by enchytraeids may serve to stimulate the activities of the soil microflora.

Snails and slugs

The phylum **Mollusca** is a group of predominantly marine animals; there are many terrestrial forms but their diversity is small compared with that shown by marine species. Truly terrestrial molluscs belong to the order Pulmonatiformes. There are two main groups of land pulmonates, the slugs and the snails; many of these live in well defined habitats. Slugs and snails are very sensitive to micro-environmental changes and therefore are not distributed uniformly throughout the soil nor over its surface but instead aggregate in favourable situations. Population densities of 10–25 m^{-2} are not uncommon, and they may exceed 50 m^{-2}.

The food of terrestrial molluscs is extremely varied: many feed on living plant material while others prefer dead or decaying vegetation. Many snails and slugs browse on fungi whilst others eat algae and lichens; some are carnivorous and prey on earthworms, and on other slugs or snails. A few species are omnivorous. Many of those which consume surface vegetation move down into the soil after feeding; it seems likely therefore that they have a role in incorporating organic matter in the soil. A number of slugs and snails appear to produce cellulases (see Ch. 2); in addition they may contain cellulase-producing bacteria in their guts. The part played by molluscs in the degradation of plant material in the soil may thus be considerable. Molluscs also produce slimy muco-proteins (as do earthworms, see above) and these products help in the formation of water-stable soil aggregates; thus molluscs may contribute to the improvement of soil structure.

1.7.6 Arthropods

The phylum **Arthropoda** contains more species than any other animal phylum. Arthropods often dominate all other groups of the meso- and macrofauna in the soil, both in terms of the number of individuals and the number of species present. They reach their greatest diversity and abundance in undisturbed habitats such as forests, woodlands or permanent grasslands. The most common soil arthropods are woodlice, centipedes and millipedes together with mites and certain insects. The more important insect groups are the springtails, termites and ants and also the adults and/or young stages of beetles and flies. Springtails and mites are often considered together under the name microarthropods, the larger forms being then termed macroarthropods.

Arachnids

Mites

The acarina (order **Acariniformes**) are the most numerous of all soil animals. Although their classification and nomenclature is subject to debate, five sub-orders are generally recognized (B. J. Main, in Marshall & Williams 1972). Four of these sub-orders are represented in the soil fauna, namely the Prostigmata, Mesostigmata, Astigmata and Cryptostigmata (Wallwork 1970, 1976). Members of the first two groups are usually active forms which range freely through the soil, and many are predatory. Most Crypostigmata are smaller, slower moving mites, which are mainly detritus feeders. The Astigmata are not usually common in soils.

The predatory Mesostigmata prey upon nematodes, small enchytraeid worms, insects or other mites; other members of the order are saprophagous, while some are mycophagous or coprophagous. The Prostigmata is a large, diverse group of mites. Many are predators and their prey is similar to that of the Mesostigmata; some species are phytophagous but little is known about the feeding habits of many of the smaller forms.

The group Cryptostigmata contains most of the typical soil mites. They are found in greatest numbers in mor soils under forest, and are most abundant in the F-layer. The distribution of Cryptostigmata, like that of the collembola (see below), is governed by the need for habitats with a saturated atmosphere. Many Cryptostigmata are relatively unspecialized feeders not restricted to any particular type of food, ingesting organic matter in various stages of decomposition. Many of them however prefer soil fungi and bacteria, while others eat woody tissues and some are coprophagous.

Mites probably contribute very little to the chemical decomposition of plant litter although many of them are associated with the later stages of decay. They do however have a significant role in the fragmentation of litter; this is especially true of the Cryptostigmata. The small litter fragments expose a large surface area to microbial activity. Evidence suggests that the faeces of detritivorous mites are particularly susceptible to invasion by microbes. Mites are also important transporters: they may disseminate fungal spores and many of them (especially the Cryptostigmata) carry decomposing organic matter from the surface to the deeper soil layers.

Other arachnids

Also counted among the soil fauna are the scorpions and pseudoscorpions, and some spiders and harvestmen. Scorpions (order **Scorpioidiformes**) are widely distributed in warm, dry tropical and subtropical habitats (Wallwork 1970). Like isopods (q.v.), they are nocturnal, sheltering by day and preying by night on a wide variety of surface-dwelling invertebrates and small vertebrates. Spiders (**Araneiformes**) mostly live above ground and the majority therefore cannot be considered to be part of the soil fauna. Representatives of several families, however, inhabit the leaf litter of forests where they act as predators of smaller soil animals

(Fig. 1.14) Pseudoscorpions (**Pseudoscorpioidiformes**) have a pantropical distribution (Wallwork 1976) though they occur in temperate regions as well. They may number hundreds per square metre in decomposing leaf litter but are usually not so abundant. Pseudoscorpions prey on a variety of detritus-feeding arthropods. Finally, there is the relatively small group of harvestmen (**Opilioniformes**), some of which are regular soil inhabitants, widely distributed in temperate regions. They also are predacious though their ecological role is thought to be slight.

Isopods

The **Isopodiformes** (Isopoda) is a terrestrial order of crustaceans (class Crustacea) which flourishes in wet, decomposing organic matter such as deciduous woodland leaf litter and compost. Commonly known as woodlice, they tend to congregate beneath stones or bark by day and emerge to feed at night on plant debris or carrion. Their clumped distribution makes it difficult to estimate population densities with any accuracy but many hundreds of individuals per square metre may be found in favourable conditions.

Myriapods

This group of arthropods comprises four sub-classes, to which some authorities assign the rank of class. All four are typical soil dwellers and three of them, the **Chilopoda, Diplopoda** and **Symphyla**, are widely distributed. Symphylids are delicate, centipede-like creatures, usually less than 1 cm long, which are often quite abundant in cultivated soils rich in organic matter. Most are saprophages but some feed on plant roots and can become serious pests of plants raised in glasshouses.

Centipedes (Chilopoda) are found in moist habitats in the temperate and tropical parts of the world (Fig. 1.14). The larger ones live under stones and logs or in cracks or crevices in the soil; the smaller forms are truly subterranean and many can burrow. Centipedes are found mainly in woodland and forest, where they are among the more important predators. They occur to a much lesser extent in grassland and arable soils. Although mainly carnivorous, there are some species which feed on decaying plant material.

Millipedes (Diplopoda) prefer calcareous soils though they occur in moderately acid environments as well; they are more common in woodland than in grassland or cultivated soils (Fig. 1.14). Some are restricted to the loose litter layer while others burrow in the soil. Some species feed on living plants, others on fungi, others on fresh litter but most millipedes prefer decaying plant material as food. There is little evidence to suggest that any chemical breakdown of plant material occurs in the guts of millipedes; their main role in the soil seems to be the fragmentation of litter which is thus made more susceptible to rapid microbial decay. Like woodlice, millipedes have a clumped distribution and occur at densities ranging from tens to hundreds per square metre.

Insects

Springtails

The collembola (order **Collemboliformes**) are small, primitive insects. The group is one of the most abundant and widely distributed of the soil arthropods, numerous with regard to both numbers and species diversity. Springtail populations in the soil commonly reach a density of 10^4 m^{-2} but these insects contribute little to the soil biomass because of their small size. They are usually less than 1 mm long and rarely exceed 3 mm (Fig. 1.14).

Collembola exist only in moist situations but some of them can resist desiccation to a certain extent. In the soil some springtails live in the surface layers while others are commonly found deeper in the soil. The surface dwellers are usually larger, have a spring-like appendage and simple eyes; the spring and the eye-spots are usually reduced or absent in the smaller species which live deeper in the soil. Populations of springtails are largest in the surface layers of the soil, especially where the macropore space is greatest. Soil-dwelling springtails, like other soil microarthropods, tend to aggregate; the reason for this is not known for certain.

The feeding habits of soil collembola are very varied: their food may be bacteria, fungal hyphae and spores, decomposing organic material, faeces, living plants or animals. It seems that the detritus feeders consume decaying plant material for the fungi and other microbes it contains rather than for the nutritive value of the detritus itself. Thus collembola do not appear to play a direct part in the turnover of soil nutrients but they are active in fragmenting plant litter and in this respect they may be a significant factor in certain soils.

Termites

Termites (order **Isopteriformes** or Isoptera) are mainly tropical or sub-tropical in distribution. They constitute one of the two major groups of soil-inhabiting social insects, the other being the ants.

Termites have a highly developed social organization and build up complex colonies which may take several years to become properly established. They may be divided into three groups on the basis of food preference, that is, those which feed exclusively on wood, those which consume humus or decaying litter and those which cultivate fungi for food. Cellulose is the main carbohydrate in the diet of all termites which consume wood or other relatively undecomposed plant tissue. Such termites do not produce their own cellulases but rely on the activities of a rich gut fauna of cellulose-producing flagellates (sect. 1.7.2). Those termites which feed on humus or on fungi appear to lack the flagellate gut fauna. Hemicelluloses as well as cellulose can serve as a food source, and some species are believed to digest lignin also. The rate of decomposition of lignin is much slower than that of the polysaccharides, and it is possible that most of the lignin breakdown which occurs is brought about by associated fungi rather than by the insects' gut fauna (see sect. 4.1).

Most litter decomposing termites forage for freshly fallen or old leaf litter and transport it to their nests, where it may be eaten immediately or stored for later consumption; these fall mostly in the family Termitidae. Other species (mainly in the family Hodotermitidae) harvest fragments of living plants, especially grasses. Most of the wood feeding termites (Kalotermitidae and Rhinotermitidae, among others) consume rotting wood but a few species (Kalotermitidae and Mastotermitidae) feed on living wood. It is thought that wood-rotting fungi are an important component of the food of some termites that utilize decaying wood. A number of species consume large amounts of mineral soil, particularly if it is rich in humus, their mouth parts being specially adapted for this purpose. They frequently mould the mineral particles and organic matter into a discrete mass, thus providing a favourable habitat for fungi, which apparently serve as a food source for the insects.

In the tropics and subtropics termites play an important part in the decomposition of cellulosic residues and in transporting and mixing organic matter with mineral soil from different horizons. In other words, their role in the soil is similar to that of earthworms in more temperate climates, and, along with earthworms, they are probably more important in litter decomposition than any other category of invertebrates. Some termites erect earth mounds of varying degrees of complexity, these being constructed of sand and clay particles cemented with saliva or faeces. Not all species build mounds: many humus feeders are entirely subterranean and some wood feeders live in galleries excavated in rotting logs or trees. Termite mounds are, however, conspicuous features of the landscape throughout much of tropical and subtropical Africa, Asia, Australia and South America.

Despite the consensus of opinion that the functional roles of termites and earthworms are similar, Lee and Wood (1971) have emphasized some important differences. The gallery systems of Australian termites, for example, are rarely extensive enough to affect soil porosity, although there is no doubt that they tend to invert soil profiles. Moreover, the sequestration of chemically altered soils in mounds places them out of reach of the general soil microflora, and in any event the residues of plant tissues degraded by termites cannot be regarded as readily available substrates for microorganisms. Notwithstanding these reservations, the fact remains that there is little accumulation of surface litter or soil organic matter in places where termites are abundant. This is especially true of arid regions where these insects probably do have an important role in organic matter decomposition and nutrient cycling.

Flies

Many soil and litter inhabiting insects belong to the order **Dipteriformes** (Diptera) or two-winged flies (Fig. 1.14). Their larvae rival the coleoptera in the number of species and in the range of feeding habits, but relatively few adults are true soil dwellers. Diptera as a whole are less well adapted to dry soil conditions than coleoptera. Dipteran larvae are generally found in damp situations especially in the L- and F-layers of soils, where these are present. Very few of them are able to burrow; the majority move through existing crevices in the soil. The more familiar dipteran larvae are carrion feeders, but many species consume plant residues or living roots, others such as the fungus-gnats eat fungal hyphae. Still others are coprophilous, while a number are predacious, preying upon a variety of other soil animals.

There is no evidence that plant material undergoes humification as it passes through the gut of dipteran larvae but these larvae, like many other arthropods, are important in the fragmentation of organic debris; their faeces may also provide favourable habitats for microorganisms.

Beetles

The **Coleopteriformes** or Coleoptera (Fig. 1.14) is the largest of the insect orders, and contains a number of forms which live in the soil as larvae and/or adults. Soil coleoptera are not particularly numerous, rarely exceeding 100 m^{-2} in number or 1 g m^{-2} in biomass; most of them live on the surface of the soil or in the upper layers.

Some soil coleoptera are predators of other soil animals, others are phytophagous; included in the latter group are the well known larvae of certain scarabs and chafers which feed upon plant roots.

Other soil coleoptera are saprophagous, feeding on decaying plant or animal matter while still others feed on dung. Some beetles consume the more refractory plant materials such as cellulose and this may involve the intermediate activity of cellulase-producing gut microbes. The main role of coleoptera in the soil seems to be that of assisting in the breakdown and incorporation of organic matter.

Ants, bees and wasps

A few species of bees and wasps of the order **Hymenopteriformes** (Hymenoptera) are permanent members of the soil fauna and their burrowing activity may help to aerate the soil. The most important soil-inhabiting hymenoptera, however, are ants. These well known social insects forage on the surface and make extensive subterranean excavations; some also build large earth mounds above ground level. They are often the first animals to become established in newly exposed habitats such as riverine deposits.

The food of ants is extremely varied. Many species are predators on the larger arthropods of the soil, many take plant food, leaves, woody tissues or seeds; others feed on the hyphae of fungi which they cultivate in underground galleries. The mound building activities of ants, as in the case of termites, may involve the transfer of large quantities of mineral matter from below ground to the soil surface; this is believed to promote the development of crumb structure in the soil. Their activities may also bring about local increases in the organic matter content of the soil.

1.8 Notes

1. This definition requires that allowance be made for that fraction of primary production which is lost from the plant biomass by the shedding of dead tissues and through the grazing of herbivores. For an analysis of ecosystem productivity, see section 4.7.
2. Under natural conditions, generation times for soil bacteria are much longer than in the optimal environment of the laboratory (see sect. 4.6). Nevertheless, they reproduce much more rapidly than most other organisms.
3. Coenocytic organisms are multinucleate but not multicellular, although at some stage of their life-cycle they may exhibit cellular structure.
4. Structurally, flagella and cilia are homologous; they differ only on the basis of relative length.
5. Gliding is a peculiar means of movement which occurs only when cells are in contact with a solid substratum; its mechanism is obscure.
6. The oxygen relationships of bacteria are described in Chapter 2.

2

Energy and nutrient sources for soil organisms

In order to comprehend the role of soil organisms in such aspects of ecosystem function as energy flow and nutrient cycling, it is necessary to have an understanding of the manner in which the organisms themselves satisfy their demands for energy and nutrients. These considerations are taken up in the present chapter.

2.1 Nutritional categories and energy-yielding processes

The concept of living organisms falling into one of two nutritional categories, autotrophic or heterotrophic, is a long standing one in biology. It was originally based on the belief that autotrophs are entirely independent of the presence of organic matter in their environment, but this is not strictly accurate since many such organisms are now known to require specific organic growth factors, such as vitamins, even though their principal nutrients are inorganic. The classification retains its value, however, provided the use of the terms is restricted to indicating the nature of an organism's **principal carbon source**, that is, whether organic (heterotrophic) or inorganic (autotrophic) (Stanier *et al.*, 1977). The nature of the **energy source** forms the basis of another fundamental division of organisms, those using radiant (solar) energy being called phototrophic (or photosynthetic), and those dependent on the energy released during chemical oxidations being termed chemotrophic.

Combining these two basic criteria leads to the recognition of four major nutritional categories (Fig. 2.1):

(a) **Photoautotrophs**, utilizing light as an energy source and CO_2 as the principal source of carbon. Examples are higher plants,

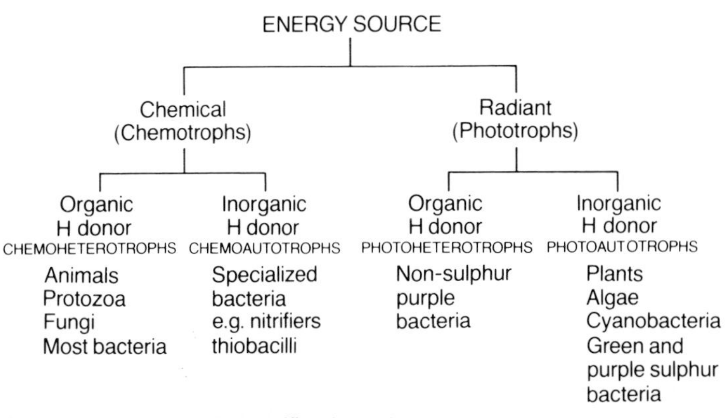

Fig. 2.1 Functional classification of organisms based on the nature of their energy and principal carbon sources.

algae, cyanobacteria and the purple and green sulphur bacteria.

(b) **Photoheterotrophs**, dependent on light as a source of energy and deriving much of their carbon from organic compounds. This category is represented by a specialized group of photosynthetic bacteria known as non-sulphur purple bacteria.

(c) **Chemoautotrophs**, deriving energy from the oxidation of inorganic compounds and using CO_2 as the principal carbon source. This category comprises several groups of specialized bacteria, including the nitrifying bacteria and thiobacilli.

(d) **Chemoheterotrophs**, utilizing organic compounds as both energy and carbon sources. Included here are animals, protozoa, fungi and most bacteria.

Chemoterotrophs may be further subdivided according to the manner in which they satisfy their carbon and energy requirements, being called **osmotrophs**, if, like bacteria and fungi, they absorb organic substances in solution, and **phagotrophs** if, like many animals and protozoans, they ingest them as solid particles. Alternatively, subdivision of chemoheterotrophs may be made on the basis of whether they meet their needs by feeding upon other living organisms, in which case they are called **biotrophs**, or on dead organic matter, when they are known as **saprotrophs**.

As will become apparent in Chapter 3, predators are obviously biotrophs and so too are some parasites, namely those which derive their energy from living cells of the host plant. However, there are some parasitic fungi which kill the cells of their hosts before utilizing them as an energy source and it is desirable, from an ecological standpoint, to distinguish between these and saprotrophs in the strict sense, which are usually found on plant and animal tissues occurring as organic residues on or in the soil. This can be done by assigning such parasites to a third category, that of **necrotrophs**.

D. H. Lewis (1973) in a discussion of fungal nutrition, pointed out that much confusion and ambiguity had arisen from the practice, formerly common among plant pathologists, of using as a basis for classification the criterion of whether or not a fungal parasite could be grown in axenic culture. By abandoning this approach, and using instead a combination of ecological and nutritional behaviour, Lewis recognized five classes of organisms (Table 2.1). The behavioural attribute on which this classification is based is whether or not an organism occurs in association with another organism in a relationship involving intimate contact and some degree of permanency, that is, in symbiosis (sect. 3.5).

The five groups are as follows:

(a) **Obligate saprotrophs**. These have no capacity for entering symbiotic association and so always occur free living in nature. Saprophytic fungi and bacteria are typical.
(b) **Faculative necrotrophs**. These include those soil organisms whose nutrition is necrotrophic when behaving as parasites but which more often than not are free living i.e. saprotrophic. The so-called 'damping-off' fungi are found here.
(c) **Obligate necrotrophs**. This group includes specialized plant parasites whose saprotrophic phase is essentially restricted to survival in dead host tissues. Examples are the vascular wilt fungi and the wood-rotting basidiomycetes.
(d) **Facultative biotrophs**. These organisms have a biotrophic mode of nutrition when living symbiotically, but otherwise

Table 2.1 Categories of chemoheterotrophs based on nutritional and ecological behaviour

Ecological behaviour	*Nutritional behaviour*		
	Saprotrophic	*Necrotrophic*	*Biotrophic*
Obligate saprotrophs	+	–	–
Symbionts			
facultative	–	+	+
obligate	–	+	+

Source: After Lewis (1973). For definitions of the five classes of organisms recognized, see text.

occur as saprotrophs. The group includes facultative mycorrhizal fungi (ch. 7).

(e) **Obligate biotrophs**. These never occur free living in nature but only in symbiosis with other organisms. Many mycorrhizal fungi including most of the ectomycorrhizal and vesicular–arbuscular forms (Ch. 7) fall into this category.

2.2 Energy-yielding processes of chemotrophic organisms

Any substance which is oxidized by an organism to provide energy for metabolic processes is termed a **substrate**. By definition, substrates of chemoautotrophs are inorganic, while those of chemoheterotrophs are organic. Many biological oxidations take the following form:

$$AH_2 + B \rightarrow BH_2 + A$$

Such a reaction is representative of the type known as dehydrogenation, and involves the transfer of hydrogen atoms (and electrons) from the hydrogen donor, or substrate (AH_2), to the hydrogen acceptor (B). While the composition of the substrate will clearly influence the kinds of end-products, it is the nature of the hydrogen acceptor which characterizes the energy-yielding reactions of chemotrophic organisms. According to their specific hydrogen acceptors, three kinds of energy-yielding metabolism may be recognized (Table 2.2), but it is emphasized that the categories listed in the table do not provide a basis for dividing the living world into mutually exclusive groups. For example, many microorganisms respire aerobically in the presence of oxygen, but turn to anaerobic respiration or fermentation as an energy source when the oxygen supply is depleted. Nevertheless, since an organism's prime requirement is for energy, a classification based on the four major energy-yielding processes – **respiration, anaerobic respir-**

Table 2.2 The energy-yielding processes of chemotrophic microbes

Energy-yielding process	*Hydrogen acceptor*
Respiration	Molecular oxygen
Anaerobic respiration	Inorganic substance other than oxygen
Fermentation	Organic compound

ation, fermentation and **photosynthesis** – provides a useful framework for examining the ecological role of soil organisms (see, for example, Ch. 5.)

It should be noted that the term substrate has a particular meaning in biochemistry, as a chemically defined substance which reacts with a specific enzyme in the course of metabolism. It has however been traditionally used by microbial ecologists to describe any organic residue such as leaves, bark, wood, dung, etc. or the chemical components thereof, for example cellulose, lignin, sugars, amino acids and so on. Thus, in the field of soil ecology, the term substrate may be used in a narrow, almost biochemical, sense but it usually has a much wider connotation, the differing usages being discerned by the context. While this can be confusing, it seems preferable to using the latin 'substratum', which elsewhere connotes an inorganic layer, or substituting a term such as 'resource' (Swift *et al.*, 1979) which has a variety of meanings outside the field of ecology. In this book, 'substrate' is used in the broad sense to indicate a substance, whether structurally and chemically heterogeneous or homogeneous, during the decomposition of which energy is released by any of the processes listed in Table 2.2, and utilized by the decomposer organism for biosynthesis and growth.

Oxygen relationships: the significance of soil aeration

So far as microorganisms are concerned, access to oxygen may or may not be beneficial, and four categories can be distinguished on the basis of their reactions to molecular oxygen:

(a) **aerobes**, which grow only in the presence of oxygen and are completely dependent on respiration as a source of energy;
(b) **anaerobes**, which are inhibited or killed on access to oxygen, and so depend on fermentation or anaerobic respiration as energy sources;
(c) **microaerophiles**, which are obligate aerobes but which develop best at low oxygen tensions;
(d) **facultative anaerobes**, which are active under either aerobic or anaerobic conditions.

The four types of oxygen relationships are not found in all taxonomic groups of microbes. Most of these, like plants and animals, are aerobes, but yeasts and a few other fungi are facultative anaerobes, while among the bacteria all degrees of sensitivity to oxygen exist. In passing, it should be noted that while most photosynthetic organisms (plants, algae and cyanobacteria) are aerobes, the photosynthetic bacteria are anaerobes. It should also be appreciated that the oxygen relationships of microorganisms form a continuum, with obligate aerobes at one end and obligate anaerobes at the other.

As indicated in Chapter 1, the question of soil aeration cannot be separated from a consideration of soil moisture status, both depending in turn on soil structure. Many microbes (especially bacteria) exist in water films surrounding the soil peds, hence their access to oxygen depends on the solubility of that gas in water and its rate of diffusion in aqueous solution. Oxygen is only slightly soluble in water and if diffuses very slowly, with the result that anaerobic microsites are likely to exist temporarily even in well aerated soils. Because it is the rate of diffusion of oxygen to active microsites that limits aerobic processes, rather than a decline in O_2 concentration itself, the change from aerobic to anaerobic metabolism in facultative anaerobes, or the cessation of activity in obligate aerobes, occurs at soil oxygen levels much higher than those at which aerobic respiration can proceed *in vitro*.

2.3 Nutrient supply

An organism requires not only a source of energy, but must also find in its environment all the materials needed to build and maintain its cellular organization, that is to say, it must have available a supply of **nutrients**. Phagotrophs obtain all their nutrients from ingested materials, but many of the nutrients absorbed by osmotrophs enter the organisms more or less independently from the ambient solution. Two factors determine whether or not a given compound can be used as a nutrient by an osmotroph: first, the ability of the compound to penetrate the cytoplasmic membrane and enter the cell, and second, the ability of the organism to metabolize the substance once it enters. Molecules too large to penetrate the plasma membrane may however be utilized as nutrient sources if the organism can hydrolyse them to their constituent moieties by enzymatic action outside the cell (see sect. 2.4). The compounds which are taken up and metabolized by microorganisms in order to satisfy their requirements for various nutrients are discussed subsequently, but first it is necessary to consider briefly the principal sources of nutrients for the ecosystem as a whole, the manner in which those nutrients are distributed throughout the system, and the mechanisms whereby they come within the sphere of influence of the soil biota.

Sources of mineral nutrients

Mineral nutrients in the soil derive from atmospheric or geological sources. Atmospheric inputs are of several kinds, the most significant being bulk precipitation (including sedimentation or dry fallout into gauges between rain events), aerosol impaction on vegetation, and the fixation of gases by biological processes. Accessions of some elements in rainfall can be quite substantial especially in coastal regions or in areas subject to industrial pollution. High sulphur inputs originating from chimney smoke may lead to 'acid rain' which has become a major cause of environmental concern in highly industrialized countries. Biological nitrogen fixation by symbiotic or free-living soil microbes is the ultimate source of nitrogen for all organisms, since rocks contain virtually no nitrogen accessible to plants or microorganisms. Carbon too must be fixed biologically (mainly by photosynthesis) for much the same reason, and the atmosphere is also the primary source of oxygen used in aerobic respiration.

Most other inorganic nutrients derive ultimately from rocks, which are the parent material of soils. The products of rock weathering cannot, however, be considered as sources of nutrients for ecosystem development unless they are generated within the root zone of plants, for only then are they capable of being incorporated in the biomass (Gorham *et al.* 1979). Primary minerals vary in their susceptibility to weathering, and some of their soluble products may be leached from the root zone. The fate of phosphorus during soil formation is of particular interest because of its central role in biochemical processes. Forming highly insoluble complexes with iron and aluminium, it is very resistant to leaching and so the amount retained in soil closely reflects its concentration in parent material. This in turn may have a marked influence on the kind of vegetation which develops: the distribution patterns of mesic and sclerophyllous elements in the Australian vegetation have been attributed by N. C. W. Beadle (1954, 1966) to phosphorus deficiencies in the underlying bedrock.

The relative significance of atmospheric and geological nutrient sources depends on a number of factors (Bormann & Likens 1979; Charley & Richards 1983), including the stage of development of the system. During primary succession, when nutrients are being sequestered in the aggrading biomass, parent material sources are generally the more important, but with the passage of time the retention of nutrients already accumulated becomes critical for the persistence of the system and so there is likely to be an increasing dependence on atmospheric sources. The older and more deeply weathered a soil, the less its residue of primary minerals and the less the ability of root systems to acquire nutrients before they are leached from the profile. Ecosystems such as tropical lowland rainforests on deep soils derived from nutrient-poor parent materials may be virtually dependent on atmospheric inputs in order to replace the small losses to drainage waters which inevitably occur.

Nutrient movement from source to soil biota

Nitrogen and carbon are the only elements to reach the soil biota directly from their primary source, and then only if microbes capable of fixing atmospheric N_2 or CO_2 are present. Most carbon and nitrogen, and all other elements, become accessible to soil microorganisms via the vegetation subsystem.[1] This is true not only of those originating from parent material sources and absorbed by plant roots, but it also applies to aerosols impacted on foliage and to that fraction of precipitation intercepted by plants. Rain striking plant surfaces either drips onto the soil surface as 'throughfall' or is channelled to the ground by 'stemflow'. Compared to rainfall, both are enriched in nutrients leaked from plant tissues or derived from particulate matter deposited on plant surfaces by sedimentation or impaction.

The principal mechanisms involved in transferring nutrients from vegetation to soil are leaf consumption by herbivores (followed by their death or defaecation, leaching from the canopy, and litter fall together with root sloughing. While the contribution of each of the three pathways may be substantial, much greater quantities of nutrients are transferred from vegetation to soil in plant residues than by any other means. With few exceptions, litter inputs are greater by an order of magnitude than inputs from animals. Litter fall is particularly significant in forest ecosystems where it is a continuous source of energy and nutrients for chemoheterotrophic soil microbes. It is relatively less important as a nutrient transport mechanism in savanna and grassland ecosystems, and the same is probably true of throughfall and stemflow. On the other hand, root sloughing is likely to be a major pathway of plant–soil nutrient transfer in a wide range of ecosystems.

Nutrient distribution in soil

One of the most consistent features of soil is its anisotropic or nonrandom distribution of nutrients. Elements such as nitrogen and sulphur, which occur mainly as organic forms, usually show decreased concentration from surface to subsoil. Similar patterns are found with organic phosphorus and exchangeable calcium and potassium (Fig. 2.2). Horizontal patterns are also evident (Zinke 1962) and while these are often most pronounced in ecosystems dominated by long-lived woody perennials, that is in forests and

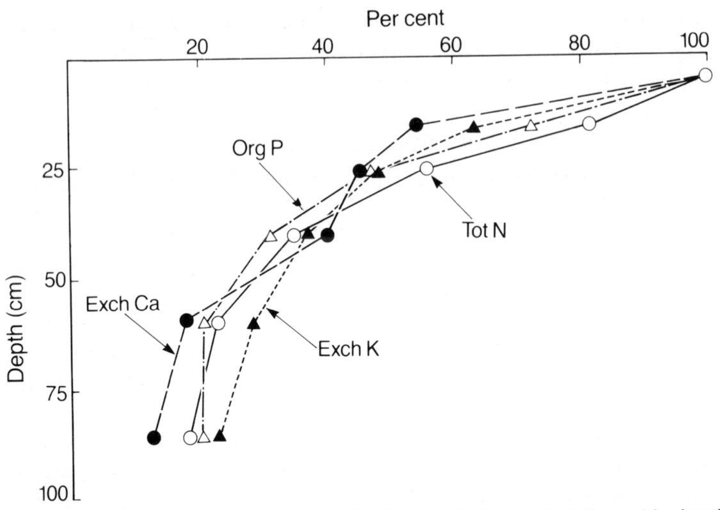

Fig. 2.2 Relative changes in soil chemical characteristics with depth in some Australian forest soils. Within profiles, values are plotted as percentages of the highest concentration recorded for the profile. (From Charley & Richards 1983)

shrublands, they have been discerned in grassland communities as well.

The patterns formed are related to the various pathways, described above, along which nutrients pass from vegetation to soil. Consequently, different plants tend to induce the formation of characteristic patterns within their respective zones of influence. Recognition and delineation of patterns in nutrient distribution have important implications for the study of functional processes in the soil ecosystem, stemming from the effects of energy and nutrient supply on microbial activity. This is treated in some detail in Chapters 4 and 5. We turn now to a discussion of where microbes obtain their main inorganic nutrients.

Inorganic nutrition of soil microbes

Carbon

Carbon is the element required by organisms in greatest amount. All photosynthetic forms can reduce atmospheric CO_2, but not all can use CO_2 as the sole source of carbon. In the photoheterotrophic non-sulphur purple bacteria, organic substances such as acetate and succinate act as hydrogen donors for CO_2 reduction. Chemoautotrophic microbes such as the nitrifying bacteria utilize inorganic substances exclusively, and must therefore use atmospheric CO_2

as their sole carbon source. Most chemotrophic forms however, oxidize organic compounds which serve not only as substrates in energy-yielding reactions, but also as sources of carbon for nutrition. The range of organic compounds from which carbon is extracted by chemoheterotrophic organisms is extremely wide. Microorganisms, as a group, are especially versatile in this respect: in fact, for every naturally occurring organic material there are presumed to be microbes capable of decomposing it. This is the reason why microorganisms play such a vital role in the geochemical cycling of the elements, especially carbon and nitrogen.

Carbohydrates are among the most readily available sources of carbon for microorganisms. Monosaccharides, especially hexoses, are widely used, but polyhydric alcohols such as mannitol and glycerol are often good carbon sources too, especially for fungi and actinomycetes. Organic acids of the tricarboxylic acid cycle cannot usually penetrate the cytoplasmic membrane, but amino acids are readily used as carbon sources by most microorganisms, and fatty acids can also be utilized by some. Hydrocarbons can serve as a carbon source for a few bacteria of the genera *Corynebacterium, Mycobacterium* and *Pseudomonas*. Utilization of aromatic compounds such as lignin may be quite extensive under aerobic conditions, but when oxygen is limiting such compounds are not decomposed and therefore accumulate, for example as peat and coal. The major lignin decomposers are fungi, notably some of the higher basidiomycetes, while among bacteria the ability to use aromatic compounds is often found in the pseudomonads and actinomycetes. A discussion of several important naturally occurring organic substrates will be found in Chapter 4.

Although the range of organic materials which can be metabolized by microorganisms in general is very great, some species have narrow preferences, others wide. Furthermore, there are certain synthetic carbon compounds which are relatively resistant to decomposition by microbes. These 'recalcitrant molecules', to use the phraseology of Martin Alexander (1971), usually derive from insecticides, herbicides or detergents, and constitute one of the major pollution problems of technological society (see Ch. 4). The study of their interactions with the soil microflora and fauna is the province of environmental microbiology.

Nitrogen

Relatively large amounts of this element are required for the synthesis of amino acids and proteins, purine and pyrimidine nucleotides, and certain vitamins. The nitrogen atom occurs in nature in a variety of oxidation states, each of which can be utilized by different microorganisms. The preferred state seems to be the ammonium ion, which is not surprising since it is in this form (NH_4^+) that nitrogen is incorporated into organic compounds. However, nitrate (NO_3^-) can be used by many algae and fungi, though not so extensively by bacteria. Certain bacteria including the cyanobacteria can use molecular nitrogen (N_2); these will be discussed in Chapter 8. Organic nitrogen compounds are used as sources of nitrogen by those organisms which decompose them to produce ammonia. Many microorganisms have a particular capacity to do this, a single amino acid often serving as their sole source of nitrogen.

Phosphorus

Phosphorus occurs in living organisms chiefly as sugar phosphates in nucleotides and nucleic acids, and as phytates and phospholipids. Phytates are phosphate esters of inositol (hexahydroxycyclohexane) which occur widely in living organisms. Myoinositol hexaphosphate (phytin) is the only isomer found in plants and animals but a variety of other forms is synthesized by microorganisms. Phospholipids are essential components of cell membranes. They are substituted glycerols related to triglycerides but differing in containing phosphorus; some also contain inositol. Phosphorus, usually as inorganic phosphate, therefore needs to be provided in considerable amount if an organism is to grow. Most of this phosphate is of mineral origin but some is derived from the enzymatic breakdown of inositol hexaphosphates in soil organic matter by phytase-producing microbes (see sect. 5.2.1).

Sulphur

Sulphur is present in organisms as the sulphydryl (–SH) group of the amino acid cysteine, and most other sulphur compounds in the cell, such as the amino acid methionine and the vitamins biotin and thiamine, derive from cysteine. Most microorganisms, like plants, absorb sulphur as sulphate (SO_4^{2-}) and must reduce it to sulphydryl. Thiosulphate ($S_2O_3^{2-}$) can also serve as the sole source of sulphur for many microorganisms. A few microbes are unable to reduce sulphate or thiosulphate, however, and require reduced sulphur compounds as nutrients, such as hydrogen sulphide or cysteine. Such sources are quite distinct from those compounds which provide sulphur for any microorganism that is capable of oxidizing them to sulphate, during the mineralization of soil organic matter (see Ch. 5).

Other mineral elements

Unlike animals and plants, which have additional requirements for certain minerals to build or repair structural tissues, microbes require the remaining elements mainly as activators of various enzymes. The macronutrient elements potassium, magnesium and calcium are required in concentrations of about 10^{-3} to 10^{-4} M. Micronutrients are needed in much lower concentrations (10^{-6} to 10^{-8} M). Because of the technical problems involved, it is extremely difficult to demonstrate a micronutrient deficiency in microorganisms, though the need has been shown for manganese, iron, zinc and copper. The micronutrient requirements of most microbes are in fact so low that it is considered unnecessary to incorporate these elements in microbiological media used for cultivating microorganisms in the laboratory; they are normally present in adequate concentration as impurities in other constituents of the media. Notwithstanding this generality, some fungi are sufficiently sensitive to the concentration of micronutrients in the environment to enable them to be used as indicators in the bioassay of micronutrient deficiencies in soils. For example, copper is needed for the production of the brownish-black melanin pigments found in the conidia of *Aspergillus niger*, so that if soil samples are added to copper-free nutrient media which are inoculated with this fungus the copper content of the soil can be determined by assessing the degree of pigmentation of these spores.

Organic nutrients: growth factors

Many organisms require certain organic compounds in addition to those needed as carbon and energy sources. Such nutrients are called **growth factors**, and include amino acids, purines, pyrimidines and vitamins. These are specific nutrients only in the sense that the organisms which require them lack the ability to produce them; an organism which requires a particular growth factor is said to be auxotrophic for that substance. Those organisms which have no requirement for growth factors must of necessity possess all the enzyme systems needed to synthesize, during the normal course of metabolism, all of the organic compounds they require for biosynthesis and growth. Some microorganisms are more demanding than others for growth factors: lactic acid bacteria and protozoa are among the most fastidious. A growth factor may be essential or stimulatory, depending on whether the organism requiring it is completely unable to synthesize it or is able to synthesize small amounts but not enough to satisfy all its metabolic needs.

Any of the twenty or so amino acids found in proteins may be required as a growth factor by microorganisms; one strain of the bacterium *Leuconostoc mesenteroides* requires no fewer than seventeen amino acids for growth. Purine and pyrimidine requirements are most commonly encountered among protozoa and in certain bacteria, especially lactobacilli. Protozoa and the lactic acid bacteria are also the most exacting microbes for vitamins, and may require up to five or six separate vitamins for growth. Some root-infecting fungi, including the mycorrhizal basidiomycetes (Ch. 7), have a requirement for thiamine (vitamin B_1) and biotin, while among the algae and phytoflagellates a requirement for vitamin B_{12} is not uncommon.

2.4 Nutrient uptake

All cells continually exchange materials with their environment yet cell composition remains relatively constant. Osmotrophs absorb nutrients from the surrounding medium and release excretory products. These exchanges are specific in that only certain substances move in and certain ones out. The reason for this selectivity lies in the nature of the plasma membranes. In bacteria, it is primarily the cytoplasmic membrane that is involved, but eukaryotic cells have intracellular selective membranes in addition to that bounding the cytoplasm. Cell membranes are principally lipoprotein, and have both hydrophilic and lipophilic properties. They have the important property of being semipermeable, that is they are able to keep out solute molecules except in certain circumstances.

There is no evidence that pores exist in the membrane, except for those small enough to permit water and a few other molecules of like size to pass, hence some mechanism of penetration other than simple diffusion must exist. In this regard the membrane contrasts with the cell wall itself, which is relatively porous and does not appreciably restrict the entry of materials. This is not to say that the cell walls of plants and microbes are freely permeable to all solutes, but their permeability is greater, and their selectivity is much less than that of the plasma membrane. Because water can pass freely through the cytoplasmic membrane, it is taken up by osmosis. In organisms with rigid cell walls it is the rigidity of the wall itself which eventually prevents further entry of water. Organisms which lack rigid cell walls are restricted to a limited osmotic environment, unless they possess some mechanism for removal of excess water, such as the contractile vacuoles found in some protozoa.

For the uptake of solutes, the plasma membrane is not only a semipermeable barrier but a vectorial one, the direction of flow of solutes being as important as their nature. Thus cells can take up K^+ faster than the rate at which it can leak out, resulting in the internal concentration of potassium ions being greatly in excess of their concentration in the external medium. In other words, the vectorial property of the membrane permits movement of ions against an electric potential gradient, or of non-electrolytes against a concentration gradient, giving rise to the phenomenon known as accumulation, that is the accumulation of solutes within cells at concentrations greatly in excess of their concentration in the surrounding medium. Not all materials accumulate against a gradient, however. Some, like urea and glycerol, pass in by **passive diffusion**, the rate of transport being proportional to the concentration gradient, the size of the molecules and their solubility in lipids. None the less, only a very small proportion of the total solutes enter thus. Most substances pass across the plasma membrane by means of specific carrier transport mechanisms. Three types of carrier-mediated solute transport are recognised. The simplest is **facilitated diffusion** in which the driving force is (for non-electrolytes) the difference in concentration or (for electrolytes) the difference in electrical potential across the membrane. With this form of transport, the uptake process necessarily ceases once the concentration of electrical potential inside the cell equilibrates with that outside; accumulation does not occur. If, however, facilitated diffusion is coupled to the release of metabolic energy, then solutes may accumulate within cells by a process known as **active transport**. The third kind of carrier-mediated transport is called **group translocation**, and it differs from the other two in that the solute is chemically modified to effect its movement across the membrane: the carrier molecules behave like enzymes in that they catalyse group transfer reactions by using the solute as a substrate. Since the product inside the cell is different from the external substrate, no concentration gradient develops across the membrane.

Various sugars, including glucose, mannose and fructose, are taken up by group translocation. Some purines, pyrimidines and fatty acids may perhaps be taken up in this way as well, but probably no amino acids. The latter, together with other sugars, and most inorganic ions, appear to enter the cell by active transport. These carrier-mediated transport processes are often the rate-limiting step in the metabolism of carbon and energy sources (Rose 1976).

Nature of the uptake process

Most hypotheses advanced to explain carrier-mediated transport involve the reversible binding of solutes to a proteinaceous constituent of the cytoplasmic membrane known as a permease. The use of this term is misleading since these transport proteins are not enzymes.

The mechanism by which a solute molecule, once it has combined with its carrier, is transported from one side of the plasma membrane to the other (a distance of 7.5–10 nm) is not known for certain. Thermal energy and the molecular deformation that occurs during binding could account for facilitated diffusion down a concentration gradient, but other mechanisms must be sought to explain active transport. Several hypotheses have been advanced. One proposes that a conformation change, brought about by the expenditure of metabolic energy, lowers the affinity of the carrier for the solute molecule at the inner surface of the membrane, resulting in its release into the interior of the cell. A second proposal involves the mediation of dehydrogenases, whereby transport proteins are subject to cyclic oxidation and reduction, during which the carrier has alternately high or low affinity for the solute. The reduced or low affinity form of the carrier releases the solute and is itself reoxidized by electron transfer in the normal course of respiratory metabolism. A third hypothesis assumes that the release of metabolic energy is linked to solute transport by the extrusion of protons into the cell from the end of a respiratory oxidation chain arranged across the membrane. This leads to a different in pH and electrical potential which causes an influx of protons accompanied by, on appropriate carriers, neutral molecules including sugars and amino acids, or anions such as phosphate. For a more detailed account of active transport mechanisms, the reader is referred to Rose (1976) and Brock (1979).

Role of exoenzymes

Molecules too large to enter microbial cells directly by diffusion or permease action may be digested extracellularly by enzymes secreted into the medium (Fig 2.3). These are known as extracellular enzymes or **exoenzymes**. They are mostly hydrolases, that is they catalyse the hydrolysis of high molecular weight molecules such as polysaccharides, proteins and lipids into their constituent subunits, which are then taken into the cell by passive or active processes. Polysaccharases include α -amylases, cellulases and chitinases, among others. The α -amylases hydrolyse the polysaccharides starch and glycogen into smaller glucose polymers known as dextrins and eventually into glucose, while the group of exoenzymes known as cellulases is responsible for the degradation of cellulose (see sect. 4.1); both enzyme complexes are commonly produced by saprotrophic fungi. Extracellular proteases and peptidases, which promote the hydrolysis of proteins (proteolysis), are typical of bacteria such as *Proteus, Clostridium* and *Pseudomonas*, and fungi such as *Aspergillus* and *Penicillium*. The decomposition of complex organic residues in soil is considered in more detail in Chapter 4.

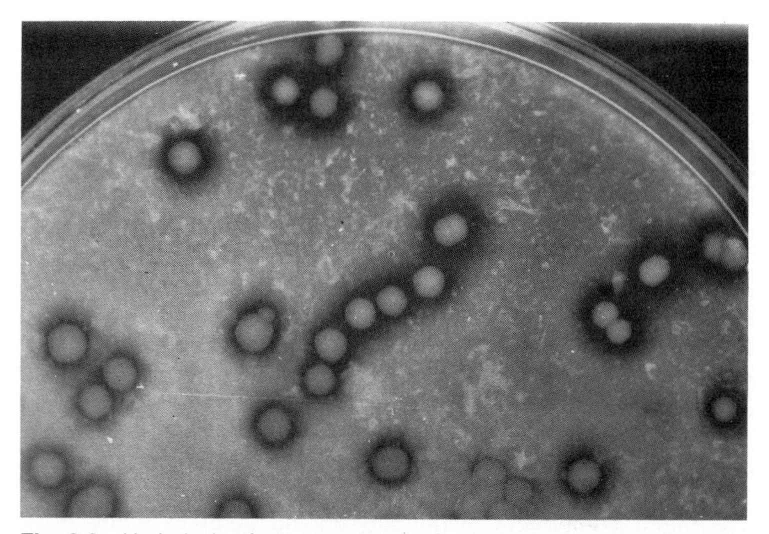

Fig. 2.3 Hydrolysis of macromolecules by exoenzymes. Milk agar plate inoculated with *Bacillus subtilis*, the clear zones around the colonies indicating hydrolysis of the milk protein, casein.

Not all exoenzymes diffuse into the environment, some remaining bound to the surface of the plasma membrane or held in the cell wall. Many are inducible, in other words they are often synthesized only in the presence of an inducer which is usually (though not necessarily) the substrate for the particular enzyme involved.

Soil enzyme activity

That soils exhibit enzymatic properties has been known since the early years of this century, and more than fifty different enzymes may be involved. R. G. Burns pointed out in 1977 that the total enzyme activity comprises:

(a) an intracellular component;
(b) an extracellular fraction which is stabilized by soil colloids;
(c) an extracellular component which comprises exoenzymes excreted by living cells together with formerly intracellular enzymes released following lysis of dead cells.

In some soils the intracellular fraction (a) is dominant while in others, components (b) and (c) account for 50–90 per cent of the total. As indicated previously, extracellular enzyme activity must of necessity be responsible for much of the initial decay of organic polymers in the soil, since these molecules are too large and too insoluble to be absorbed by microbial cells. The degradation of complex structural polysaccharides by exoenzymes is discussed in Chapter 4.

The efficacy of digestion by enzymes outside the cell is a function of their longevity and rate of replenishment. Free enzymes in the soil solution, or attached to the surfaces of living microbes, are subject to proteolysis by other extracellular enzymes, or may be used as energy and nutrient sources by microorganisms, or may be denatured by physicochemical factors in the soil environment. It appears however, that some retain their catalytic properties for long periods of time. For example, J. Skujins and A. D. McLaren (1968) reported finding active urease and phosphatase in permafrost soils which had been preserved in their present state for 9000 years. The reasons for this persistence lie in the protection afforded by the clay-organic complex (sect. 1.4). Without impairing their function, enzymes can be adsorbed both on and within clay lattices, or be trapped within colloidal organic matter – humus – during its formation, or be adsorbed on or chemically combined with humus. The humic component of the clay-organic complex seems to be primarily responsible for the stabilizing effect of soil colloids on extracellular enzymes.

Ingestion

Ingestion is the usual means of food intake among the higher animals. It also occurs in amoebae, which are thereby able to transport solutes, and even solid particles such as bacteria, into the interior of their cells. In these protists, the process involves the pinching off of small vesicles from the end of a deep invagination in the plasma membrane. The contents of the vesicle are subsequently digested by intracellular enzymes, the mechanism being known as **pinocytosis** when solutions only are taken in, and **phagocytosis** when solid particles enter; both are encompassed by the more general term **endocytosis**. It is not known whether the process occurs in other microorganisms, but it has been detected in mammalian cells. Endocytosis may be regarded as active uptake since presumably metabolic energy is expended in the process of invagination. However, it is a relatively non-selective mechanism for effecting the entry of solutes.

2.5 Notes

1. There is evidence that some bacteria and fungi can dissolve primary minerals under certain circumstances, in which event they might gain direct access to the chemical elements released. See sects. 5.2.1 and 6.5.

3

Pattern and process in the soil ecosystem

Soil organisms occur in a wide array of habitats, the diversity in habitat being brought about by variation in such soil properties as moisture, aeration, temperature, pH, and nutrient or food supply. No one species of organism is able to grow in every kind of habitat. For every environmental factor, there is a minimum level below which the organism will not grow at all, an optimum level at which growth is best, and a maximum level above which again no growth occurs. The limits between which the species makes good growth define its **ecological tolerance** for that factor, and any factor which tends to slow down the growth of the organism is referred to as a **limiting factor.**

The concept of limiting factors was first formulated in 1840 by the German chemist Justus von Liebig, who observed that the growth of crops was restricted by whatever essential nutrient was in short supply. Liebig's 'law of the minimum', as it became known, was later extended by ecologists to include the limiting effect of the maximum, that is to say, an excess of any particular factor can limit growth as well as a deficiency. Since to make good growth, an organism requires a source of energy together with a supply of nutrients and an appropriate physical environment, any of these things is a potential limiting factor. It should also be recognized that the factor in short supply may affect the requirement for another which is not itself limiting. E. P. Odum (1971) has therefore restated Liebig's concept of limiting factors as follows: the success of a population or community depends on a complex of interacting factors; any factor which approaches or exceeds the limits of tolerance for the organism or group concerned may be regarded as a limiting factor.

Figure 3.1 illustrates the interaction of two limiting nutrients on the growth of a coniferous tree. Yield, which is represented by the surface of the three-dimensional figure, depends on an optimum

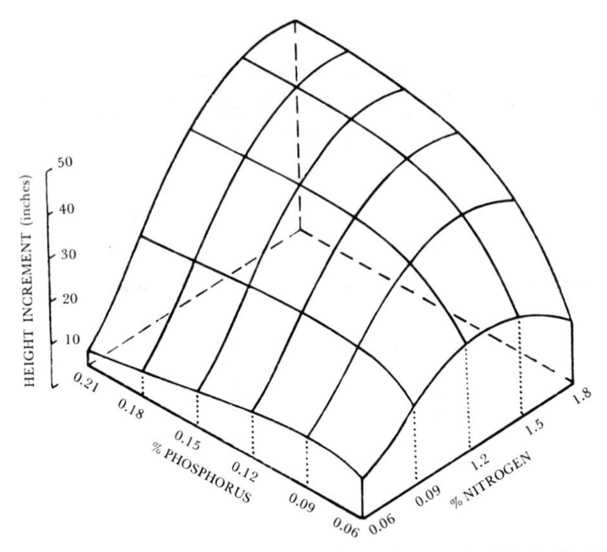

Fig. 3.1 The interaction of limiting factors, illustrated by the effect of nitrogen and phosphorus on the growth of *Araucaria cunninghamii*. Note how the response to increasing foliar concentrations of either nutrient depends on the level of the other. (From Richards & Bevege 1969).

combination of the two nutrients: when either is grossly limiting, increases in the other have little or no effect. One could, in theory, depict the effect of all limiting factors as a multidimensional figure, the several factors varying along horizontal axes radiating from a common origin, and the yield (or some other measure of productivity) along a vertical axis. The surface of this figure would be a gently sloping plateau representing an optimum combination of all factors, and falling increasingly steeply in every direction as the various factors become more and more limiting.

As indicated in Chapter 1, the action of a limiting factor may be depicted in energy circuit diagrams by the work gate module. The interaction of limiting factors can also be illustrated by this means. In the statistical sense, two factors may be said to interact when the response to both together is significantly different from the algebraic sum of their individual effects: if the combined response is greater than the sum of the separate responses then the interaction is said to be positive; if less then it is termed negative. Figure 3.2 illustrates a positive interaction of two soil amendments on the population size of heterotrophic bacteria. The native population density was 5.2×10^6 cells g^{-1}. Supplying additional substrate in the form of finely macerated pine needles caused an increase of

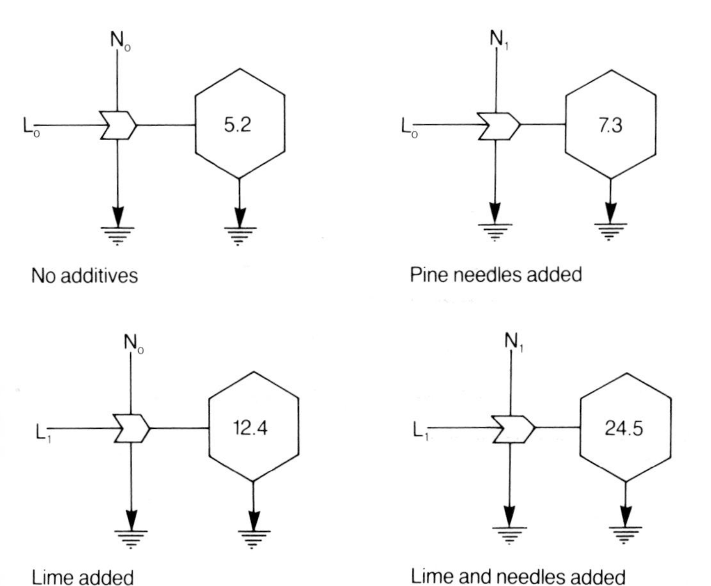

Fig. 3.2 The interaction of limiting factors, illustrated by energy circuit modules. The diagrams show the effect of lime and/or macerated pine needles on the population of heterotrophic bacteria in a sandy podzolic soil incubated in the laboratory. Values given are numbers of bacteria per gram $\times 10^6$ as estimated by plate counts. (Based on the data of Jones & Richards 1977.)

2.1×10^6 cells g^{-1}, while adding lime to give a more favourable (less acid) environment allowed the population to rise by 7.2×10^6 g^{-1}. When both limiting factors were removed by incorporating lime and pine needles together however, the population density rose by 19.3×10^6. This is more than twice what might have been expected from a consideration of the additive effect of the individual factors. A comprehensive account of limiting factors for soil microbes is given by Dommergues *et al.* (1978).

3.1 Limiting factors in relation to ecosystem development

One way of analysing ecosystems, and of studying their evolution and development, is the state factor approach of Hans Jenny (1961), which has its origins in the work of the Russian pedologist, V. V. Dokuchaev, and which is based on the principle of interacting limiting factors. According to this viewpoint, the state of the system at any point in time is defined by a certain number of variables, or **state factors.**

Since ecosystems are open systems, throughout the course of their development they are subject to gains and losses of matter and energy in the form of light, heat, gases, water, solutes, particulates and migrating organisms. Starting at time zero, the value of any attribute of an ecosystem, l, at time t is:

$$l_t = l_o + \sum_{1}^{t} (l_{in} - l_{out}) \qquad (3.1)$$

where $(l_{1n} - l_{out})$ is the annual net change in the property which may vary from year to year.

According to Jenny (1961), an understanding of ecosystem development may be derived by considering potential flux gradients across the system boundary. The flux of matter and energy, i.e. l_{in} and l_{out}, per unit area of boundary layer of thickness $\triangle x$ permeability m, is given by:

$$l_{in}, l_{out} = \frac{(P_{outside} - P_{inside})}{\triangle x} m \qquad (3.2)$$

where $(P_{outside} - P_{inside})$ is the potential gradient or difference in intensity of the factor over the distance $\triangle x$, that is outside and inside the system boundary. P_{inside} and m are characteristics of the system itself and vary with the state of the system, whereas $P_{outside}$ is an environmental property. Designating the latter as the external potential, P_x we may write:

$$l_{in}, l_{out} = f\ (P_x, \textit{state of system}). \qquad (3.3)$$

Hence $(l_{in} - l_{out})$ is also a function of P_x and the state of the system. If P_x is constant, or a function of time, then the state of the system at any time is uniquely determined by its age, so that we may rewrite equation (3.1) as follows:

$$l_t = l_o + \sum_{1}^{t} f\ (P_x, t). \qquad (3.4)$$

Thus, in general terms, any measurable property, l, of the system is a function of (a) the amount of that property or material present at time zero, l_0, i.e. when system development begins; (b) the external flux potentials, P_x, which determine gains from and losses to the environment of energy and matter; and (c) the time, t, or the age of the system from time zero. Expressed mathematically, this becomes:

$$l = f(l_o, P_x, t). \qquad (3.5)$$

When analysing ecosystem development, the state factors l_o and P_x are usually partitioned into more specific components. The factor l_o is made up of soil parent material, p, and topography or relief, r. External potentials, P_x, are the properties of the environment, such as the regional climate, cl, and the biotic factor, o, the latter consisting of all the organisms which were available for colonization of the site, or already present, at time zero. Having partitioned the state factors we can rewrite eq. (3.5) as follows:

$$l = f(cl, o, r, p, t). \qquad (3.6)$$

Equation (3.6) is the general **state factor equation**, and it applies not only to the total system but to various subsystems as well. Thus we may regard l as a composite entity being the product of three interacting subsystems – the soil, s, vegetation, v, and animal, z, subsystems. These three components of the total system are mutually interdependent, so that a particular property of the soil subsystem, for example, is determined not only by its own external potentials but also indirectly by many others which directly affect the vegetation or animal subsystems. We may, therefore, extend the general state factor equation as follows:

$$l, v, z, s = f(cl, o, r, p, t). \qquad (3.7)$$

The influence of any state factor, F, on the development of a given ecosystem property, for example soil nitrogen, N, is given by the expression

$$\int_a^b \frac{\partial N}{\partial F} \, dF \qquad (3.8)$$

where a and b are the extremes of the range of F, and the slope of the curve ($\partial N / \partial F$) a measure of its effectiveness. Thus the contribution of any of the state factors to the property under consideration will be small or negligible if either the range of the factor is narrow or its effectiveness small. When these conditions hold for all but one of the state factors, the property in question is essentially a function of that factor. In this way, Jenny was able to show that soil nitrogen content was a function of climate by studying 'climofunctions', this is, a range of climates in situations where parent material, topography and the biotic factor were relatively constant. In this instance, eq. (3.7) becomes $s = f(cl)_{o,r,p,t}$. In the same way, we could determine the effect of parent material on any soil property by studying a 'lithofunction' such that $s = f(p)_{cl,o,r,t}$, or the effect of relief by examining a 'topofunction' that is $s = f(r)_{cl,o,p,t}$.

One conceptual difficulty inherent in state factor analysis is the fact that a term for organisms appears on both sides of eq. (3.7).

This difficulty may be resolved by regarding those biotic factors which existed prior to time zero as independent components, while those which have developed since are considered to be dependent variables. In practice, the regional flora and fauna is assumed to have preceded time zero. For a further treatment of the organism factor in state factor analysis, the reader is referred to a paper by R. L. Crocker (1952).

Some ecologists question the validity of this 'functional, factorial' approach to ecosystem analysis, arguing that it is impossible to distinguish between dependent and independent variables when most, if not all, are in any event interrelated. It is further argued that the response of an ecosystem to variations in environmental factors is difficult, if not impossible, to interpret since the environment itself is part of the ecosystem. Such argument fails to take account of the fact that some environmental factors, namely those within the system boundary, can be clearly recognized as dependent variables, as distinct from those which operate at or beyond the boundary and which are therefore independent. It also fails to recognize the full extent of the potential contribution which the general systems approach can make to the study of ecosystems. The techniques of systems analysis will permit the identification of those variables which are important in ecosystem function, and lead to the building of models which may be subjected to rigorous mathematical testing.

3.2 The biotic factor: succession

The development of ecosystems involves the phenomenon of **ecological succession**. Succession is one of the oldest, most studied, yet still widely debated ecological concepts. The traditional view, which owes its origin to F. E. Clements (Clements 1936), is that of an orderly process of community change resulting from modification of the physical environment by organisms, and culminating in the system attaining a steady state, or **climax**. This concept is fundamental to the strategy of ecosystem development as envisaged by E. P. Odum (1969). It emphasizes the importance of biological attributes in controlling succession, rather than physical factors. The physical environment sets limits as to how far development may proceed, and also influences the pattern and rate of change, but it is the biotic factor which is generally seen to be of overriding significance.

Although adherents of this view of succession see autogenic (within system) processes predominating in succession, the interplay between biotic and environmental factors is such that external (allogenic) processes sometimes become dominant. As indicated by Vitousek and Reiners (1975), many observations point to the importance of recurrent disturbance in determining ecosystem development which is reflected in the structure and floristic composition of plant communities. In a detailed study of temporal and spatial variation in the subalpine fir forests of New England in the United States, Reiners and Lang (1979) found evidence to support A. S. Watt's classic concept of pattern and process in plant communities, leading to a situation which Watt (1947) described as the 'phasic equilibrium', and which Bormann and Likens (1979) have termed the **shifting mosaic steady state** (see Ch. 1). The corollary to this view of community dynamics is that succession must be seen as part of a cyclic or repetitive process, rather than the unidirectional one of the Clementsian model.

Reiners and Lang recognized an hierarchical set of overlapping patterns in subalpine vegetation which were the result of exogenous and endogenous factors. Environmental gradients established first order patterns in the landscape, and superimposed on these were second order patterns, produced by hurricanes and avalanches. Still more subtle patterns are generated by the endogenous processes of growth, senescence, death and decay. Disturbance, according to P. S. White (1979), should not be seen simply as an allogenic or exogenous process of vegetation change. Some disturbances, such as tree fall following death or senescence, are effectively generated from within the system, that is, they are autogenic, and the distinction between these and external events such as storm blowdown is one of scale rather than kind. Disturbance and 'cyclic successions' are seen as complementary facets of community development and, in this context, the widespread nature of disturbance makes the distinction between allogenic and autogenic models of succession arbitrary, and the concept of climax nebulous.

The failure of classical succession theory to describe vegetation development adequately in all situations has led to more widespread acceptance of a population-oriented approach to the study of succession. This approach emphasizes the importance of individual rather than community properties. It sees the availability of particular species at an exposed site – the organism factor of Jenny (1961) – and their ability to establish and grow at that site as the key features of either primary or secondary succession (Noble & Slatyer, 1981). Replacement strategies are determined by the life-history characteristics of component species, and J. H. Connell and R. O. Slatyer (1977) envisage three such successional pathways:

(a) the facilitation pathway (succession *sensu* Clements) in which early occupants facilitate the entry of successive suites of species;
(b) the tolerance pathway, where late successional species with the ability to tolerate competition, and to survive at a low level of resources, are successful irrespective of whether or not other species have preceded them;
(c) the inhibition pathway, where later species are inhibited from developing by early successional species.

Mechanisms of microbial succession

When considering microbial population dynamics, the distinction between allogenic and autogenic successions, though arbitary, may still be useful. Thus allogenic succession of soil microbes may be observed as occurring in close association with the early developmental stages of an apparently autogenic plant succession. In this situation, external factors such as plant roots and the deposition of litter (plant residues) on the soil surface, control the pattern of microbial succession. Nevertheless, the biotic factor plays a major role in many successions involving chemoheterotrophic soil organisms and to this extent, these successions might be seen as primarily autogenic. Such is the case during the colonization of specific substrates, be they simple or heterogeneous.

Succession in the soil subsystem differs in one important respect from that in the vegetation subsystem. In autogenic plant succession, there is a progressive increase in organic matter and the amount of energy stored in the system as it matures from the pioneer stages towards the climax. In contrast, the succession of heterotrophic microbes in the soil ecosystem leads to a progressive depletion of energy sources (plant and animal residues and exudates), so that as S. D. Garrett (1946) has expressed it, 'the end-point of the microbial succession is not a climax association, but zero'. In so far as the final product of organic matter decomposition in soil, namely humus, is a more or less permanent constituent of the soil ecosystem (see Ch. 4), the theoretical zero end-point will never be reached. Rather, the climax microbial community in the soil is represented by what is termed the 'autochthonous microflora'; this is discussed later (sect. 4.1) in the context of organic matter decomposition.

Heterotrophic succession may also be differentiated from autotrophic succession on the basis of system energetics. In the developmental stages of an autogenic succession, the gross production/community respiration (*P*/*R*) ratio exceeds one, whereas in a heterotrophic succession the (P/R) ratio is much less than one initially. In both kinds of succession however, the P/R ratio approaches unity as the systems mature.

The apparent differences between microbial and plant successions do not necessarily reflect different mechanisms. As pointed out by J. C. Frankland (1981), any of the three pathways described by Connell and Slatyer (1977) may apply. The **facilitation** model is appropriate when physical access to the substrate is improved by early colonizers, as for instance when comminution and processing of plant residues by the macrofauna and mesofauna facilitate colonization by the soil microflora, or when early colonizers break down heterogeneous materials to provide more homogeneous substrates for other microorganisms. Further examples are discussed under the heading of commensalism (sect. 3.5). The **tolerance** model could apply to microbes capable of growing and reproducing at low levels of substrate availability, such as the wood-decomposing basidiomycetes. These fungi can grow in substrates deficient in available nitrogen because they are able to translocate and reuse the nitrogen released by autolysis of older parts of their own mycelia. The tolerance model might also explain many replacement sequences culminating in the autochthonous microflora, that is in populations which utilize humic matter highly resistant to decomposition, but evidence for this hypothesis is lacking. The **inhibition** pathway would account for successional patterns observed when antibiotic-producing microbes are among the early arrivals. While this situation frequently applies when readily decomposable substrates become available for colonization, there remains some doubt about the efficacy of antibiotics in nature. A discussion of the potential role of antibiosis in determining pattern and process in the soil ecosystem will be found later in this chapter, under the heading of antagonism.

Finally, it is pertinent to ask how the state factor approach applies to microbial successions. The answer depends on the level of analysis used. The general state factor equation (3.7) applies to the soil ecosystem as a whole. This equation however, is based on the premise that energy is not limiting, which is true for the general case since the soil–plant–animal system as a whole derives its energy from the sun. However, as we have seen in Chapter 2, energy supply is normally the most important limiting factor for the chemoheterotrophic soil microflora, so that if one is considering the development of microbial subsystems on particular substrates within the soil, then an energy factor must be included in the general equation before it can be applied. We must also allow for the fact that metabolic processes other than respiration are used by many microorganisms for biosynthesis and growth, so that oxygen supply may have a marked effect on microbial succession. On the other hand, at this level of analysis, regional climate, relief and parent material are constant, so that the development of a microbial ecosystem on any given substrate may be depicted, in functional, factorial terms, thus:

$$m = f(substrate,\ oxygen,\ o,t)_{cl,r,p} \qquad (3.8)$$

The same replacement sequence may occur repeatedly on identical substrates, and in this sense it resembles the cyclical process of development envisaged by A. S. Watt.

3.3 Microbial succession in soil

Succession in the soil ecosystem is epitomized in a schema proposed by S. D. Garrett (1951) for the colonization of plant tissues by soil fungi. According to this hypothetical sequence, moribund tissues of plants are normally colonized by weak parasites before they fall to the ground. Upon reaching the ground, they are invaded first by saprophytic 'sugar fungi' which utilize sugars and other carbon compounds simpler than cellulose. These fungi, typically certain phycomycetes and imperfect fungi, are characterized by the capacity for rapid spore germination and a high mycelial growth rate, which give them a competitive advantage over slower growing species. This advantage is further enhanced by the fact that some of them produce metabolic waste products and antibiotics (see sect. 3.5) which have an inhibitory effect on other fungi. Following on after the primary sugar fungi are the cellulose decomposers together with associated secondary sugar fungi. The cellulose decomposers, mainly ascomycetes, imperfect fungi and some basidiomycetes, are quite capable of utilizing simpler carbon compounds such as sugars, but are usually denied the opportunity of developing in nature until such substrates have been exhausted. The secondary sugar fungi likewise cannot compete successfully with the primary colonizers for the sugars present in recently fallen leaves, and must therefore share the breakdown products of cellulose hydrolysis with the cellulose decomposers. The final stage of Garrett's hypothetical fungal succession is represented by the lignin-decomposing basidiomycetes.

H. J. Hudson (1968) refined Garrett's generalized scheme of fungal succession on plant residues lying on the soil surface. Initial colonization of senescent leaves is by a group of ascomycetes and imperfect fungi (*Alternaria, Aerobasidium, Botrytis, Cladosporium* and *Epicoccum* spp.) many of which may already have been present on mature leaves as regular constituents of the phylloplane (leaf surface) microflora or as facultative necrotrophs. These 'common primary saprophytes' are ubiquitous in distribution but they may be accompanied by 'restricted primary saprophytes' which are not so cosmopolitan but tend to be confined to particular hosts. Subsequently, a group of 'secondary saprophytes', comprising ascomycetes, imperfect fungi and basidiomycetes, invades the decomposing litter, followed by the 'soil inhabitants' such as the zygomycetes *Mucor, Absidia, Rhizopus* and *Zygorhynchus*, and the ascomycete *Penicillium*. The soil inhabitants participate late in the succession as secondary sugar fungi (*sensu* Garrett) associated with cellulose and lignin decomposers.

A somewhat similar succession has been postulated by W. A. Kreutzer (1965) for the reinfestation of soil which has been 'partially sterilized' by treatment with a strong chemical fumigant, such as chloropicrin. Four intergrading seral, i.e. successional stages are visualized: *Stage 1* – Fast growing fungi (phycomycetes and imperfect species) invade the treated zone from without, while actinomycetes and various bacteria arise from resistant residues (conidia, endospores) within the treated zone. These initial invaders utilize simple substrates such as sugars and amino acids. *Stage 2* – With the simpler substrates largely exhausted, micro-organisms capable of hydrolysing complex carbohydrates make their appearance. Imperfect fungi still dominate, but ascomycetes and basidiomycetes also develop, together with other cellulose-degrading organisms such as actinomycetes, bacteria and myxobacteria. *Stage 3* – When the only remaining energy sources in non-rhizosphere soil are lignin and other complex materials of biological origin, basidiomycetes begin to dominate together with a few autochthonous bacteria. Most of the simple 'sugar-organisms' and cellulose decomposers are now limited to the rhizospheres of developing plants. *Stage 4* – After most of the potential energy sources originally available have been depleted, the micro-organisms in non-rhizosphere soil either grow very slowly, utilizing humic materials, attacking and parasitizing one another, or else enter a dormant phase. The fumigated soil is now ecologically indistinct from the surrounding untreated soil. The only regions of available free energy left are in the rhizospheres of young plants, in sloughed-off root cells, or in decaying rootlets.

Experimental studies of heterotrophic succession

Garrett's generalized hypothesis of a nutritional sequence of fungi on plant residues was first put forward in 1951, and dominated mycological thinking on fungal succession for two decades. As evidence accumulated, however, it became apparent that the relationship between substrate composition and physiological groups of fungi is not nearly so clear cut. To illustrate, many of the common primary saprophytes are cellulose decomposers, so that depletion of sugars, that is exhaustion of substrate, is unlikely to be the cause of their replacement by secondary saprophytes (Hudson 1968). Moreover, many members of the latter group are also cellulose decomposers, yet they too are often replaced by other species long before the residues are depleted of cellulose. Attractive as they may be, therefore, the hypothetical successions described above are far too simple and generalized, and fail to take account of the fact that successional development is influenced not only by the kind of energy source or substrate, but also by the nature of the microflora and mesofauna (the organism factor), and by environmental factors as well.

Two experimental approaches have been used to study microbial succession. One is to follow the colonization of specific, chemically defined substrates introduced into the soil, the other is to follow succession during the breakdown of complex natural substrates such as leaf litter.

Colonization of specific substrates

Using the former approach, the primary colonizers of cellulose film buried in English soils were found by H. T. Tribe (1957) to be fungi, the actual species present varying from soil to soil. The imperfect genera *Botryotrichum* and *Humicola*, and the ascomycete *Chaetomium*, were dominant in neutral to alkaline arable soils, whereas the imperfect fungus *Oidiodendron* was common in acid soil under coniferous forest. The absence of phycomycetes among the primary colonizers is explained by the complexity of the substrate. After several weeks, bacteria grew profusely around the fungal mycelium and over the cellulose film, but they were not usually prominent before fungal growth occurred. The bacteria invariably supported a population of nematodes and sometimes colonies of amoebae, and the nematodes were often parasitized by predacious fungi. Frequently the microbial tissue and cellulose were consumed by soil animals: in acid sand, mites dominated the fauna but in neutral to alkaline soils, springtails and enchytraeid worms were more common. These observations demonstrate clearly that decomposition of substrates in soil is a complex process in which many members of the mesofauna participate along with microbes.

A similar study was made by D. M. Griffin (1960) of fungal

succession on sterile human hair in three soils from the environs of Sydney, Australia; the particular substrate of interest here was keratin, the major protein of hair, wool, horn and feathers. Aqueous extracts of hair however are known to contain, in addition to keratin, appreciable quantities of readily available nitrogen and energy sources including pentoses and amino acids, so that many fungi that are not keratinolytic can develop. The first colonizers were the imperfect genera *Fusarium* and *Penicillium*, and various phycomycetes. These overlapped with or were followed by a second group, of which the imperfect genera *Gliocladium* and *Humicola*, and the ascomycete *Chaetomium*, were characteristic; other *Penicillium* spp. were also common at this stage. The third and final group were typified by the slow-growing keratinolytic fungi *Keratinomyces* and *Microsporium*. The pattern of succession was broadly similar on all three soils, but there were differences in detail which reflected initial differences in the native microflora.

Fungal succession in the litter layer

The second approach to the experimental study of microbial succession in the soil ecosystem is illustrated by the data of W. B. Kendrick and A. Burges (1962) in Fig. 3.3, which slows the temporal and spatial changes that take place in the fungal populations colonizing pine needles in the litter layer of a Scots pine forest in England. In this forest, needle fall occurs towards the end of summer but the needles were found to be infected by parasites, such as the 'leaf-cast' ascomycete *Lophodermium*, several months before they were shed. *Lophodermium* remained active in the fallen needles of the L-layer throughout the autumn and winter, sporulating in late winter and spring to provide an inoculum for re-infecting living needles. Shortly before leaf fall, the saprophytic imperfect fungi, *Aureobasidium* and *Fusicoccum*, invaded senescent needles, and remained prominent on the fallen needles while these were in the L-layer. In this layer, the needles were invaded by rather uncommon species, some of which, the imperfect fungi *Sympodiella* and *Helicoma* for example, formed a network of hyphae on the needle surface while others, such as the ascomycete *Desmazierella*, attacked the internal tissues. The needles remained in the L-layer for about six months, on the average, and as they passed into the F_1-layer, the saprophytic fungi sporulated. During the ensuing summer, many of the spores and much of the superficial mycelium were eaten by mites and springtails. A year after the first sporulation, a second one occurred and was again followed by renewed grazing by the mesofauna. As more needles fell they compressed the older needles, and the micro-environment became

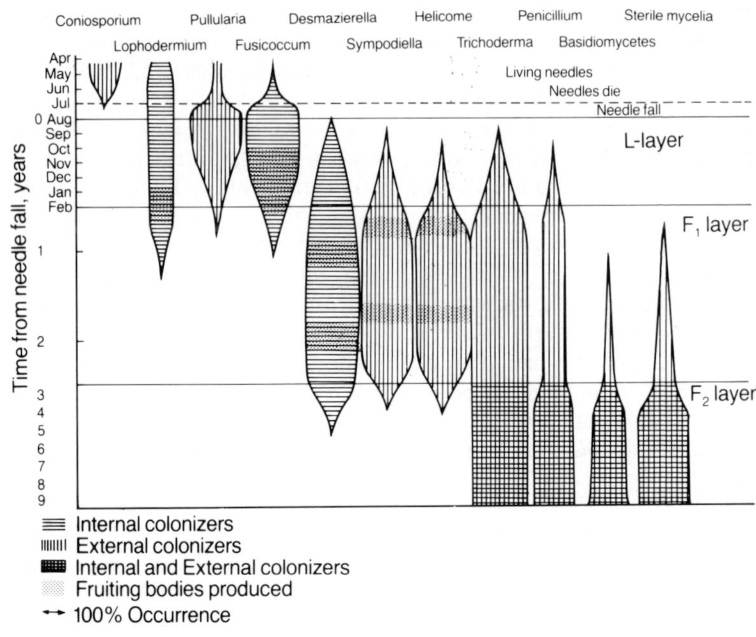

Fig. 3.3 Fungal succession in the leaf litter of *Pinus sylvestris* (From Kendrick and Burges 1962 – for explanation, see text.)

much moister. During this phase much of the mesophyll was destroyed, leaving the outer shell of the needle with its thick cuticle and the vascular strands plus associated lignified tissues. About $2\frac{1}{2}$ years after being shed, the needles had entered the F_2-layer, and this is the stage when basidiomycetes, which first gained entry in the F_1-layer, caused active decomposition of cellulose and lignin. Also prominent in all but the earliest stages of the succession were the common soil saprophytes, *Trichoderma viride* and *Penicillium* spp. Basidiomycete activity was terminated by a considerable increase in the activity of the soil fauna, and the remains of the needles and the fungi were converted to a dark mass of faecal pellets which became part of the H-layer at the junction of the litter and the mineral soil. In this zone, chitin-decomposing fungi were active. Needles took about seven years to progress through the F_2-layer, after which they passed into the H-layer where biological activity was reduced to a low level.

In this succession on pine needle litter the dominant organisms were fungi and small animals; bacteria played relatively little part. This is perhaps because the pH of the litter is low (3–4), and perhaps because some of the phenolic compounds released during decomposition of pine needles are bacteriostatic. A few actinomycetes were present, but their role in the succession is unknown. In other successions, for example in straw composts with a pH of 6 or higher, bacteria play a very active part and large populations develop. Nor is the decomposition of fungal mycelium always restricted to the activity of other fungi or to small animals, since many bacteria are chitin decomposers and have been shown to lyse fungal hyphae.

Another study of fungal succession on leaves was carried out by B. J. McCauley and L. B. Thrower (1966) in the *Eucalyptus regnans* forests of southeastern Australia. As with pines in England, some of the early colonizers of leaf litter were imperfect fungi which invaded the leaves before they were shed, *Protostegia* and *Readeriella* for example, while others such as *Piggotia* and *Hormiscium* became established as saprophytes on freshly fallen leaves. None of these genera is particularly widespread in soils, but well known soil fungi such as *Mucor* and *Penicillium* were prominent later in the succession.

Several other studies have confirmed that the fungal colonizers of freshly fallen leaves are often not typical soil saprophytes, but are rather common airborne ascomycetes and imperfect fungi. As noted previously, some of these primary colonizers are weak parasites, and others infect senescent leaf tissues, but still others may be a part of the normal surface microflora of leaves. Members of the latter group possibly have some affinities with the sugar fungi, living as they do on the relatively simple organic compounds found in leaf exudates. Many of those fungi which invade mature or senescent leaves, however, are not considered to be sugar fungi, and the same is true of many of the species which colonize freshly fallen leaves. Perhaps this is not so surprising when one considers that leaves may be already largely depleted of simple carbohydrates before abscission, or become so very rapidly after reaching the ground. The extent to which the processes of autolysis and translocation within the leaf influence primary colonization is an unknown factor, but both leaf physiology and anatomy are certain to have some effect on the pattern of succession in decomposing leaf litter (see sect. 4.3). In a steady state ecosystem, one might speculate that the whole decomposition process is 'programmed' by the leaf in such a way as to regulate the release of nutrients to the producers, and so govern the rates of nutrient recycling. This is consistent with E. P. Odum's views on the strategy of ecosystem development (Odum 1969), culminating in maximum conservation of nutrient elements at system maturity. As will be explained in

Chapter 5, however, there is little definitive evidence to support such an hypothesis; rather it would seem that maximum control over biogeochemical cycling is exerted at an earlier stage of succession.

Succession of coprophilous fungi

It would seem that the 'nutritional hypothesis' of fungal succession, inherent in the sequences proposed by Garrett and Kreutzer, is not fully substantiated by the results of experimental studies such as those just described. As pointed out by Webster (1970), the hypothesis was derived in part from observations on the sequential development of fruiting bodies of coprophilous fungi. When fresh herbivore dung is incubated in a moist chamber in the laboratory, a succession of fruiting bodies occurs, with phycomycetes such as *Mucor* and *Pilobolus* appearing first, followed by the ascomycetes *Ascobolus* and *Sordaria*, basidiomycetes such as *Coprinus* developing last; various imperfect fungi are found at all seral stages. Few if any terrestrial phycomycetes have the ability to utilize cellulose and lignin, whereas cellulose decomposers are common in the ascomycetes and fungi imperfecti, and many basidiomycetes can decompose both cellulose and lignin. Furthermore, at least some ascomycetes and basidiomycetes are known to germinate and grow more slowly than many phycomycetes. Thus the physiology and nutrition of these major taxonomic groups, combined with evidence that, as manures and composts decompose, sugars, starches and proteins disappear first, followed by hemicelluloses, cellulose and lignin, in that order, provided a basis for the interpretation of fungal successions as nutritional sequences.

An experimental analysis of coprophilous succession by J. E. Harper and J. Webster in 1964 revealed however, that the sequential appearance of fruit bodies could be explained by the minimum time taken by the various species to produce them, and was not primarily the result of competition for substrates. Furthermore, it is likely that many ascomycetes and imperfect fungi in soil successions germinate and grow as rapidly as phycomycetes. Again, there is no doubt a differential depletion of other nutrients, nitrogenous compounds for example, concomitantly with the progressive utilization of more and more complex carbohydrates. Hence fungal successions might perhaps be interpreted as readily in terms of nitrogen availability as on the basis of carbohydrate nutrition. It must be concluded that while the nutritional hypothesis explains many of the observed facts, it does not account for all. Not only this, but all of the stages postulated by Garrett and Kreutzer will not be found in every microbial succession in the soil–litter system.

In particular, the stage of dominance by the so-called primary saprophytic sugar fungi, and typified by the phycomycetes, may be greatly curtailed or even omitted altogether.

The role of soil fauna in microbial succession

As will be apparent from the foregoing account, experimental studies of heterotrophic succession have revealed a significant involvement of the soil fauna, and there is much circumstantial evidence that the activities of the meso- and microfauna help to determine the sequence of microbial species replacement during organic matter decomposition. It is not always easy to demonstrate successional patterns among soil and litter animals themselves however, because they are usually not organized into well defined communities. This has been emphasized by J. M. Anderson (1977) who pointed out that, while the distribution of larger forms is determined by macro-environmental gradients of rainfall, temperature, organic matter content, pH and cation exchange capacity, the smaller members of the mesofauna, and the microfauna, occur in a mosaic of microsites where their relationships with the factors of the environment are obscured.

Despite these difficulties, patterns of succession involving both animals and microorganisms may be discerned, especially during the decomposition of woody residues. Few if any experimental studies have covered all stages of this process, but M. J. Swift (1977a) has integrated the results of a number of investigations to provide a comprehensive picture of wood decay (Fig. 3.4).

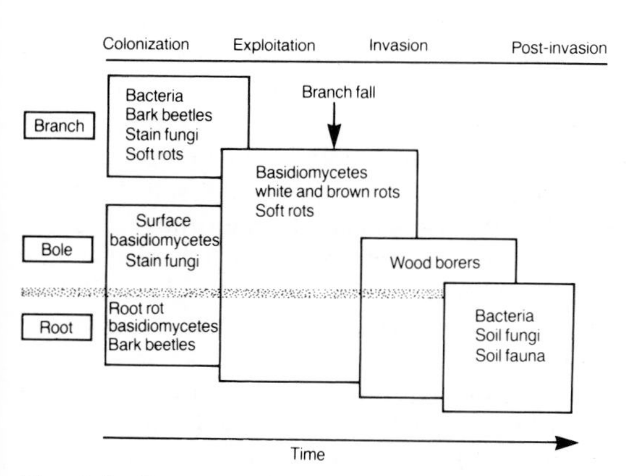

Fig. 3.4 Patterns of succession in decaying branch and stump wood (From Swift 1977a.)

Colonization of woody residues

With the proviso that the type of wood (gymnosperm or angiosperm, sapwood or heartwood), the species' propagules available for colonization, and variable environmental factors will all influence the detailed pattern of succession, three overlapping stages can generally be recognized:

(a) the initial colonization, which begins on the death of cambium and inner bark (phloem) cells, often as a result of parasitism;
(b) a decomposition phase, when major degradation of cell walls occurs and which is dominated by basidiomycetes and sapstain fungi, along with beetle larvae and other specialized wood-boring insects such as bark beetles, fly larvae and termites;
(c) the terminal stage, characterized by generalized soil and litter animals and microorganisms which gain entry through holes made previously by the specialized wood-inhabiting insects.

The role of parasitism in determining the initial pattern of colonization is most evident on stumps left after tree-felling. The freshly cut surfaces of stumps are normally invaded by parasitic, root-rotting basidiomycetes such as *Fomes annosus* and *Armillariella* (*Armillaria*) *mellea*, which are also able to initiate decay in living roots. Sapstain fungi, together with some common soil saprophytes, compete for sites on the stumps, and can prevent invasion by the root rot organisms. These observations form the basis for a successful method of biological control of *Fomes annosus* (Rishbeth 1951, 1963), which involves inoculating the surface of stumps left by thinning operations with spores of the saprophytic basidiomycete, *Peniophora gigantea*, in order to prevent their colonization by the pathogen (see also sect. 3.6).

Unlike decay in stemwood, branchwood decay is rarely initiated by basidiomycetes, the earliest colonizers being bacteria. sapstain fungi, and soft rot fungi. Sapstain in wood results from the presence of pigmented hyphae of so-called 'microfungi' – zygomycetes, fungi imperfecti and ascomycetes – which utilize the relatively simple carbon sources in sap (cf. sugar fungi). Cellulose decomposers among the microfungi cause structural damage known as soft rot. Brown rot is the result of cellulose decomposition by a small group of basidiomycetes, while white rot results from the utilization of both cellulose and lignin by a wide range of basidiomycetes and a few ascomycetes (see also sect. 4.1).

During the second stage, when most decomposition takes place, a complex succession of specialized wood borers is sometimes evident, and there is circumstantial evidence that their invasion may be facilitated by fungal decay. Many wood-boring insects are coprophagous (sect. 4.2) and this invites speculation that microbial degradation of faeces improves its nutritive value for the insects; unequivocal proof of this hypothesis has not yet been advanced, however.

The change from the specialized fauna and microflora of the main decomposition phase to the more generalized decomposer community of the terminal stage may be quite rapid. Common litter inhabitants such as springtails, mites and enchytraeid worms develop first, followed by larger animals including millipedes and earthworms, and predators such as spiders and centipedes. The animals speed the comminution process and inoculate the interior of the wood with a wide range of soil-inhabitating bacteria and fungi such as zygomycetes and penicillia.

Successional interactions in microcosms

Controlled experiments on the role of soil and litter fauna as potential regulators of microbial successions are few. D. Parkinson and his colleagues (Parkinson *et al.* 1979) used simple microcosms to simulate the surface litter layer of an aspen forest in Canada, and showed that selective grazing by collembola could influence the competitive ability of two fungi to colonize leaf litter. The fungi chosen for study were common inhabitants of the L-layer, a hyaline basidiomycete and a sterile dark mycelium, and showed the same growth rate in pure culture. A species of springtail, *Onychiurus subtenuis*, also abundant in the L-layer where it preferentially grazes the sterile dark hyphae, was the animal used in the investigation. When sterilized macerated leaf litter was inoculated with either fungus, the presence of *O. subtenuis* did not affect fungal development. When a mixed inoculum of both fungi was used however, the basidiomycete colonized more macerate particles than the dark fungus when no grazing was permitted, and its competitive advantage was enhanced by grazing. Further studies using sterile 2.5 cm discs of leaf litter placed on the surface of inoculated macerates, together with examination of the gut contents of *O. subtenuis*, confirmed that selective feeding by the springtail greatly reduced the colonizing ability of the sterile dark fungus.

R. D. G. Hanlon and J. M. Anderson also showed in 1979 that the feeding activity of collembola in laboratory microcosms exerted a strong differential effect on fungal and bacterial populations. Microbial respiration was stimulated at a low grazing intensity but the effect was reversed at higher intensities. Grazing by springtails increased bacterial and reduced fungal biomass in proportion to the grazing intensity.

Further research of this kind is needed to elucidate the regulatory role of soil and litter animals on pattern and process in the microbiotic community.

3.4 Reproduction, dispersal and survival of soil microbes

Availability of substrates is the most important single limiting factor for the heterotrophic microflora, and furthermore, substrates are discontinuous in space and time (see Ch. 4). It follows that soil microorganisms must have a mechanism for reaching new substrates once existing ones are depleted. **Dispersal** is thus a key factor in the continuity of the microflora. Soil microbes can move to new substrates, or fresh substrates can be brought to them by mechanisms such as leaf fall and root decay of plants, and by death or defaecation of soil animals. In other words, they can disperse either in space or time. Dispersal in space is linked to reproduction, while dispersal in time requires the existence of survival structures or organs; all three processes are interrelated, however (Fig. 3.5).

Dispersal in space involves both active and passive mechanisms. Motility and growth are active means of dispersal. Motile structures such as the zoospores of lower fungi may be important in aquatic habitats but their efficacy in the soil environment is limited to situations where seasonally high water tables develop. Likewise it is doubtful whether motile bacteria move very far in soil, though in

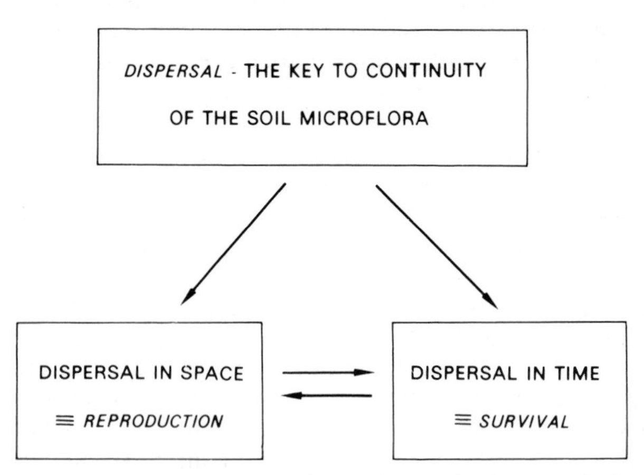

Fig. 3.5 The interrelationships of reproduction, survival and dispersal of soil microorganisms.

relation to the size of their habitat even small distances could be significant. Of all soil microbes, only fungi and perhaps actinomycetes can disperse by growing towards new substrates. The usefulness of growth as a means of dispersal is difficult to evaluate because it is hard to follow hyphal extension through soil. Only occasionally does it become evident, as in the familiar 'fairy rings' of mushrooms or by the presence of macroscopic aggregations of hyphae into strands or rhizomorphs. Passive dispersal embraces transport by air, water and animal vectors. Dispersal on dust particles is one of the most important means available to bacteria. These organisms may also provide the solid phase of aerosols, as too, may fungal spores. Basidiospores can be effective in dispersing root disease fungi such as *Fomes annosus* and some mycorrhizal fungi.

Where no means exist for dispersal of soil microbes in space, survival mechanisms become of paramount importance. Various structures are implicated in survival. Many of them, especially those of fungi, are often bulky. Possibly great bulk is necessary to provide sufficient energy reserves for the colonization of fresh substrates. Certainly survival structures cannot be effective in the dispersal of soil microbes unless their activation is followed by successful establishment on a new substrate. Sclerotia, the largest of fungal resting structures, are frequently regarded as storage organs but they undoubtedly have considerable survival value also. Chlamydospores are well adapted for long term survival, and many other reproductive structures of fungi can be converted into chlamydospores, conidia for example. The phycomycetes which form endomycorrhizas with plants (see Ch. 7) seem to persist in soil as thick-walled resting cells resembling chlamydospores. Even unmodified conidia have some survival value. In addition, viable mycelium is not uncommon in soil and some of it appears to be in a resting condition. Hyphal strands and rhizomorphs are believed to function in part as survival structures too.

Actinomycetes seem to survive in soil as spores rather than as mycelium or mycelial fragments. Some species of bacteria persist as active cells. A few genera produce endospores that are extremely resistant to adverse conditions, but this is thought to be an adaptation to aerial dispersal rather than to survival. Resting cells, rather than endospores, seem to be the main agents of bacterial survival. Furthermore, bacteria rarely occur in soil as single cells but survive as microcolonies each of which may be surrounded by mucilage (possibly originating as extracellular capsules or slime layers) and embedded in humic materials.

The concept of soil as a matrix of discrete and transient microhabitats provides a rationale for the growth pattern of heterotrophic microorganisms, which is one of successive cycles of substrate colonization, exploitation and exhaustion, separated by periods of migration or quiescence. The frequency or brevity of the cycles does not diminish the significance of the dispersal process, although it may disguise it.

3.5 Microbial interactions

Organisms respond to the factors of the environment in such a way that overall homeostasis of the ecosystem is achieved. In other words, the conditions of existence (physical state factors) act as regulatory agents. As indicated previously, however (Ch. 1), homeostatic controls may be self-regulatory processes not involving the physical state factors at all. As the functional – factorial approach to ecosystem development illustrates, the concept of limiting factors is not confined to physical parameters, since biological interactions may be equally or even more important in controlling the development of ecosystems. These interactions frequently prevent some organisms from taking full advantage of optimum physical conditions, and at times may prevent an organism from developing even though all physical factors are well within its range of tolerance. In other instances, biological interactions may result in better utilization of the conditions of existence than would otherwise have been the case.

Interactions between plants and microorganisms, and their regulatory function in terrestrial ecosystems, are the subject matter of the last three chapters of this book. Microbe × microbe interactions in the soil also have a regulatory role in the total system, albeit indirectly. The existence of purely microbial interactions provides the rationale for the biological control of plant pathogens, that is the control of pathogenic microbes through encouraging the growth of non-pathogenic species which inhibit the development of the pathogens.

Interactions between individuals or between populations may be either positive or negative, according to whether one or both parties are stimulated or inhibited. Most negative interactions can be described as examples of either competition or antagonism, while positive interactions comprise commensalism, protocooperation and mutualism. Other more specific terms can be used when describing particular interactions, but this simplified scheme is satisfactory for the present purpose. It should be mentioned that the term 'symbiosis' is often used by botanists as a synonym for mutualism. Since its literal meaning is 'living together' is it perhaps best used in that general sense, without regard to the effect of the symbiosis on either partner. According to D. H. Lewis (1973), this is the sense in which it was first used by Anton de Bary in 1887.

Competition and antagonism

Competition

Broadly speaking, competition involves active demand by two or more organisms for a material or condition in short supply. This definition was given by F. E. Clark in 1965 and is based on an earlier one of Clements and Shelford (1939). Botanists began to study competition many years before microbiologists. The classical botanical concept is that plants compete for light, water, nutrients, and possibly space. Soil microbiologists have frequently assumed that with the exception of light, these same factors are involved in competition among heterotrophic microorganisms, but many now subscribe to the view that microbes compete basically for substrate, that is energy supply, and that competition for water, nutrients and space is insignificant or nonexistent. There is no doubt that water is essential for microbial activity, but it appears that the metabolic reactions of microbes frequently produce water rather than consume it: provided sufficient water is available to initiate decomposition of a substrate, the process becomes autocatalytic in that metabolic water is produced. This should be apparent from the general equation for the aerobic respiration of carbohydrates:

$$C(H_2O)_n + O_2 \rightarrow CO_2 + nH_2O.$$

In an attempt to distinguish competition *per se* from the phenomenon of antagonism (q.v.) it is sometimes referred to as **exploitation competition**, its scope being then confined to the depletion of a resource by one species without restricting the access of another species to that resource (Lockwood 1981). Other forms of competition, which involve denial of a species' access to a resource by the behavioural or chemical characteristics of a competitor (antagonism for example), may then be referred to as **interference competition**. Numerous studies have confirmed the importance of substrate, that is, carbon or energy source, as the primary object of exploitation competition. The annual input of substrates to the soil subsystem is usually insufficient to permit the turnover of microbial populations more than three or four times each year. For example, Clark and Paul (1970) showed that one-fifth of the substrate available annually was required for each new generation of cells produced. Moreover, maintenance energy requirements, that is the energy needed to keep the existing population alive without growing or reproducing, may consume as much as one-third of the available substrate. That the growth and metabolic activity of microorganisms in the soil is severely constrained may also be inferred from the observation that the amount of CO_2 evolved from soil is one or two orders of magnitude less than would be expected from the respiration rate of microbial populations of similar biomass in the laboratory. This failure to achieve full metabolic potential in the field is most likely to result from competition for limiting substrates.

Habitat and niche. While accepting the qualitative significance of substrates in exploitation competition, it is still pertinent to ask, how intensively do microbes compete for these resources? The answer to this question depends on whether one is considering intraspecific competition – that which occurs between individuals or groups within a species – or interspecific competition – that which takes place between different species. A close spatial association is undoubtedly important in determining the degree of intraspecific competition: two individuals will compete more strongly if contiguous than if separated. This does not necessarily hold for interspecific competition, however, because of the operation of a concept known as the Volterra-Gause principle. This states that as a general rule only one species may occupy any specific niche in a habitat. A corollary to this is that no two species in a stable community are in direct competition: they must have different niche requirements or they could not continue to coexist in the same habitat (Fig. 3.6). Because one species would be excluded as a result of competition, the Volterra-Gause principle is also referred to as the **competitive exclusion principle**. To understand it fully, it is necessary to know what is meant by the terms 'habitat' and 'niche'.

Habitat is a word in common usage outside the field of ecology, and is generally understood to mean simply the place where an organism lives. It is defined by the physicochemical factors of the environment and the communities which that set of factors supports. Thus habitats vary along environmental gradients such as soil fertility or elevation, and with the different communities that develop in response to those gradients. The concept of **ecological niche** has a wider connotation, embracing not only the physical space occupied by an organism but also its activities, that is its role in the community by virtue of its contribution to functional processes such as energy flow and nutrient cycling. Two species may occupy the same habitat but different niches if their activities are separated in time or if they use different resources. Like many broad concepts, the idea of niche is difficult to define and quantify, and confusion

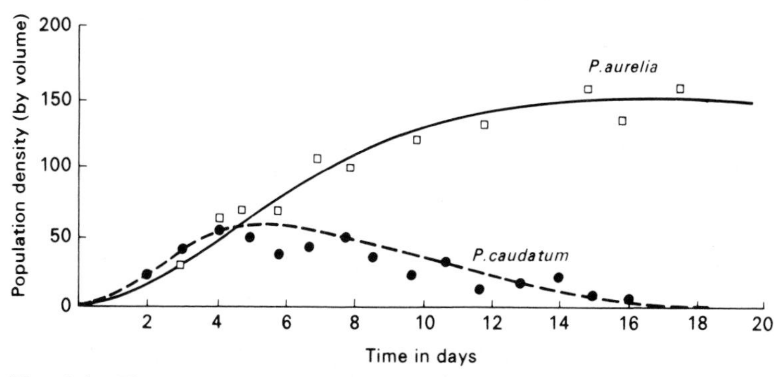

Fig. 3.6 The competitive exclusion principle, illustrated by the experiments of G. F. Gause. When grown separately in controlled cultures with constant food supply, both *Paramecium caudatum* and *P. aurelia* exhibit normal sigmoid growth curves. The diagram illustrates that, when in competition, *P. caudatum* is eliminated while *P. aurelia* maintains its usual growth pattern. (From Whittaker 1975.)

has arisen because it has sometimes been used interchangeably with habitat. In discussing the two terms, R. H. Whittaker and his colleagues (Whittaker *et al.* 1973) distinguished three senses in which the word 'niche' is used: (a) to denote the role of a species within a community, (b) to indicate its distribution between communities relative to a range of environments, and (c) for a combination of the two, based on both intra- and intercommunity factors. The first usage covers the functional meaning of niche, the second the place or habitat concept, and the third embraces both ideas. These authorities argue that the usage of 'niche' should be restricted to the intracommunity role of species, and that 'habitat' should be used for the definitive set of attributes at the intercommunity level, that is for environmental variables with an extensive spatial component, such as elevation, slope (exposure), soil type etc., and to the community gradients consequent on these.[1]

G. E. Hutchinson (1958) proposed that a species' niche be defined as a volume in n-dimensional space, where n is the number of environmental factors affecting its fitness to exist indefinitely. Any organism must be able to survive and reproduce over a certain range of environmental variables, both physical and biological, and Hutchinson emphasized the multifactorial nature of the niche. He also differentiated the **fundamental niche** – an abstract, n-dimensional hypervolume determined by an organism's genetic properties – and the **realized niche**, which is defined by the expression of those properties in the presence of competitors. Where the fundamental niches of two species overlap and the efficiency of their resource utilization is unequal in the overlap zone, then the less efficient species will be excluded, that is, the realized niche of the poorer competitor is only part of its fundamental niche. It should now be apparent that the Volterra–Gause principle applies only to interacting realized niches: if the fundamental niches were identical the implication would be that two distinct species could have precisely the same physiological and ecological tolerances, a proposition which is untenable.

Soil organic matter contains a great many different chemical compounds which act as substrates for soil microorganisms, and the species involved in organic matter decomposition are often in separate niches and not in direct competition. In many instances these niches are complementary, as for example with the nitrifying bacteria where nitrite produced by *Nitrosomonas* forms the substrate for *Nitrobacter*. A close spatial association does not therefore increase competition between these two genera and in fact *Nitrobacter* benefits from it. The degree of physiological and biochemical specialization among microorganisms, combined with the complexity and variety of organic substrates, makes possible the existence of a large number of realized niches in a given habitat. In spite of the overwhelming significance of energy source in determining the tempo of microbial activity in soil we should not jump to the conclusion that soil contains a teeming mass of microorganisms that are fiercely competitive for common substrates. In many situations, soil microbes are as likely to be cooperative as competitive.

While competition for substrate is undoubtedly the primary determinant of community development, other resources may become limiting in certain circumstances. The production of metabolic water does not necessarily mean that microorganisms do not have a water requirement nor does it imply that they are not influenced by changes in the soil water potential; indeed there is evidence (reviewed by D. M. Griffin 1969) that the composition of the fungal flora of the soil is affected by the differential response of different fungi to changes in the availability of soil water. Likewise, oxygen could become a factor in competition when the supply is limited, as in flooded soils or in microsites of intense oxygen demand; the competitive superiority of facultative anaerobes in such situations should be readily apparent. Also, competition for nitrogen can occur in substrates where the N concentration is low, such as wood and water. Some fungi seem to be physiologically well adapted for growth in such habitats (Ch. 4) and may therefore have a competitive advantage in them. Perhaps one should not even dismiss the idea of competition for space without more definitive evidence, even though the space occupied by microorganisms represents but a small fraction – usually less than 1 per cent – of the volume of substrate apparently available for colonization.

Finally, it is well to remember that exploitation competition is not the only kind of population interaction involved in determining a species' place in the organization and functioning of microbial communities. The parts played by tolerance and inhibition in guiding development of the soil ecosystem have been discussed previously when considering the phenomenon of succession (sect. 3.2), and there is good evidence that interactions other than exploitation competition are important in determining just how much of a species' fundamental niche is realized in nature. For example, many lichen algae have light-sensitive pigments that are bleached at intensities much lower than full sunlight, and the shading effect of the associated fungi expands the range over which the lichens can function effectively. In general, it might be thought that positive interactions such as mutualism would increase niche breadth, while negative interactions such as parasitism and predation would reduce it. The complexity of trophic relationships makes it difficult to generalize, however. To illustrate, the results of Elliott *et al.* (1980) indicate that predation of bacteria by the protozoan *Acanthamoeba* enlarges the realized niche of the nematode *Mesodiplogaster* in fine-textured soils by foraging in soil pores inaccessible to the latter, then re-emerging to be preyed upon by the nematode, whose significance in the mineralization of organic matter is thereby increased.

Antagonism

Competition as described above is competition for a scarce, common resource. Another form of competition (interference competition) involves the antagonism of one species towards another which results in the inhibition of the second.[2] The growth-regulating properties of metabolites excreted into the environment by microorganisms exert an important influence on the composition of the soil microflora. Some microbial excretions are stimulatory because they contain substances which may be used by other organisms as energy substrates, nutrients or growth factors, others are inhibitory or antagonistic even in very low concentration; the latter are called **antibiotics**. Often one substance acts in two capacities, depending *inter alia* on its concentration and the identity of the species affected. Antagonism may be readily demonstrated in the laboratory (Fig. 3.7).

Antibiosis. The production of antibiotics is considered by some microbiologists to be one of the most important mechanisms

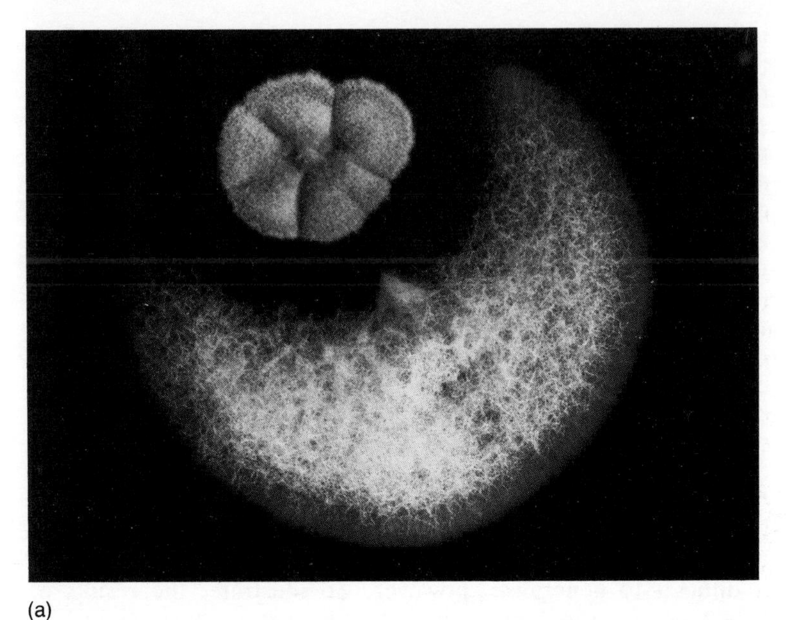

(a)

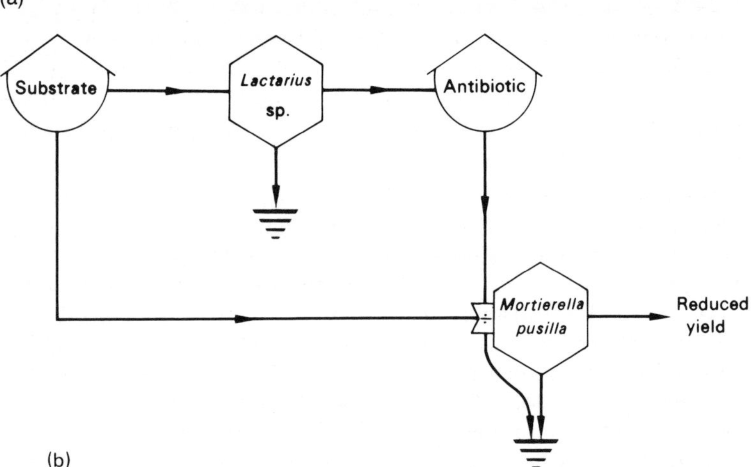

(b)

Fig. 3.7 (a) Microbial antagonism, illustrated by the inhibition of the saprophytic fungus *Mortierella pusilla* (right) by *Lactarius* sp. (left), a fungus isolated from ectomycorrhizas of *Tilia americana*. (From Park 1970) Photograph kindly supplied by Mr. Douglas C. Anderson. (b) Action of the antibiotic, illustrated diagrammatically in energy circuit language. The work gate symbol, which is normally used for multiplicative effects, is shown containing a division sign, indicating that the output is some fraction of the interacting inputs.

operative in interference competition. Not all antagonistic effects seen in soil can be attributed to antibiosis, however. Soil contains many substances of biological origin which when present in sufficiently high concentrations are toxic to certain microorganisms. Thus organic acids liberated during microbial metabolism can inhibit the growth of fungi, especially in acid soils. Carbon dioxide produced by the respiration of one group of microorganisms may restrict the growth of others. Ammonia formed by mineralization markedly inhibits *Nitrobacter* in alkaline soils (sect. 5.2.1), and the consequent high level of nitrite may adversely affect other microbes, and even plants. Decomposition products of certain plant constituents, such as resins, tannins and phenolic compounds, may also prove toxic to microorganisms. All of these products or by-products of microbial metabolism differ from antibiotics in that they must be present in relatively high concentrations to be effective.

Many microorganisms isolated from soil exhibit antibiotic effects on laboratory media. Actinomycetes are particularly active, especially *Streptomyces* isolates. In fact many actinomycetes are utilized for the commercial production of antibiotics including streptomycin, chloramphenicol, and the tetracyclines aureomycin and terramycin. Many fungi imperfecti are prominent antibiotic producers, for example *Penicillium* (the commercial source of penicillin), *Trichoderma, Aspergillus* and *Fusarium*. Aerobic spore formers (*Bacillus* spp.) and some pseudomonads (*Pseudomonas* spp.) are among the more common bacterial antibiotic producers. Antibiosis has also been recorded in algae and protozoa.

While antibiosis is easy to demonstrate in the laboratory there has been a great deal of argument and much speculation about its possible regulatory role in microbial ecosystems in the soil. Evidence that antibiotics are active in nature was summarized by P. W. Brian in 1957: they can be extracted from non-sterile soil; they are produced in sterile soil inoculated with antibiotic-producing microbes and supplemented with organic amendments; antibiotic-producing microorganisms can be readily isolated from natural soils and are known to antagonize other microorganisms *in situ*; and conditions unfavourable for the accumulation of antibiotics reduce the antagonistic behaviour of the antibiotic producer. This evidence must be supplemented by a demonstration that the antibiotic can be purified after extraction, and the purified material shown to be active against the test organism in pure culture.

The potential significance of antagonism in microbial succession has been mentioned previously (sect 3.2). There is considerable interest in the phenomenon as a means of controlling parasitic symbioses in managed ecosystems. It is well known that many saprophytic rhizosphere fungi, which use as substrates the sugars and amino acids in root exudates, can produce antibiotics and consequently, it is thought that such organisms might provide a barrier against pathogenic, root-infecting fungi. Despite strong circumstantial evidence, it has not yet been shown conclusively that antibiotics are produced in sufficient quantities to inhibit fungal pathogens in nature. Large amounts may not be necessary however, and indeed could prove to be disadvantageous. As pointed out by D. T. Wicklow (1981), were an antibiotic to cause cell lysis and consequent nutrient release beyond the zone from which the antagonist could retrieve the nutrients, any advantage conferred by antibiotic synthesis would be lost; moreover, the nutrients released could become available to a potential competitor.

A particular kind of fungal antagonism known as **hyphal interference** was first observed to occur by F. E. O. Ikediugwu and J. Webster in 1970. When the hyphal tips of *Coprinus heptemerus* come into contact with the hyphae of *Ascobolus viridulus*, the cells of the latter lose turgor, become vacuolate, and suffer granulation of the protoplasm. *C. heptemerus* has the same effect on other coprophilous fungi. It represents a form of antagonism which differs from others in that it comes into play only after hyphal contact has been made, and it does not involve the production of any known antibiotic. It appears to be quite a common attribute of many slow growing basidiomycetes, and presumably increases their competitiveness. More than just interference competition may be involved however, since the antagonistic fungi is thought to derive nutrients directly from the affected hyphae, and does not merely deny the antagonized fungus access to a common substrate. Hyphal interference has also been proposed by Dickinson *et al.* (1981) as the mechanism by which streptomycetes isolated from spruce litter were able to antagonize, *in vitro*, fungi from the same habitat.

Mycostasis. The germination of spores and other fungal propagules is characteristically restricted in soils, except in nutrient-rich microsites where germination and growth occur. This long term and widespread restrictive phenomenon is known as mycostasis, and is not to be confused with antibiosis which operates on a smaller scale and is more limited in duration. Mycostasis was first described by C. G. Dobbs and W. H. Hinson in 1953, who termed it fungistasis. J. L. Lockwood (1977) has argued that it would be more appropriate to consider the phenomenon more generally as microbiostasis, because there is strong evidence for imposed quiescence of bacteria and actinomycetes under the same conditions and with the same general characteristics as those affecting fungi.

The experiments of Dobbs and Hinson (1953) clearly demonstrated the nature of mycostasis. They placed spores of *Penicillium frequentans* in folds of cellophane, buried them in soil, and found spore germination to be inhibited, thus indicating that the suppressive factor was water soluble and diffusible. Subsequent research has shown that mycostasis, though extremely common, is by no means universal, and that a range of sensitivity exists, some fungi being unaffected. Neutral or slightly alkaline soils are more suppressive than acid soils, perhaps because they support larger and more active populations of bacteria, and are generally less favourable for the growth of fungi. Where it occurs, the fungistatic effect can be overcome by adding an appropriate energy and carbon source such as glucose. It can also be eliminated by 'partial sterilization' of the soil by steam or chemical fumigants, and reinstated by inoculating with various bacteria, fungi or actinomycetes (which do not have to be antibiotic producers), or with a suspension of non-sterile soil.

All this evidence points to mycostasis as being microbial in origin, and it was formerly thought to be due to the accumulation of antibiotics in sufficient quantities to give the soil as a whole antibiotic properties. If this were so, one would not expect mycostasis to be overcome by the addition of sugars, or by inoculation with non-antibiotic producing microorganisms. In any event, antibiotics released into the soil are subject to enzymatic degradation, or inactivation by adsorption on clays, so it is improbable that they could produce such a persistent and widespread effect. Other explanations must be sought.

J. L. Lockwood's comprehensive reviews (1964, 1977) provide a good understanding of the phenomenon. Species with small spores, which usually germinate slowly and require an exogenous energy supply for germination, are very sensitive to fungistasis in soils. In contrast, species with large spores or sclerotia are often independent of exogenous energy sources, the larger spores germinating rapidly and being generally less sensitive. The most effective amendments for annulling fungistasis are complex organic materials such as plant residues, and only those substances which can be utilized as energy sources by chemoheterotrophs are suitable.

It is generally believed that fungistasis results from microbial competition for substrates in the immediate vicinity of the spore, but there are some aspects of the phenomenon that require a more detailed explanation. Thus the energy deprivation hypothesis seems at first sight unable to explain why many sclerotia and some larger spores, while germinating independently of external energy sources *in vitro*, that is in water alone, fail to germinate in soil unless organic amendments are supplied. This can be explained, however, as due to enhanced loss of endogenous reserves through microbial competition, since exudates from spores and sclerotia which are known to stimulate germination can also serve as substrates for other microorganisms. Evidence for this view comes from experiments in model systems, by leaching energy-rich exudates from ^{14}C-labelled spores, thus simulating microbial competition and thereby imposing fungistasis on otherwise energy-independent propagules. Lockwood (1981) has suggested that, at low levels of competition, exudation losses from fungal propagules are small and endogenous reserves are nearly all available for germination, but when competition is intense, exudation losses increase to the point where germination is restricted. This may be seen as an adaptation for conserving the energy reserves of the propagule, in the short term at least.

Once germination occurs, the thin walled and delicate germ tube is very vulnerable to the action of lytic enzymes. Fungi with melanized hyphae are less subject to lysis than those with hyaline hyphae. The main features of this 'mycolysis' are similar to those of mycostasis – removal by partial sterilization, reinstatement by microbial inoculants, annulment by organic amendments – and are therefore consistent with autolysis also being the result of energy deprivation. Competition for substrates by microorganisms in the micro-environment of the hypha is believed to activate autolytic enzymes, leading to its death. The parent spore may survive the death of its germ tube and germinate again, perhaps several times. Alternatively, the fungus may form resistant structures such as chlamydospores or sclerotia, and this characteristic may have adaptive significance in an environment such as the soil, which is chronically short of energy to support the microbial populations it contains.

In summary, 'the evidence is strong for the causal role of substrate depletion, through microbial competition for energy resources, in imposition of exogenous dormancy, induction of lysis, and formation of resistant structures' (Lockwood 1981). It should be kept in mind, however, that mycostasis has also been attributed to volatile and non-volatile inhibitory substances in soil, lysis to the extracellular enzymes of other microbes, and the formation of survival organs to morphogenetic initiators of microbial origin. Lockwood envisages these factors as being 'superimposed upon the more pervasive substrate deficiency caused by microbial competition ...'

Predation and parasitism

In these cases of antagonism one organism is actually consumed by the other as a source of energy and materials for biosynthesis. A **parasite** lives either in or on its host, which is therefore both substrate and habitat to it whereas a **predator** tends to be freeliving and uses its prey as a source of energy and nutrients but not as a habitat. There is, however, no clear line of demarcation between the two and they have a similar role in ecosystem regulation.

While predation is of common occurrence among soil animals, in the microbial world it is restricted to phagotrophic organisms such as protozoa, which are built in such a way that they can engulf particles or even whole organisms. Microoganisms with cell walls, such as bacteria, fungi and algae, are not usually regarded as predators because the cell wall prevents entry of solid particles. Nevertheless the fungi which trap and digest nematodes are commonly referred to as predacious fungi. Bacteria form the staple food of soil protozoans and it is conceivable that the protozoa have a role in soil as regulators of the bacterial population.

While perhaps not so dramatic in its effects as predation, parasitism is also seen among soil microbes. Fungi may parasitize other fungi and this phenomenon is known as mycoparasitism. It is however difficult to distinguish this from the colonization and lysis of hyphae which have died from other causes. Nor is there conclusive evidence that bacteria can parasitize fungi, that is by actually penetrating fungal hyphae and destroying them from within. However, lytic enzymes produced by bacteria on the surface of fungi can digest living hyphae, and this can be seen to happen in soil using appropriate microscopic techniques. Lysis of fungi by extracellular enzymes is clearly a form of antagonism yet it cannot properly be described as parasitism. The infection and lysis of bacteria by phages is however clearly parasitism, although it is not known to what extent bacteriophages regulate bacterial populations in the soil. A little studied but apparently widespread soil organism is *Bdellovibrio*, a very small vibrio-like bacterium with a single polar flagellum which is highly motile and which parasitizes other bacteria, especially Gram-negative forms. After attaching to the surface of its host, the parasite penetrates the wall and replicates between this and the cell membrane, deriving energy by oxidizing amino acids and acetate. The ecological role of bdellovibrios may be to regulate bacterial populations, though this is not proven.

Commensalism and mutualism

Commensalism. Two broad groups of beneficial interactions may be recognized: one is termed commensalism, the other mutualism. In commensalistic relationships one organism benefits while the other is unaffected. Microbial succession provides numerous examples of commensalism. Among the pioneer organisms colonizing plant remains are some which decompose complex organic compounds and in so doing release simpler substances which serve as substrates for secondary colonizers. The relationship between cellulose-decomposing fungi and secondary sugar fungi (sect. 3.3) is a case in point. Again, there are many soil bacteria which will not grow in pure culture unless the medium is supplemented with water-soluble B vitamins and amino acids. In nature, they are of necessity dependent on exudates of other microorganisms or plant roots for a supply of these growth substances. The widespread occurrence of such bacteria in soil is evidence that commensalism is among the major biological determinants of the composition of the soil microflora.

It has already been pointed out that two or more kinds of microbe are frequently associated in complementary niches and so are cooperative rather than competitive with respect to substrate. Complementarity of ecological niches is a form of commensalism and many examples can be cited. The provision of nitrite as a substrate for *Nitrobacter* or *Nitrosomonas* is one, as previously indicated, so too is the release of ammonium ions by bacterial deamination of amino acids to serve as a source of energy for *Nitrosomonas*. In fact, most of the transformations of sulphur, carbon and nitrogen in soil involve the production by some microbe of substances which can be used by others as sources of energy and materials.

Other examples of commensalism involve a change wrought by one organism in the physical environment which makes it favourable for the growth of another. Aerobes frequently reduce the redox potential of the environment to the point where anaerobes can grow, thus explaining why it is possible for obligate anaerobes such as clostridia to persist and grow in an essentially aerobic environment like soil. Similarly, osmophilic yeasts can grow in concentrated sugar solutions, and by utilizing the sugar they reduce the osmotic concentration to the point where less tolerant organisms can grow. Again, T. D. Brock (1971) has observed that hyphae of the zygomycete *Rhizopus nigricans* (syn. *R. stolonifer*) grow rapidly across the surface of an agar plate and provide a transportation route along which motile bacilli are able to travel, whereas the bacteria are unable to move across the surface of agar in the absence of the fungus. He has speculated that filamentous fungi might perform a similar function for motile bacteria in soil.

Mutualism. When an interaction between two organisms confers benefit on both, but is not obligatory for either, it is spoken of as protocooperation. Where the association is so close as to be obligatory for one or both partners it is referred to as mutualism. It is, however, often difficult to distinguish between the two. Among the best known cases of mutualism are those involving microorganisms and green plants, for example mycorrhizas and root nodules, and these will be discussed in subsequent chapters.

Bacteria of the genus *Azotobacter* assimilate molecular nitrogen but only simple organic compounds are suitable substrates for these bacteria in pure culture. H. L. Jensen and R. J. Swaby (1941) have shown, however, that N_2 could be fixed by *Azotobacter* using cellulose as an energy source provided a cellulose decomposer was present to convert the polysaccharide into simple sugars or organic acids. The cellulose decomposer benefited in turn from an increased supply of nitrogen. Since the association is not obligatory for either partner, the relationship is one of protocooperation. Further examples of increased growth of N_2-fixing bacteria, brought about by associated microorganisms partially decomposing otherwise unavailable substrates, were given by V. Jensen and E. Holm in 1975. These authors also produced convincing evidence of stimulation of bacterial N_2-fixation due to growth-promoting substances produced by associative fungi.

Mutualistic associations between two microbes also exist, permitting the two organisms to grow together where neither could grow alone. Thus in a medium deficient in phenylalanine and folic acid neither *Streptococcus faecalis* nor *Lactobacillus arabinosis* can develop, since the former requires folic acid as a growth factor and the latter phenylalanine. Nevertheless, if the same medium is inoculated with a mixed culture, both organisms will grow because *Streptococcus* synthesizes and excretes phenylalanine, and *Lactobacillus* synthesizes and excretes folic acid (Nurmikko 1956).

A widespread mutualistic symbiosis is the association between an alga and a fungus which is known as a lichen. The interaction between the two organisms results in a wide range of morphological adaptations but the functional relationship remains more or less constant, the alga providing carbon compounds as energy sources and vitamins for the fungus and the fungus apparently aiding the alga by the provision of mineral nutrients and water; the fungus also affords the alga some protection from the elements, especially from the deleterious effects of desiccation and high light intensity. Some lichens contain a blue-green alga (cyanobacterium) that fixes atmospheric nitrogen and presumably some of this nitrogen is eventually passed on to the fungus (see Ch. 8). Of the 17 000 known species of lichen, only a few contain more than one alga and one fungus. The algae are generally greens or blue-greens, the fungi nearly all ascomycetes. Lichen fungi rarely if ever occur freeliving in nature; the lichen algae may do so, but not usually under the exacting environmental conditions where lichens occur. In most lichens, the alga occurs in a narrow, clearly demarcated zone just beneath the surface of the thallus, with a thin, tough fungal cortex above and a thicker, looser hyphal medulla below. The symbionts are in intimate contact with each other, the algal cells being penetrated by fungal haustoria or, more usually, surrounded by closely appressed, modified hyphae.

The growth of lichens in nature is extremely slow, and crustose lichens, for example, increase in diameter at the rate of a few millimetres per year, at most. They are extremely sensitive to air pollutants and in consequence are rarely found in urban areas. Their great sensitivity to a polluted atmosphere is thought to be due to the fact that they absorb and concentrate nutrients from rainwater and possess no means of excreting them, so that toxic substances may rapidly build up to lethal levels. Many lichens live in habitats exposed to direct sunlight where they are subject to intense variations in environmental factors such as moisture and temperature. Under these conditions, organic compounds dissolved in rainwater may supplement the carbohydrates produced by algal photosynthesis although it is generally believed that net photosynthetic assimilation of carbon dioxide occurs in lichens, despite the adverse physical environment.

The algal component of the lichen excretes carbohydrate when in symbiosis with the fungus, but loses the ability to do so when grown in pure culture; it may also lose the capacity to synthesize the excreted carbohydrate altogether. In those lichens which contain cyanobacteria the carbohydrate is transferred to the fungus as glucose, while those containing green algae transfer polyols (sugar alcohols). In both cases the lichen fungi convert the excreted carbohydrate into mannitol and thus maintain a concentration gradient from autotroph to heterotroph (Smith *et al.* 1969).

3.6 Biotic interactions in microbial succession and ecosystem development

D. T. Wicklow (1981) discussed in some detail the potential significance of antagonism in fungal successions. As a means of limiting competitor access to a common substrate, it provides many examples of inhibition competition. There is considerable evidence to show that antibiotics produced by early fungal colonists of natural substrates prevent the establishment of subsequent invaders or inhibit those already present. Much of this evidence derives from observations on soil saprophyte–root pathogen interactions. In 1956, S. D. Garrett proposed that because root pathogens can escape competition from soil saprophytes by entering into parasitic symbioses, then they might be less competitive in the presence of such saprotrophs in the soil or rhizosphere; conversely, saprotrophic fungi in the rhizosphere ought to be able to contend with the antibiotics produced by other microbes.

Support for the hypothesis that competition from soil saprophytes can inhibit the development of pathogenic fungi in roots comes from experience with the biological control of wood-rotting fungi, such as *Armillariella* (*Armillaria*) *mellea* and *Fomes annosus*, which cause considerable economic losses in horticultural and forest tree crops. Corke and Rishbeth (1981) have briefly outlined the background to biological control methods for *F. annosus*, a widespread cause of butt-rot in man-made coniferous forests in Britain. Airborne spores colonize the freshly cut surfaces of stumps which then become foci for the infection of roots of nearby trees, the inoculum potential of the pathogen increasing as more stumps are created by successive thinnings. Furthermore, the fungus may persist in stumps left after the final harvest, and thereby reinfect the trees of the subsequent rotation. The first attempts at control involved killing the stumps with 40 per cent ammonium sulphamate, which encouraged their colonization by the common soil saprotroph *Trichoderma viride* and so inhibited the establishment of *F. annosus*. The antagonism of *T. viride* was supplemented by hyphal inference from the wood-rotting fungus *Peniophora gigantea*, which also colonizes the treated stumps and prevents *F. annosus* entering them from infected roots below. Stump inoculation with *P. gigantea* is now a routine procedure following thinning in many British pine plantations, though it is not so effective following the clear-felling of older, heavily infected plantations.

Other fungal successions seem to fit the concept of a 'competitive hierarchy' such as that proposed for vascular plants by H. S. Horn (1976), whereby a directional sequence of species replacement permits certain plants to achieve dominance by virtue of their ability to tolerate a low level of resources. This is the tolerance model of succession (sect. 3.2) and reflects the 'competitive saprophytic ability' of interacting species. As Wicklow points out, fungal communities in a forest soil profile become increasingly depauperate with depth, as substrate availability and microclimatic heterogeneity decrease. Fungal dominants in the F, H and A_1 horizons would need to be especially competitive since, as primary substrates (fresh litter components) are converted to terminal ones (humus), total niche space is progressively reduced and niche overlap among fungi increases. Because litter falling onto the soil surface is already colonized, species replacement must take place, with the fungi dominant in one layer ousting those of the horizon immediately above. Since the attributes that make for a high competitive saprophytic ability include the production of antibiotics and a tolerance of antibiotics produced by others, as well as a high growth rate (Garrett 1956), it is likely that inhibition plays a part in this succession along with tolerance. It is therefore not surprising to find that well known antagonists such as *Trichoderma* spp. and *Penicillium nigricans* become increasingly dominant in the H and A_1 horizons.

The foregoing illustrations have dwelt on the significance of negative interactions. Positive interactions also influence the course of microbial succession, via the facilitation pathway (sect. 3.2), and several examples have already been cited under the heading of commensalism. According to E. P. Odum (1971), 'in the evolution and development of ecosystems negative interactions tend to be minimized in favour of positive symbiosis that enhances the survival of the interacting species'. Whether this statement applies to the soil–litter subsystem is not entirely clear. While many of the early colonizers of organic substrates are involved in negative interactions such as inhibition by antibiosis, for example penicillia and actinomycetes, these organisms are by no means confined to the pioneer stages of the sere. Nor is the production of antibiotics the sole province of those species which do not enter into mutualistic symbioses. D. H. Marx (1969a, b) showed, for example, that the mycorrhizal fungus *Leucopaxillus cerealis* var. *piceina* inhibits the root pathogen *Phytophthora cinnamomi* by producing the antibiotic diatretyne nitrile, and subsequently concluded (Marx 1973) that many basidiomycetes which show antibiosis in pure culture can form ectomycorrhizas (see, for example, Fig. 3.7). Furthermore, other positive symbioses such as commensalism seem also to occur at all stages of microbial succession.

The complexity, diversity and ephemeral nature of organic substrates, combined with the microscopic scale on which succession occurs in the soil ecosystem, make generalization difficult. At one and the same time, adjacent microhabitats may exhibit the two extremes of a successional series. Such a situation is not however incompatible with the achievement of overall homeostasis, in the total system of which the soil–litter subsystem is a part, but whether this results from the dominance of positive over negative interactions cannot be stated with certainty.

3.7 Notes

1. Whittaker *et al.* (1973) advocated use of the term 'ecotope' to cover the broader concept entailing both intra- and intercommunity variables. Thus defined, the ecotope represents the relationship of a species to the whole range of environmental and biotic factors affecting it.
2. The term 'amensalism' is sometimes used to describe a situation in which one species of an interacting pair is inhibited while the other is not, the use of the word 'competition' being then restricted to cases of mutual inhibition.

4

Carbon mineralization and energy flow

The flow of energy through the soil subsystem is bound up with th
process of organic matter decomposition or carbon mineralizatio
both occurring concurrently and both being the province of th
decomposer group of soil organisms, in particular the chem
heterotrophic microorganisms. Since for these microbes, the suppl
of substrates is a major limiting factor, a knowledge of the manne
in which their activities are influenced by substrate availability
essential to an understanding of ecosystem function. There ar
other microbial groups, the phototrophs and chemoautotroph
which are producer organisms but, as has already been noted i
Chapter 1, their role in terrestrial ecosystems as organic matte
accumulators is insignificant beside that of the seed plants. Furthe
more, when compared with the chemoheterotrophs, their part i
energy and carbon transfers in such systems is slight also, so th
they need not be considered further in this context.

4.1 Distribution of chemoheterotrophic microbe in relation to substrate

Chemoheterotrophic microorganisms are not distributed uniforml
throughout the soil: on a microscopic scale there are large volume
of soil which appear to be unoccupied even though available fo
colonization (Fig. 4.1). Knowledge of the actual extent of micro
bial habitats is very slight, despite the advent of the scanning elec
tron microscope. One point which must be emphasized is that th
scale is minute. As R. Y. Stanier indicated in 1953, 'a single cellu
lose fibre provides a specialized environment with its own charac
teristic microflora, yet may occupy a volume of not more than
cubic millimetre'. While it is difficult to determine the individu
factors which control the distribution of microorganisms, ther

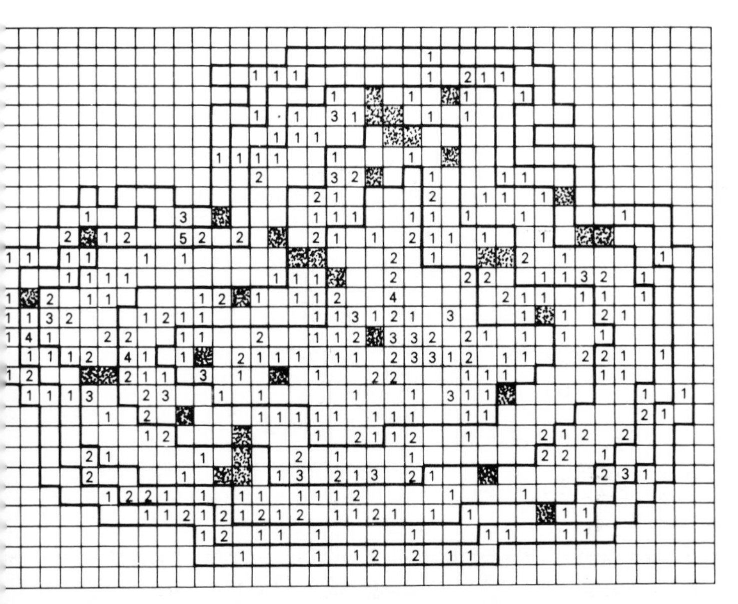

g. 4.1 Distribution of microorganisms in soil. Plan of a section through soil crumb, with figures indicating the number of bacterial colonies r 70 μm square; shaded squares are empty or obscured by debris. rom Jones & Griffiths 1964.)

ems little doubt that the primary limiting factor for chemo-terotrophs is the availability of substrate. Substrates for these ganisms have been defined by S. D. Garrett (1951) with respect soil fungi, as 'living or dead, virgin or partially decomposed ant or animal tissues lying in or upon the soil, or soluble products ffusing therefrom'. The addition to soil of carbonaceous materials ch as these greatly increases the number and activity of chemo-terotrophs, and in particular it activates the so-called zymo-nous fraction of the population, that is those organisms that pend upon an external source of energy and nutrients, in the sence of which they remain quiescent. Among the fungi, for ample, genera such as *Penicillium, Aspergillus, Trichoderma, sarium* and *Mucor* become dominant.

Within the soil, chemoheterotrophs are characteristically con-ntrated in the upper part of the profile. In podzols and podzolic ils, most are found in the A_1 horizon (Table 4.1). This pattern of stribution is not unexpected, since most of the organic matter in dzols and podzolics is found in the A_1 horizon. However, a simi-r pattern exists in soils where the organic matter is distributed

Table 4.1 Distribution of chemoheterotrophic microbes throughout th profile of a podzol soil

Horizon	*Depth* (cm)	*Numbers of microbes per gram* ($\times 10^3$)			
		Aerobic bacteria	*Anaerobic bacteria*	*Actino-mycetes*	*Fungi*
A_1	3–8	7800	1950	2080	119
A_2	20–25	1804	379	245	50
A_2/B	35–40	472	98	49	14
B_1	65–75	10	1	5	6
B_2	135–145	1	0.4	—	3

Source: After Starc (1942)

more uniformly throughout the profile, so that factors other th energy supply must be limiting also. The most likely explanati of this phenomenon lies in the change which occurs in the s atmosphere with depth. In general, as one proceeds farther fro the surface, oxygen becomes depleted and the carbon dioxi concentration rises, and either or both of these factors may lin the development of chemoheterotrophic microorganisms.

Some substrates for chemoheterotrophs in soil originate fro the roots of vascular plants. These substrates may be orga materials sloughed off from the roots or exuded by them. T presence of such substrates markedly affects the kinds a abundance of soil microorganisms in the immediate neighbou hood of roots, and gives rise to what is known as the 'rhizosphe effect'. This will be discussed further in Chapter 6.

When the soil ecosystem approaches a steady state, the gro pattern of distribution of substrates does not change appreciab with the passage of time. On a microscopic scale however, the are still vast changes, and these changes occur from place to pla at any one time, and from time to time in any one place. Of par mount importance is the fact that substrates for chemoheterotrop in soil are ephemeral. Consequently the distribution of su microbes or their propagules is determined not only by the prese but also by the past, disposition of their substrates.

Types of substrates in the soil: soil organic matter

Soil organic matter derives from several different sources, bei ıe remains of dead plants and animals and their excretory products ı various stages of decomposition, the final stage being known as umus. The cells of microbes themselves also serve as substrates ɔr succeeding generations of microbes. In fact, any organic compound that is synthesized biologically is subject to decomposition y soil microorganisms, provided the environmental conditions re appropriate. Many kinds of compounds are therefore available s microbial substrates. Since each individual organism has a range f enzymes which enables it to attack only a certain number of ıese compounds, a knowledge of those which can be used as substrates by a particular microbe is useful in studying its ecology. uch information can be obtained from pure culture studies in the ıboratory, but care is needed in transposing laboratory results to ıe field, for it cannot be assumed, just because an organism is apable of utilizing a specific substance in a test tube or petri dish, ıat it will utilize that compound as a substrate in the soil. Thus ellulose-decomposing fungi will grow very well on simple sugars ı pure culture, but in the highly competitive environment of the ɔil they may seldom have the opportunity of attacking such ubstrates.

While small amounts of simple organic compounds, sugars and mino acids for example, do occur in soil, particularly during the ecomposition of organic residues, or as exudates of plant roots, ıost natural substrates for microbial growth are complex, rarely eing composed of a single substance. Although soil organic matter therefore far from homogeneous, we may conveniently divide it ıto two broad types for the purpose of simplifying its description Hayes & Swift 1978):

a) unaltered, recently incorporated residues of plants, animals and microorganisms together with the non-transformed components of older residues;
b) transformed products (humic and non-humic) having no morphological resemblance to the tissues from which they derive. The non-humic component comprises recognizable compounds such as polysaccharides, polypeptides, lignin etc., or derivatives of these. The humic substances (humus) can be further subdivided into three forms, based on their solubilities in acid and alkali, namely humin, humic acid and fulvic acid.

Jnaltered substrates

'he cell walls of vascular plants are constructed mainly of microbrils of cellulose, embedded in a continuous matrix of lignin, ectin and hemicelluloses. Cellulose is present in all cell wall layers

though it is concentrated in the primary wall and the first layer
the secondary wall. Pectin occurs mainly in the primary wa
Hemicelluloses are found in both primary and secondary walls
mixture with lignin and cellulose; they also occur in the midd
lamella in association with lignin (Towle & Whistler 1973). Th
biodegradation of lignified tissues is, in a quantitative sense, th
most important process in the recycling of photosynthetically fixe
carbon in terrestrial ecosystems.

The utilization of structural polysaccharides as substrates pr
sents a variety of problems to microorganisms (Burns 1983):

(a) the relative inaccessibility of stable crystalline regions fibrillar molecules such as cellulose and chitin;
(b) the degree of polymerization and the steric rigidity of th primary units;
(c) the extent of hydration of the polymer which affects both th water requirements for hydrolysis and the surface area avai able for enzymatic degradation;
(d) the association in natural substrates of a number of differe polymers – celluloses, hemicelluloses, pectins and lignins which calls for the concerted action of a range of enzym and often the presence of two or more synergistic micr organisms before decomposition can proceed;
(e) the physical barriers imposed by a meshwork of cell wall pol saccharides and their protective substances (cuti and suberin), which highlights the importance of filamento organisms such as fungi and actinomycetes in the initi stages of polysaccharide decomposition because of the penetrative powers (see sect. 1.1) and their ability to gro through regions containing unsuitable substrates;
(f) the bacteriostatic and fungistatic properties of some pla products, especially lignin and its constituents.

Before considering these major potential microbial substrate mention should be made of those vital cytoplasmic constituent namely proteins, lipids and nucleic acids, that are present in a biological residues entering the soil (Jenkinson 1981). Activel metabolizing tissues contain a high proportion of protein, up to 5 per cent of dry weight of bacteria, but the percentage decline dramatically during senescence. Water-soluble cell constituent (mono- and disaccharides, amino acids, peptides etc.) form a signi ficant part of the tissues of all organisms but the proportion c these will also be much less in the senescent or dead material tha makes up the bulk of organic matter input. It should also b remembered that minor amounts of specific plant products – pig ıents, resins, terpenes, alkaloids, tannins and other polyphenols – nter the soil, and some of these may affect the rate of decomposiion of plant residues (see sect. 4.5).

Iemicelluloses. Once thought to be precursors of cellulose (hence ıeir name), hemicelluloses have long been recognized as a distinct roup of structurally and functionally unrelated polysaccharides. 'hey can be extracted from plant tissues by dilute alkali, and have o part in the biosynthesis of cellulose. Most hemicelluloses are eteroglycans containing two to four different sugar units, and are lassified according to the identity of the sugars present (Towle & Vhistler 1973). Xylans, which are polymers of xylose, are among ıe most prominent polysaccharides of seed plants comprising up ɔ 30 per cent of the weight of hardwoods and up to 12 per cent of oftwoods; they are structural components of the cell wall. Iannans are mannose polymers and function as reserve food but ıe functions of galactans, derived from galactose, are unclear: ıey occur in the seeds of several annuals and also in so-called 'eaction wood', that is wood formed in response to stresses induced ı the stems of leaning trees.

Hemicelluloses not only occur in woody shrubs and trees but lso in cereals, grasses and ferns, and consequently are among the ıajor plant products added to soil. They are subject to decompoition by a wide range of microorganisms, and their breakdown roducts include uronic acids, as well as hexoses and pentoses.

'ellulose. This long chain polymer of glucose is the principal tructural component of the cell walls of vascular plants and algae. 'he hyphae of some fungi are also constructed to cellulose (though ıost are chitinous) while a few bacteria can synthesize cellulose s an extracellular polysaccharide.[1]

The linear nature of the cellulose molecule permits long segıents to be held in alignment by intermolecular secondary bonds, esulting in a strongly interlocked framework of great physical trength. These highly oriented crystalline segments are separated y amorphous regions where the molecules are not so strongly ligned nor so firmly bonded (Teng & Whistler 1973). Chemically, ellulose is a β-(1→4)-linked glucan made up of uniformly linked -glucopyranosyl units:

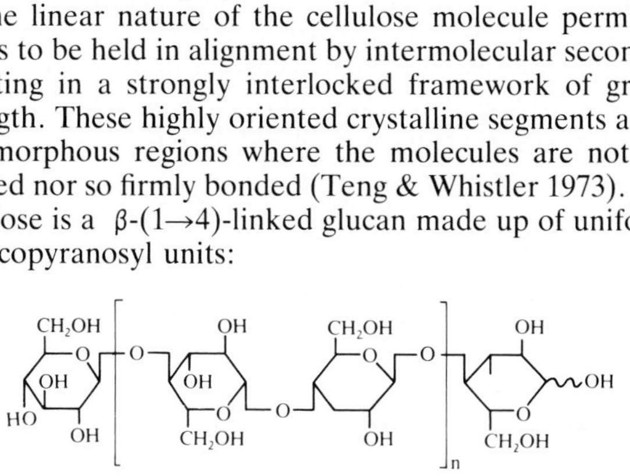

Cellulose makes up about one-third of the biomass of annua plants and about half that of perennials, and consequently is th most abundant substrate in soils. Decomposition is slow: relativel few animals are able to decompose it and most of those which d depend on the cellulolytic ability of symbiotic microorganisms such as bacteria in the rumen of cattle and protozoa in the gut o termites. Cellulose-decomposing microbes in soil include man fungi, and some aerobic and anaerobic bacteria, actinomycetes protozoa and myxobacteria. The heterogeneous nature of th cellulolytic microflora permits decomposition over a wide range o environmental conditions. It should be noted however that, whil many microbes have the ability to metabolize isolated cellulos (and hemicelluloses), most are unable to degrade lignin and there fore, because the wood polysaccharides are physically protecte by lignin (q.v.), they cannot decompose intact wood.

Cellulose biodegradation is thought to be brought about by th combined and synergistic activities of at least three different group of hydrolytic enzymes (Burns 1983), endoglucanase, exoglucanas and cellobiase (β-glucosidase). Microorganisms differ widely i the way in which they attack cellulosic substrates, both in th numbers and molecular weights of the enzymes comprising th cellulase complex and their location within the cell or whether the are secreted or not. Penetration of the enzymes into the cellulos molecule, and hence its susceptibility to hydrolysis, is influence by the relative amounts of crystalline and amorphous segment present.

Chitin. In most fungi, the usual structural element is not cellu lose but a closely related polymer of glucosamine known as chitin This is also the major structural component of the exoskeletons o invertebrate animals such as crustaceans and arthropods, and minor constituent of the cell walls of green algae. Chemically chitin may be regarded as cellulose in which the hydroxyl on carbo atom C-2 of each D-glucopyranosyl unit has been replaced by a acetylated amino group (Teng & Whistler 1973):

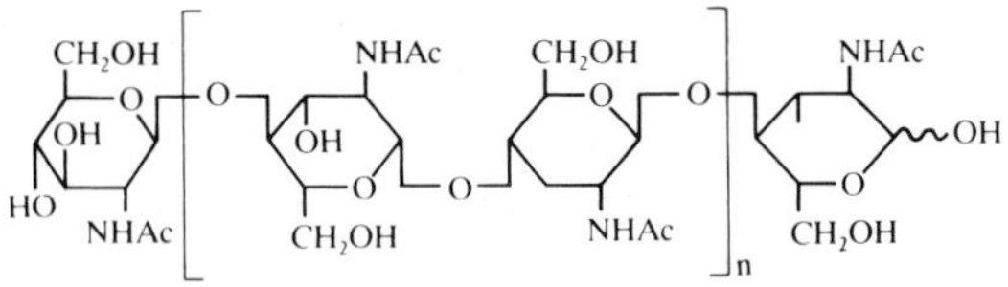

Naturally occurring chitin is always found linked to protein as glycoprotein. Because of its widespread occurrence, substantia quantities become incorporated in soils. Chitin-utilizing microbes include bacteria and actinomycetes, and some fungi.

Lignin. This is the generic name for complex aromatic polymers consisting of phenyl propane building blocks held together by irregular carbon-ether and diaryl-ether linkages. A major structural component of seed plants, lignin makes up between a fifth and a third of woody tissues and occupies a central position in the global carbon cycle. It functions in wood as a cementing agent, binding the polysaccharide components of cell walls both physically and chemically, thereby greatly increasing the strength of this natural material.

Lignins are formed by dehydrogenative polymerization of *p*-hydroxycinnamyl alcohols, and are generally classified into three major groups according to the nature of their monomer units (Table 4.2). Gymnosperm lignin is composed mainly of coniferyl alcohol but angiosperm and grass lignins are mixed polymers of two or three monomers, respectively (Higuchi 1980). The cinnamyl alcohol precursors are derived from phenylalanine or (in grasses) from both this and tyrosine, and these in turn are synthesised from sugars via the shikimic acid pathway (Kirk 1984). The lignin content of normal wood of temperate zone hardwoods is 15–25 per cent, and of conifers 25–33 per cent; springwood is usually higher in lignin than latewood in both groups. Lignification is initiated in the differentiated wood cells from the primary walls adjacent to cell corners, and extends to the intercellular middle lamella and throughout the primary and secondary walls.

While the central features of lignin biosynthesis are fairly clear, the biology and biochemistry of its degradation are not yet fully understood. Its large molecular size, poor solubility and complex cross-linked structure make it relatively inaccessible to micro-organisms and their extracellular enzymes. Earlier reports of the anaerobic metabolism of lignin have not been confirmed, so that it must be depolymerized or otherwise chemically altered before substantial decomposition can be effected under anaerobic conditions. The decomposition process is thus oxidative but the precise nature and role of the phenol oxidase enzymes involved remain to be elucidated. Degraded polymers contain α-carboxyl groups and both aromatic and aliphatic carboxyl groups, and vanillic acid is a major product (Table 4.2). The lignin monomers can be absorbed and metabolized by numerous microbes. but 'a complicated and incompletely understood series of extracellular events must occur in order to release the low molecular weight fragments of native lignin' (Burns 1983). Several different metabolic pathways are involved and many intermediate products are produced.

Table 4.2 Types of lignins and their structural sub-units

Lignin type	*Alcohol monomers*	*Aldehyde oxidation product*[†]
Guaiacyl lignin (gymosperms)	Coniferyl alcohol 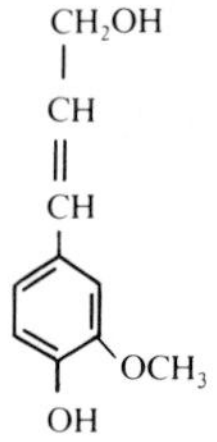	Vanillin 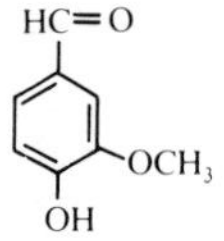
Guaiacyl-syringyl lignin (angiosperms)	Coniferyl alcohol Sinapyl alcohol 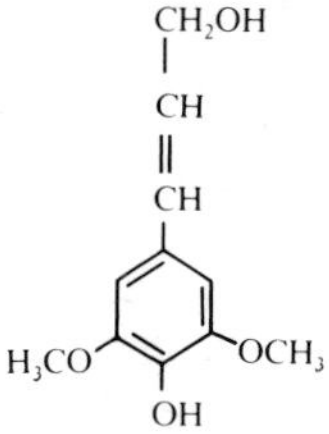	Vanillin Syringaldehyde 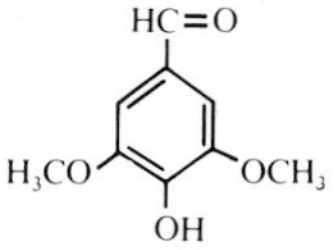
Guaiacyl-syringyl-*p*-hydroxyphenyl lignin (grasses)	Coniferyl alcohol Sinapyl alcohol *p*-Coumaryl alcohol 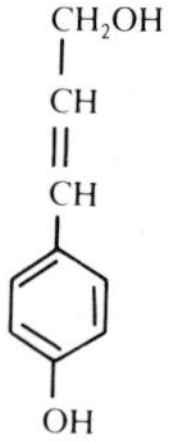	Vanillin Syringaldehyde *p*-Hydroxybenzal-dehyde 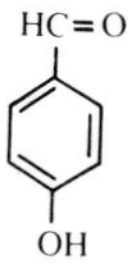

Source: Based on data of Higuchi (1980).
[†] Formed by alkaline nitrobenzene oxidation of lignin for 2–3 h at 160–170 °C.

In 1980, T. K. Kirk reported that the white rot fungus, *Phanaerochaete chrysosporium*, began to metabolize lignin in pure culture at the end of the linear growth phase, long before substrate glucose was exhausted. From this and other evidence he concluded that lignolytic activity is not induced by the presence of lignin itself but accompanies a physiological shift to secondary metabolism which is brought about by nitrogen starvation. That lignin decomposition is, in some circumstances at least, a function of secondary rather than primary metabolism, may explain why some white rot fungi apparently do not use lignin as a carbon source but derive their energy from cellulose and simple sugars instead.

Although relatively resistant to chemical and biological degradation, lignin is not an unusually stable polymer. Indeed, a certain amount of long term non-biological fragmentation can occur, leading to the formation of low molecular weight substances which are readily metabolized (Kirk *et al.* 1980). Knowledge of the biology of lignin decomposition advanced rapidly during the last decade, following the introduction of methods based on ^{14}C-labelled lignins in the 1970s. Bacteria, especially actinomycetes, are now known to degrade lignin to some extent, though according to Kirk (1984), they do not play a primary role in the breakdown of mature xylem tissues of woody plants. The common imperfect fungi *Fusarium*, *Aspergillus* and *Penicillium* have no capacity to degrade intact lignin although these genera, along with many soil-inhabiting bacteria and actinomycetes, probably participate in the further decomposition of low molecular weight breakdown products of the lignin polymers.

There are three kinds of fungal wood decay, viz. white rots, soft rots and brown rots. White and brown rot fungi are basidiomycetes while soft rots are caused by ascomycetes and fungi imperfecti. The white rot basidiomycete, *P. chrysosporium* was recently classified as a species of the imperfect genus *Sporotrichum* or *Chrysosporium*. Soft rot and brown rot fungi are mostly cellulolytic, but among the former a *Chaetomium* sp. has been confirmed as a lignin decomposer, causing complete oxidation of the molecule to CO_2. Brown rot fungi have also been shown to oxidize ^{14}C-lignins to $^{14}CO_2$ in the laboratory, but their activities in the field seem to be limited to demethylation and incomplete oxidation. A white rot fungus frequently occurs as the sole organism in decaying wood, and in axenic culture these fungi can degrade lignin to CO_2 more effectively than any other group of microbes. Under natural conditions however, two or more white rot fungi tend to act synergistically, often in association with brown rot fungi. Finally, it should be noted that lignin decomposition is not the sole province of soil microorganisms, the termite *Nasutitermes exitiosus* being able to degrade parts of ^{14}C-lignin to $^{14}CO_2$ (Butler & Buckerfield 1979). Tissues having a relatively low lignin content, including leaf litter, are probably degraded by a wider range of microbes than is wood. 'In concert these microbes no doubt account for extensive, albeit perhaps slow metabolism of lignin in some environments' (Kirk 1984).'

Transformed products

'*Soil polysaccharide*'. About 10 per cent of the organic carbon in soils is in the form of carbohydrate, the proportion being slightly higher in virgin than in cultivated soils (Cheshire 1979). It occurs mainly as polysaccharide, and serves an important function by binding soil particles into water-stable aggregates. It derives from plant, animal and microbial detritus, and from extracellular gums of microbial origin, and is not easily extracted. Soil polysaccharides from different soils are remarkably similar in composition, and typically contain hexoses, pentoses, deoxyhexoses, uronic acids and hexosamines. Whether they are mixtures of a variety of poly-saccarides, or very complex single polysaccharides, is not known for certain. The polysaccharide fraction is potentially the most readily available microbial substrate in soil organic matter but, despite its abundance, utilization by microorganisms is severely curtailed. Soil polysaccharide was once thought to be a precursor of humus but, according to Cheshire (*loc. cit.*), there is no firm evidence that it contributes specifically to the formation of humic substances.

Humus. As indicated earlier, humic substances are of three kinds. The black insoluble residue which is left when humus is dissolved in dilute alkali is called **humin**. If this is removed and the solution made very acid, the main fraction, known as **humic acid**, precipitates. The remaining solution contains the second major component, **fulvic acid**, which can be concentrated and recovered by dialysis. The various humic fractions are heterogeneous mixtures of polymers resulting from numerous biological and chemical syn-thesis reactions. About half their bulk comprises bound amino acids, amino sugars and heterocyclic compounds, but the chemical nature of the other half is unknown, as is a detailed knowledge of their chemical structures (Swift *et al.* 1979). The presence of amino sugars (hexosamines) and diaminopimelic acid suggests that bacterial cell residues are a significant constituent of humic acid. A major component of humic acid is aromatic however, and may consist of plant polysaccharide residues protected from further degradation by the 'tanning' action of phenolic-protein complexes. Its real chemical nature remains obscure. Fulvic acid generally has a lower molecular weight and a higher carbon content (46% compared to 36%). Humin seems to differ from humic acid principally in having a closer association with the inorganic fraction of soils, which may account for its more ready precipitation from alkaline solution.

Although humus represents the final stage of organic matter decomposition in soil, it would be a mistake to regard it simply as the resistant residue of plant and animal detritus. It is also in fact, a product of microbial biosynthesis, and furthermore it remains subject to microbial decomposition, albeit slowly.

Microbial growth in relation to substrate

Substrates for microbial growth in soils vary not only in kind but also in amount, and this variability affects some groups of microbes more than others. The zymogenous microflora, normally present in the resting stage, becomes active whenever fresh organic materials are added. The autochthonous or indigenous microflora, which characteristically exists in the vegetative phase and is more or less continuously active, is believed to utilize as substrates humic materials and the fatty residues from cuticular waxes and cutins. Knowledge of the actual species of microbes that break down humus is limited, but actinomycetes are known to be involved. Both bacteria and actinomycetes are able to decompose cutins and waxes. Whether any fungi are truly autochthonous is not certain although some species are characteristically present in soil as active hyphae. It can only be inferred that such fungi form part of the autochthonous microflora, since little is known of their nutrition. There is no doubt however that soil contains a large number of inactive fungal propagules of various kinds: 'resting hyphae', sclerotia, and various kinds of spores, particularly chlamydospores, sporangiospores and conidia. These spores and other resting fungal structures in soil, with the possible exception of sclerotia, appear to germinate only when in contact with an external source of nutrients, such as roots, germinating seeds, and decomposing plant and animal remains (see sect. 3.5). After germination, growth of fungi depends on a continuing supply of energy, together with the necessary nutrients and a suitable physical environment. As J. H. Warcup (1965) has pointed out, a source of energy and nutrients sufficient for the growth and multiplication of a bacterium may be barely adequate for a germinating fungal spore and quite inadequate to sustain extensive mycelial development. Unless the external supply is maintained, lysis of the hyphae takes place or else the fungus forms a resting structure such as a chlamydospore.

It is not uncommon for some fungi to be closely associated with specific substrates and to make little free growth in the soil itself. Typical are the sugar fungi (sect. 3.3), whose capacity to germinate and grow rapidly on simple carbohydrates permits them to burst into activity whenever suitable substrates become available. Between the exhaustion of one substrate and the appearance of another, most sugar fungi lie dormant as spores. Other fungi, especially higher ascomycetes and basidiomycetes, which have longer-lived mycelia, may make more extensive growth through soil, travelling from substrate to substrate by means of rhizomorphs or mycelial strands. Such behaviour is typical of the lignin fungi which utilize, in addition to lignin, cellulose and simpler compounds if they are available.

Distribution of substrates in space and time

The fact that certain kinds of natural substrate tend to be decomposed by particular suites of microoganisms is not the only determinant of decomposition rate. Environmental conditions are also important, and these interact with substrate type and soil biota in the decomposition process. The influence of both extrinsic and intrinsic factors on litter decomposition is discussed further, later in this chapter. We turn now to a consideration of the spatial and temporal distribution of substrates.

The soil–litter subsystem is not a uniform milieu for microorganisms but varies considerably from place to place and from time to time. The minute dimensions of the microbial habitat were emphasized in Chapter 3, where the soil was visualized as a matrix of discrete and transient microhabitats. Notwithstanding the conceptual significance of this view of the soil environment, there are readily discernible and more or less permanent gross patterns of decomposer activity related to substrate distribution. For example, an important class of substrates in soil emanates from the roots of vascular plants, in the form of sloughed-off roots or root exudates. These substrates markedly affect the kinds and abundance of microorganisms in the immediate vicinity of plant roots, and give rise to the so-called 'rhizosphere effect', which forms the subject of Chapter 6.

In some forest ecosystems, such as those dominated by *Pseudotsuga menziesii* and *Tsuga heterophylla* in western North America, a surprisingly large part of the organic matter on the forest floor is found in logs, especially in old-growth communities where it may constitute as much as 50 per cent of above-ground biomass (Grier &

Logan 1977). Logs are thus a substantial sink for carbon in these coniferous forests; there is, however, great variability in the amounts present, which is partly a reflection of stand age. In dense, young regrowth stands where growth is rapid there is a slow but regular input of small stems as a result of competition-induced mortality following canopy closure. As the stand ages, and individual trees approach senescence, the rate of input of logs is probabilistic, reflecting catastrophic events such as storms, disease and insect epidemics. In a 121-year old *Picea sitchensis-T. heterophylla* stand in Oregon where storm damage is very evident, most of the fallen logs seem to have originated from windthrow or stem breakage during windstorms (Grier 1978). Windthrown trees also bring large quantities of green foliage to the forest floor. This is supplemented by the wind-pruning of foliage-bearing branches and branchlets, which is a major factor in returning organic matter to the soil subsystem in storm-prone forest ecosystems. Grier (*loc. cit.*) found that, during a 3-year period, the dry weight of green leaves removed from trees by wind was equivalent to more than one third of annual foliage production in this hemlock–spruce stand. Similarly, green branches, presumably broken by wind and snow, make a large contribution to litterfall in an *Abies amabilis* forest in Washington (Turner & Singer 1976).

These random influences on litterfall are superimposed on the more predictable, seasonal variation in the rate of deposition of litter. Seasonality of litter input is most apparent in deciduous forests, but is recognizable in other communities as well. Understorey plants add substantially to litterfall in many forest ecosystems, at rates which vary according to the stage of development of the system. Accumulation of understorey litter also contributes to the spatial heterogeneity of forest floors. This is illustrated by the observations of Birk (1979) who found that, in a mixed eucalypt forest in eastern Australia, 63 per cent of the litter layer was derived from wood and overstorey canopy and these components were spread continuously across the ground; the remaining 37 per cent was understorey litter which was concentrated in localized areas beneath individual shrubs. Another characteristic of Australian forests is the marked accumulation of bark around the bases of smooth-barked eucalypts, which may be greater downslope from such trees than upslope or along contours (Richards *et al.* 1985). Litter accumulation around individual shrubs and trees creates variation in soil chemical and microbiological properties (Zinke 1962; Charley & West 1975). Consequently, spatial and temporal patterns in substrate distribution and availability greatly complicate the measurement of decomposition rates on an areal basis. The variability must be taken into account in any discussion of the techniques of measuring soil metabolism (see sect. 4.4).

4.2 Effect of food supply on the soil fauna

Osmotrophs, as typified by fungi and bacteria, may or may not utilize the same substance both as a substrate for energy-yielding oxidations and as a source of nutrients. In contrast to this, the food of phagotrophs must provide both energy and nutrients. The feeding habits of large animals can be readily determined by observation, but many members of the soil fauna are too small to be observed directly, and consequently our knowledge of the food supply of soil animals is incomplete.

Food source as a basis of classification

Following the classification of chemoheterotrophs given in Chapter 2, they may be divided into biotrophs and saprotrophs, but this classification is too broad to be of much value in determining their food relationships, and each category is normally subdivided further. Saprotrophs include those animals which feed on dead vegetable matter (detritivores), carrion (cadavericoles) or dung (coprophages). Biotrophs include species which feed on living plants, especially roots or on freshly fallen green leaves (phytophages), or on microorganisms (microbivores, fungivores), or on living animals as predators (carnivores) and parasites. These various classes of soil animals will be discussed briefly in turn. Further details may be found in Wallwork (1970).

Phytophages

Many different kinds of sucking insects, for example cicada nymphs and aphids, and numerous chewing insects, molecrickets, beetle, fly and moth larvae for example, feed on plant roots. Millipedes, woodlice, slugs and snails may also feed on living plants, though normally their diet is restricted to decaying material. Termites and ants may sometimes attack surface vegetation. Freshly fallen leaves are devoured by a wide variety of soil animals, many of which may occasionally attack live plants. Soil algae are ingested by certain amoebae.

Detritivores

Representatives of the majority of taxonomic groups of soil-inhabiting invertebrates consume dead organic matter in varying stages of decomposition. For many of these animals, it is not certain whether their food supply is the detritus itself or the associated microorganisms of decay. When organic matter is ingested, compounds such as hemicelluloses and cellulose are usually digested through the agency of an internal symbiotic microflora, although there is evidence that a few animals synthesize their own cellulases, for example termites, earthworms and molluscs, and possibly some beetles.

Carrion feeders

The common decomposers of the carcasses of vertebrates are the larvae of flies and beetles which are not regarded as members of the soil fauna. Many soil animals do not appear to possess the enzymes necessary to digest animal protein, and hence do not feed on carrion. The bodies of many invertebrates decay rapidly in the soil, but the chitinous exoskeletons of arthropods are subject to decomposition by few animals. Fungi are the most important chitin-decomposing soil organisms, though chitinase is found in the alimentary tract of some nematodes, earthworms and molluscs.

Coprophages

Apart from the dung of surface-dwelling vertebrates, which is inhabited by many, often specialized, invertebrate animals, large quantities of excrement are produced by the soil fauna itself. Faecal pellets are utilized by bacteria and fungi, together with a great variety of small arthropods, especially springtails and nematodes. Russian workers have found that enchytraeids reproduce rapidly in forest litter, provided it contains caterpillar faeces. Enchytraeid excrement is in its turn eaten by earthworms, the casts of which provide a favourable habitat for certain beetles. The reingestion of microbially enriched faeces, that is, coprophagy, may prove to be of some significance in the functioning of the soil ecosystem (see sect. 4.3).

Microbivores

Fungi and other protists constitute an important part of the diet of many soil animals which might otherwise be classified as detritivores. In fact, the food relationships of these two trophic groups (microbivores and detritivores) are often so complex and intertwined that it is impossible to separate saprotroph from biotroph: many soil animals can apparently be either or both, as the occasion demands. Nevertheless, there are some forms which are noted for their habit of grazing upon fungal mycelia, and fungivores may provide over half the soil faunal biomass. Included here are certain families of beetles and the larvae of some flies. Many oribatid mites are fungivorous and seem to ingest plant detritus for the decay fungi associated with it. These mites, and some springtails, may be quite selective in their choice of fungi and bacteria for food.

Carnivores
In this group also it is frequently difficult to distinguish between true carnivores (biotrophs) and carrion feeders (saprotrophs). Exclusively predatory and parasitic forms do exist however, in many taxa. Arthropods and small annelid worms are preyed upon by various centipedes, spiders and beetle larvae. Predators of the smaller members of the mesofauna include many species of mite, and numerous fly and beetle larvae. Nematodes are voracious predators of other nematodes, small enchytraeid worms, and protozoa. Many rotifers and all turbellarians are predacious, and are relatively unselective in their choice of prey. In addition to these predatory forms, there are some which are parasitic on other animals, including many species of endozoic protozoa and nematodes, and numerous ectoparasitic mites.

Population regulation

As indicated above, restriction to one particular source of food is rare among soil animals. Predators may select from a variety of prey animals, detritivores may occasionally graze on fungal mycelia, and fungivores may change their diet and become detritus feeders should the need arise. In any soil with a well developed organic horizon, therefore, adult members of the soil fauna are unlikely to die from starvation. Nevertheless, there is some evidence that food shortages lead to increased mortality of some immature forms of soil animals, for example the larvae of some mites. In general however, the influence of food supply on fecundity seems to be more important in population regulation than is any effect it might have on mortality. The size of animal populations is a function of the interaction between natality and immigration on the one hand, and mortality and emigration on the other. Migration is not an important factor in the biology of most soil animals, as it is for example with leaf-eating insects. This means that the balance between natality and mortality determines population size, so that any factor which influences either the birth rate or death rate may act in a regulatory fashion to stabilize populations.

Diet may affect reproductive performance in some soil animals, although this matter has been little studied. It is known however, that a diet high in protein stimulates reproduction in earthworms, and that when protein supply is limiting, competition for food intensifies. Competition for available prey might well therefore regulate the populations of predators, though again there is little experimental evidence that such a mechanism is operative in the soil ecosystem. There are, however, data to show that predators themselves have a strong regulating influence on the populations of other soil animals, notably beetles, nematodes, and possibly fly larvae.

Biotic factors such as competition and predation often act in a 'density-dependent' manner, the effect on population size being a function of population density. Populations may also be regulated by environmental factors, which tend to be 'density-independent', in other words their effect is more or less constant over a wide range of population densities. Both kinds of factors are probably operative in most ecosystems, the density-independent mechanisms predominating in ecosystems of low diversity under conditions of environmental stress, the density-dependent agents in high-diversity ecosystems not subject to environmental extremes. In systems of the first kind, population fluctuations are usually large and irregular, whereas the second type of system displays relatively minor oscillations of population density around some steady state value. Furthermore, regulation appears to be achieved, in some populations, by the interaction of density-dependent and density-independent factors. For example, climatic conditions determine the length of the breeding system in many soil animals, and this independent variable probably acts in conjunction with predation to control the size of faunal populations in the soil.

4.3 The process of organic matter decomposition

Much discussion revolves around the relative contributions of the several components of the soil biota to organic matter decomposition and energy flow in the soil subsystem. The softer tissues of plants and small animals are believed to be decomposed by microbes alone, but the breakdown of more refractory tissues involves the combined activities of the soil flora and fauna. Some tissues decompose at the soil surface, releasing soluble products which diffuse into the soil. On the other hand, fresh organic matter may be incorporated directly into soil, by earthworms for example, and sometimes even by vertebrates, before any decomposition takes place. The effects of invertebrates on the decomposition of woody substrates may be one of the more significant examples of fauna–microflora interactions, according to Witkamp and Ausmus (1976). They suggest that the microbial succession which follows the tunnelling activities of termites and wood borers in fallen tree trunks (see sect. 5.4) enhances the incorporation of wood into soil organic matter by translocating nitrogen to the sites of decay through fungal mycelia and perhaps by stimulating bacterial nitrogen fixation *in situ*.

Chemical analyses of the food and faeces of arthropods reveal little difference in the composition of ingested and egested materials (Edwards *et al.* 1970): net assimilation of ingested litter by soil animals is less than 10 per cent, easily digestible sugars and proteins being the major constituents utilized. In view of this, the main contribution of the soil fauna to organic matter decomposition is thought to be an increase in the surface area available for microbial attack as a result of their comminution of detritus.

The importance of physical breakdown by the macrofauna and the larger members of the mesofauna is clearly shown when discs cut from leaves of plants are placed in nylon mesh bags and buried in the soil or litter. The results of one such experiment are shown in Fig. 4.2. Bags made from mesh with a 7 mm opening, large enough to allow entry of all microbes and most invertebrates (see Fig.1.13), permitted more rapid decomposition of the leaf discs than those made from mesh of 0.5 mm, through which only microorganisms, mites, springtails and other small arthropods, small

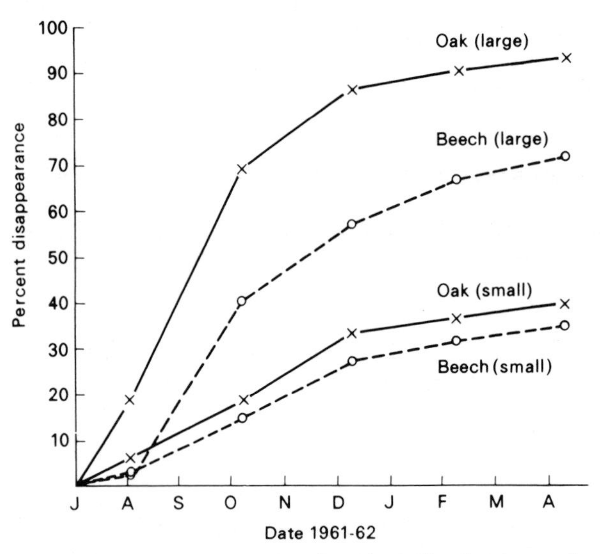

Fig. 4.2 Breakdown of plant litter by soil animals, as shown by the effect of mesh size on the decomposition of oak and beech leaves in nylon bags. The larger mesh (7 mm) admitted most invertebrates while the smaller (0.5 mm) excluded all but small nematodes and small arthropods such as mites and springtails. (From Edwards & Heath 1963)

nematodes and some tardigrades and rotifers, could pass. Leaf discs placed in bags made from mesh of 0.003 mm size, to which only micoorganisms gained access, remained intact throughout the duration of the experiment, a period of about nine months.

These results illustrate unequivocally how the mesofauna in particular disintegrates plant litter in forest ecosystems and in the process makes it more susceptible to microbial attack. In grassland ecosystems, leaves are generally softer and more readily decomposable, and breakdown by the soil fauna has a less conspicuous role. The surface area of plant detritus in forests may be increased by up to several orders of magnitude due to the action of the meso- and macrofauna. The faecal pellets which result maintain a more favourable moisture status than does unaltered plant litter, and are more readily moved down into the litter layer by percolating water or gravitational forces. Whether any significant chemical changes take place in the organic matter during its passage through the animal gut is a more contentious issue than the extent to which physical changes occur. While the composition of faecal matter is generally very similar to that of food prior to ingestion some enzymatic degradation no doubt takes place (see below), and a slight rise in pH occurs, during passage through the gut. These changes, however small, when combined with better moisture relationships are believed to make faecal material a very suitable substrate for bacterial, though not perhaps fungal, exploitation. This conclusion has been questioned by Satchell (1974) who drew attention to evidence for the resistance of faecal pellets to decomposition and their consequent persistence in soil and litter.

J. E. Satchell also emphasized an aspect of litter comminution by arthropods that is frequently overlooked, namely its effect on the topography of the litter layer. Freshly fallen litter is moist for only short periods after rain because of its looseness and consequent rapid drying, whereas on fragmentation the litter becomes more closely appressed to the soil surface, creating a zone where the humidity is higher and more stable. The increase in water-holding capacity and decrease in evaporation brought about by flattening the litter layer may prove to be a more important outcome of litter fragmentation by invertebrates than the greater surface area it provides for microbial colonization.

Despite the strength of argument in support of physical interpretations of the effect of soil and litter arthropods, there has recently been a reassertion of earlier views that the fauna influences decomposition directly (Crossley 1977). Although the elemental concentrations of ingested and egested material differ only slightly, the molecular forms of the elements may be quite distinct

(Witkamp & Ausmus 1976) and this may consequently modify the rate of decomposition. Thus the presence of invertebrates doubled the loss of organically bound ^{134}Cs during litter breakdown in the field (Crossley & Witkamp 1966) even though the litter fauna accounted for only 1 per cent of the energy expended by the decomposer group as a whole. A small initial energy expenditure by invertebrates may therefore greatly increase mineralization of litter.

Mycophagy

One of the more widespread interactions between the soil fauna and microflora, and which occurs regularly in decomposer successions such as those outlined in Chapter 3, is mycophagy. This is the term used to describe the consumption of fungal tissue by microbivores. It is common among arthropods, especially insects and mites. Some insects not only consume fungi but enter into a symbiotic relationship from which the fungus benefits by improved growth and dispersal. Thus some ants and termites culture fungi on plant material brought into their nests, the fungal tissues being the primary food of the former and an essential though minor food component of the latter. Larvae and adults of a variety of insects found in decaying wood feed either on fungal mycelium or consume both wood and fungus. The significance of this mode of nutrition has recently been reviewed by M. M. Martin (1979), from whose account the following aspects relevant to the process of organic matter decomposition are taken.

Fungal tissue is a potentially rich source of carbohydrate for an insect possessing the appropriate enzymes, but it is not known for certain whether such enzymes are widely distributed among insects. The major structural polysaccharides of fungi are chitin and non-cellulosic β-(1,3)- and β-(1,6)-glucans; the cellulose, lignin and pectin of vascular plant tissues are not found in the normal soil mycoflora. Fungi can also provide the sterol requirements of many insects, which they need for normal growth and reproduction: the main fungal sterol, ergosterol, can be utilized even by phytophagous insects which normally do not consume fungal sterols.

The nutrition of xylophagous (wood-eating) insects has excited the curiosity of entomologists for many years. The largest number of xylophagous species are found in the orders Isoptera (termites) and Coleoptera (beetles). Early workers studying cellulose digestion by wood-boring beetles attributed this to endosymbiotic microbes but it is now believed that the endophytes of xylophagous beetle larvae provide the insects with vitamins and perhaps essential amino acids, and not digestive enzymes. The source of insect cellulases is still uncertain and it is often presumed that, if they do not derive from gut microorganisms, then they are secreted by the beetles themselves. Martin (1979) has suggested an alternative hypothesis, namely that many of the enzymes present in the gut of wood- and litter-feeding arthropods are fungal enzymes acquired while feeding on mycelia and/or on a substrate into which fungal enzymes have been secreted. Thus some termites, by grazing on fungi which they grow in their combs, acquire an essential component of the cellulase complex, and it is this that enables them to digest cellulose in their guts. Such acquired enzymes, of course, must be able to function effectively in the biochemical environment of the insect gut, and not be subject to deactivation or denaturation. This certainly happens in the attine ants which feed upon the fungi they cultivate in their nests: they are entirely dependent on fungal proteinases which are consequently not subject to proteolytic enzyme attack themselves. Whether the absence of digestive proteinases is a general characteristic of insects that exploit acquired enzymes is not known for certain, however.

Regulation of decomposition processes

It is quite clear that organic matter decomposition in the soil depends on the integrated activities of microbes and small animals. What is less certain is the extent to which the interaction between the two groups of decomposers regulates the process overall. This will depend, *inter alia*, on successional patterns which develop in accord with the state factor equation (Ch. 3, eq. 3.8). There is evidence that leaves of woody perennials become depleted of simple substrates before abscission and that nutrient withdrawal occurs during leaf senescence. As indicated in section 3.3, we do not know to what degree these processes of autolysis and translocation within the leaf influence primary colonization by microorganisms, but it can be safely assumed that both leaf physiology and anatomy will have some effect on the direction of microbial succession in decomposing leaf litter. In addition, the degree of physical breakdown by the mesofauna is known to depend not only on leaf anatomy but also on the kinds of animals involved. The various litter-feeding and phytophagous invertebrates have their own distinctive patterns of leaf consumption, some mining the internal parenchyma tissues, others consuming the cuticle and epidermis, and still others reducing the leaf to a skeleton of vascular tissue. In mor soils, where the fauna consists principally of oribatid mites and collembola, plant residues with well preserved cellular structure may be found in the uppermost mineral horizon. In mull soils, where the fauna is more varied and numerous, all plant structures are completely destroyed; furthermore, many of the larger animals ingest mineral particles along with their food, thus promoting the formation of soil aggregates.

In Chapter 3, the suggestion was made that decomposition might be determined by leaf physiology and anatomy in such a way as to regulate the recycling of nutrients to vascular plants. Certainly root uptake, exchange reactions in the soil, and recycling by decomposers are processes which utilize the products of organic matter mineralization and 'minimize losses from the ecosystem of elements whose absence would severely limit primary production' (Witkamp & Ausmus 1976). This viewpoint is readily acceptable, but it is a large step from here to postulate, as some ecologists have done, that mature ecosystems are so closely integrated that decomposition processes are 'programmed' by plant physiology and anatomy in order to conserve nutrients. The regulation of nutrient loss from the soil ecosystem is discussed again in Chapter 5.

4.4 Measurement of carbon mineralization and energy flow

Qualitative studies of microbial growth in relation to substrate, for example successional studies (Ch. 3), or of soil animals in relation to food supply, do not throw much light on the contribution of decomposer organisms to energy flow in ecosystems, nor on the relative importance of various taxonomic groups. Unfortunately, there are considerable technical difficulties involved in working out energy budgets for the soil biota, and as a result the data on energy release and carbon mineralization by the decomposers are not nearly so extensive, nor as reliable, as those relating to energy capture and CO_2 fixation by the producers. An analysis of energy flow, or of the carbon cycle, in the soil ecosystem requires a population energy budget or carbon budget for each recognizable trophic group, to provide the basis for constructing a detailed food chain or food web. It is customary to recognize two kinds of food chains, namely the grazing food chain and the detritus food chain (see Fig. 4.3). Such a model is however inadequate to represent the complexity of the decomposer group. Little more can be deduced from it than an appreciation of the overall importance of the soil biota in the energetics of terrestrial ecosystems, relative to that of the other biotic components.

Before studying how energy is channelled along the various pathways and food chains, it is necessary to measure the total input to the soil subsystem. In other words, we must have a knowledge of total soil metabolism.

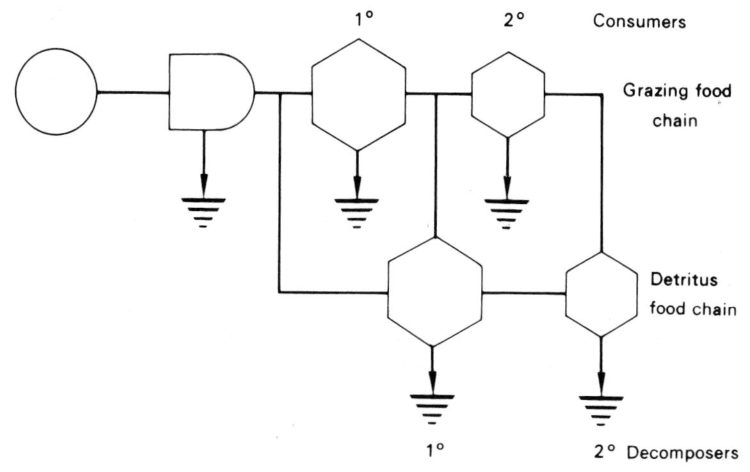

Fig. 4.3 A two-channel food chain – the standard trophic model – depicted in energy circuit language.

Methods of measuring soil metabolism

The realization that a large part of the net production of any terrestrial ecosystem passes directly into the soil–litter subsystem, and is decomposed there, is fairly recent. The study of organic matter decomposition in soil has therefore assumed great significance, not only because so much of the energy captured by photosynthesis is released by it, but also (and perhaps more importantly) because it is only during this dissipation of energy that nutrients bound in the organic form are 'mineralized', and so made available for recycling through the vegetation subsystem. The cycling of carbon is discussed later in this chapter, while the mineralization of other important plant nutrients is treated in detail in Chapter 5.

Soil metabolism reflects the energy demands of the soil biota. There are several ways in which it may be determined, namely by difference, by litter fall, by soil respiration, by litter decomposition, and by soil mineralization. As indicated previously, spatial and temporal heterogeneity in substrate distribution and availability greatly complicate the measurement of decomposition rates. If there were no pattern to the variability, then unrestricted random sampling could (subject to certain provisos related to the effect of sampling *per se* on microbiological processes) provide a reliable estimate of the rate of organic matter decomposition. Given the existence of spatial patterns however, some form of stratified random sampling is necessary, and this presupposes a knowledge of substrate distribution and decomposer activity patterns. This key problem in the study of soil metabolism has not yet been studied systematically. With this general reservation on methodology therefore, the means of measuring the rate at which substrates are decomposed will be described in turn.

Difference method

If the energy consumed by above-ground herbivores can be determined, and subtracted from net primary production, then the remainder (in a steady state system) represents energy flowing through the decomposer component. Several estimates of soil metabolism have been made in this way. While sound in theory, it results in practice in estimates which are often inaccurate and imprecise, because it compounds errors made in the determination of net primary production and herbivore consumption. Though useful in certain grassland ecosystems, especially those grazed by domestic livestock, its applicability to forest ecosystems is limited by the difficulty of obtaining an accurate estimate of the foliage consumed by phytophagous insects. Furthermore, the method takes no account of losses to soil-inhabiting biotrophs, which may be considerable. For example, Ausmus *et al.* (1976) estimated that 8.5 per cent of the annual root production in a *Liriodendron* (tulip tree) forest was consumed by nematodes.

Litterfall method

In a steady state system, the annual input of litter and the amount which is decomposed annually should be equal. Litterfall, consisting of leaves and other plant parts, can be measured directly by catching the litter in suitably designed litter traps. Excrement of herbivores and carnivorous animals is not usually caught in these traps, and should be determined separately; an appreciable amount of readily decomposable organic matter is added to some soil ecosystems from this source.

In forests, the annual rate of litterfall increases directly with decreasing latitude and altitude from about 2.5 Mg ha^{-1} in arctic and alpine forests to 25 Mg ha^{-1} in equatorial lowland forests (Rodin & Basilevich 1967). The litterfall method is not however readily applicable to grasslands because of the difficulty of collecting litter and measuring its production accurately in grassland communities. Another defect of this method is that it fails to take into account the input of detritus from root systems, so that estimates of energy input from litter fall are underestimates. They also ignore the respiration of plant roots and their associated microflora (see Ch. 6). Furthermore, the contribution of any herbaceous ground flora is usually neglected by this method.

The decomposition constant. The amount of litter that accumulates on the surface of the mineral soil depends on the balance between the rate of litterfall and the speed with which it decays once it reaches the soil surface. Different kinds of forest ecosystems differ in the rates at which organic matter decomposition proceeds. The overall rate will obviously depend on such factors as the degree of lignification of the tissues being decomposed, on the prevailing climate and on weather patterns. Decomposition is generally more rapid in hot climates than in cold. This is illustrated by the fact that less litter accumulates on the soil surface in tropical forest ecosystems than in cool temperate forests, despite the fact that, as indicated above, the rate of litter production is higher in the former.

J. S. Olson (1963) discussed the relationships involved in some detail, unifying the earlier approaches of Jenny *et al.* (1949) and Greenland and Nye (1959). If X represents the accumulated surface litter, and L the amount of new litter added each year by the process of leaf fall, then the rate of change in X is given by the expression:

$$dX/dt = L - kX \qquad (4.1)$$

where k is a constant which may be called the decomposition or decay constant.

For an ecosystem in steady state, X is constant, i.e. $dX/dt = 0$. At this point, the following relationships hold:

$$L = kX, k = L/X. \qquad (4.2)$$

The decomposition constant k can thus be estimated by measuring L and X in evergreen communities assumed to be in steady state. Some values of k obtained in this fashion are: 0.025 for Ponderosa pine forests in the Sierra Nevada mountains of the western United States, 0.25 for southern pine forests in the southeastern United States, and 4.0 for tropical lowland rainforest in Africa (Fig. 4.4). For forests with a marked autumn peak in litterfall approximating the ideal deciduous forest, the decomposition constant may be determined by measuring surface litter biomass just after (X_0), and just before (X), the annual litterfall. In this instance $k' = (X_0 - X)/X_0$ and since $X = X_0e^{-kt}$, $k' = 1 - e^{-kt}$, thus providing an indirect measure of the 'instantaneous' decay rate, k.

The decomposition constant may be used to calculate the time needed for a given fraction of the accumulated litter to decompose, assuming no further additions of litter are made. This is a special case of eq. (4.1) which becomes:

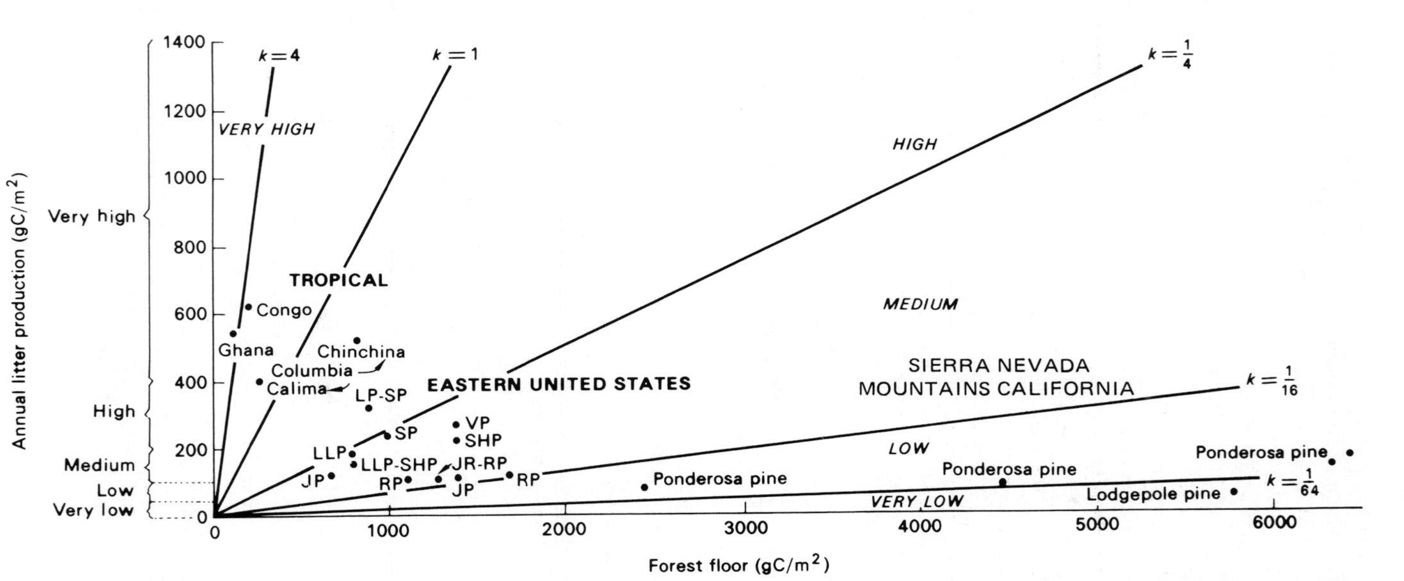

Fig. 4.4 The decomposition constant k, estimated for various evergreen forests from the ratio of annual litter production to the amount of litter accumulated on the forest floor. (From Olson 1963 by permission of the Duke Press.)

$$dX/dt = kX. \tag{4.3}$$

According to this expression, the decay rate (dX/dt) is proportional to the amount of litter present. This situation is analogous to the phenomenon of radioactive decay, and we may calculate a theoretical 'half-life' for the litter analogous to the half-life of a radionuclide, that is the time taken for the total amount present at time zero to be reduced by half.

Solving eq. (4.3),

$$X = X_0 e^{-kt} \tag{4.4}$$

where X_0 is the amount of litter present at time zero. Then, by putting $X = X_0/2$ in eq. (4.4), we may derive the following expression for litter half-life:

$$t_{\frac{1}{2}} = -\ln 0.5/k = 0.693/k. \tag{4.5}$$

Substituting the values of k given previously for the several ecosystems, half-times for litter breakdown are found to be over twenty years in the subalpine Ponderosa pine forests, two or three years in the subtropical southern pine communities, and less than one year in tropical rainforests.

Similarly, it may be shown that 95 per cent of the litter would be decomposed in $3/k$ years, and 99 per cent in $5/k$ years.

Inaccuracies in calculating decay constants arise from the methodological difficulties inherent in measuring both litterfall and surface litter: since decomposition constants are calculated from the ratio of litter input to litter standing crop, errors in the estimation of both may be compounded. Apart from seasonal variability in litterfall, which is widely recognized, spatial variability is such that the method is generally too imprecise to serve as a basis for comparing short-term (e.g. seasonal) fluxes of energy from individual litter components. Notwithstanding, provided litter trap design is appropriate and traps are adequately replicated, the method gives a reasonably precise estimate of total annual litterfall; year-to-year variation however remains high, so that measurement of litterfall may need to be continued for five years or longer in order to provide reliable data on annual inputs.

Seasonal variation in the standing crop of surface litter is most evident in deciduous hardwood forests and has only rarely been examined in other ecosystems. The difficulties in obtaining reliable estimates are such that a very high degree of replication is necessary to demonstrate any significant differences. Additionally, a substantial proportion of litter may be incorporated in the topsoil by animal activity, and can be recovered as recognizable components by wet sieving (Richards & Charley 1977). If this buried material is included in the calculation of *k* values the apparent decomposition rates will be much less: by so doing, estimates of litter half-life in two different eastern Australian forests were raised by 28 and 36 per cent.

Soil respiration methods

It is possible to determine the respiratory activity of the soil biota by the techniques of respirometry and/or gas analysis. Either or both O_2 uptake and CO_2 output may be measured, although most variants of the method are based on the latter. They all tend to overestimate the contribution of decomposers to total soil metabolism because it is impossible to distinguish the CO_2 produced by roots from that respired by the soil fauna and microflora. Root respiration is generally considered to encompass any or all of the following: metabolic respiration by living roots; respiration of root exudates by microorganisms inhabiting the rhizosphere including the special case of mycorrhizal respiration; and microbial respiration of sloughed-off root tissues. A redefinition of the term root respiration seems necessary. Richards (1981) has proposed that it include only that proportion of the total soil respiration derived from autotrophic as opposed to heterotrophic components of the system. On this basis, root respiration should only encompass metabolic respiration by live roots including any associated mycorrhiza and nodule respiration. Heterotrophic respiration of root exudates within the rhizosphere and of detritus derived from below-ground sources, as well as respiration of organic material from above-ground sources, comprises the balance of total soil respiration. This places symbiotic associations, such as mycorrhizas and root nodules, with the autotrophic component, despite the fact that the microbial symbionts are themselves heterotrophs. The rationale for this lies in the fact that mycorrhizas and root nodules are nutrient acquiring organs of the autotrophic host plants. Even if this definition of root respiration is accepted, the problem of estimating soil respiration in the presence of living rootsystems remains.

Early methods of measuring soil respiration *in situ* frequently involved the pumping of air over the soil, and gave estimates which were far too high, because CO_2 trapped in the soil pores was included along with the respiratory output. Improved techniques involve the use of boxes or cylinders inverted over representative portions of the soil surface, these small chambers being sealed for only short periods during which measurements are recorded. Even so, sources of error remain. Root respiration is still included, for example. In addition, mechanical disturbance of the soil during installation of the chambers can stimulate metabolic activity. Finally, temperature and humidity changes within the small chambers can change the rate of CO_2 evolution from the soil–litter system. For logistical reasons, the use of the infrared gas analyser is normally restricted to the laboratory, and field estimates of soil respiration usually rely on the KOH technique. The apparent simplicity of the inverted box–KOH absorption method has led many workers to overlook its experimental limitations, chief of which is chamber geometry, that is, the size and shape of the absorption vessel and its position within the inverted box. This determines the area of alkali presented for absorption and the length of the diffusion pathway from CO_2 source to KOH sink, factors which were found by R. J. Hartigan (1981) to be critical for complete recovery of CO_2.

Although soil respiration determinations must be made in the field to be of any value in ecosystem analysis, laboratory studies with the Warburg respirometer can provide realistic comparisons between various soils, even when air-dried and rewetted soil samples are used. They also provide a means of studying the effect of environmental factors on organic matter decomposition, under controlled and reproducible conditions. Other laboratory studies, involving terrestrial microcosms (terraria), may prove of value in the testing of trophic models of real soil ecosystems.

Litter decomposition methods

If weighted samples of litter are placed in mesh bags made of nylon or fibreglass, the decomposition constant can be determined by periodically reweighing them (Shanks & Olson 1961). In practice, a large number of bags is usually placed in the field, and replicate groups are destructively sub-sampled at predetermined intervals of time. By using bags of different mesh size, particular groups of organisms can be excluded, while those entering the bags can be readily extracted, and some idea of the relative contribution of various taxa obtained (see Fig. 4.2). Several variants of the method exist. Thus branches (or parts thereof), and even individual leaves (Birk 1979), may be tethered to points on the soil surface and their weight loss determined from time to time. Log decomposition studies (Grier 1978) represent a special case of this approach.

If this technique is to give a valid estimate of decomposition on the soil surface, then it is imperative that the material used (leaves etc.) be collected as it falls from the canopy in the weeks or days immediately preceding the commencement of the study, so that it closely approximates the energy status of natural litter at the time the bags are set out. Unfortunately not all workers have observed this simple precaution. In view of the great significance of substrate quality as a determinant of decomposition rate, studies which neglect to take this into account must give false estimates of litter metabolism. Confining litter in mesh bags also creates somewhat artificial conditions, and may result in reduced decay rates (Witkamp & Olson 1963). Furthermore, leaf fragments may be lost during recovery of the bags, and foreign material may enter them, so that the variability of weight loss increases greatly with the passage of time. While valuable for comparative purposes, litter decomposition methods are generally too crude to provide reliable data on energy flux in the soil–litter subsystem.

Soil mineralization methods

These are designed specifically to study the rate of release of inorganic nutrients, especially nitrogen, from decomposing soil organic matter, but may be regarded as providing an indirect measure of soil metabolism. They will not be considered further here; details may be found in the papers of Bremner (1965) and Runge (1983).

Summarizing the four major methods described above, soil respiration techniques are likely to overestimate soil metabolism, while the other three methods in general underestimate the activities of the soil and litter microbiota. This is because none takes into account the contribution of root detritus to the soil subsystem, and this may be a significant source of substrates for decomposer organisms (Ausmus & Witkamp 1974; Waid 1974; Coleman 1976; Newman 1978).

4.5 Extrinsic and intrinsic factors affecting decomposition

Any environmental factor which affects the activities of the soil biota may influence the decomposition of organic residues. These include moisture, temperature, soil pH, oxygen supply, inorganic nutrients, and clay content, any of which may act as limiting factor (see sect. 3.1). Many of the physical factors of the environment are determined by the regional climate and, in general, decomposition rates reflect the combined influence of climate and biological factors, chief of which is the quality of the substrate, that is its inherent susceptibility to microbial degradation.

Substrate quality

When crop residues, either green or cured, are intimately mixed with soil, about two-thirds of their carbon is lost during the first year but thereafter decomposition slows dramatically, so that five or ten years later some 10–15 per cent of the incorporated carbon remains in the soil (Jenkinson 1981). The marked change in the decay rate after about one year is due to the synthesis of humic substances during the initial stages of decomposition and to the formation of resting structures by the zymogenous microflora. In this regard, it should be noted that the hyphae, conidia and sclerotia of many fungi contain melanins which make them particularly resistant to microbial attack.

Decomposition results in the conversion or mineralization of organically bound nutrients to inorganic forms but, at the same time, inorganic elements are sequestered or immobilized in microbial tissues. The supply of readily decomposable substrates generally determines which process, mineralization or immobilization, dominates at any particular time (see Ch. 5). For example, a fresh input of substrates will result in immobilization of nitrogen only if these are high in energy in relation to their N status: largely carbonaceous litter will cause greater immobilization of this element than highly proteinaceous materials. The period of net immobilization depends on the relative proportions of the two factors energy and nitrogen, and this is roughly approximated by the C/N ratio of the residues undergoing decomposition.

Indices of substrate quality

C/N ratio. Cellulose is the major constituent of plant detritus and it differs from other insoluble substrates of plant or animal origin (chitin, keratin and lignin) in containing no mineral elements. According to D. Park (1976) this is why the importance of nitrogen supply has often been stressed and why the significance of the C/N ratio was recognized early. Because nitrogen is important as a limiting factor in crop production, much agronomically oriented research has centred on the C/N ratio as an index of substrate quality. If the C/N ratio is greater than 20/1 in soils under cultivation, immobilization dominates throughout the growing season and cereal crops may suffer nitrogen deficiency as a result (see sect. 5.2.1). In natural communities, plants can obtain sufficient mineral nitrogen for their requirements from decomposing residues having much greater C/N ratios. For example, the C/N ratio of leaf litter in forest communities is often greater than 100/1 when it reaches the forest floor, and may still exceed 40/1 before the comminuted fragments enter the soil, yet 5 to 10 tonnes of such litter are decomposed in productive forests each year. Clearly factors other than C/N ratio affect substrate decomposability.

Physiological adaptation of decomposer organisms is one such factor (Levi & Cowling 1969). As Griffin (1972) has emphasized, the growth of fungi in heartwoods, which have C/N ratios of the order of 1000/1, requires great efficiency in nitrogen metabolism. This is achieved in a number of ways: reuse of mycelial nitrogen by internal translocation from young to old cells of cytoplasmic contents, or their autolytic products; production of extracellular lytic enzymes which render cell walls available for assimilation by younger cells; and preferential allocation of nitrogen to metabolically active systems. Griffin noted, however, that wood-decaying fungi can only reuse nitrogen in the presence of exogenous carbohydrate, and in fact most cellulolytic fungi show greatly reduced production of cellulases as substrate C/N ratio increases. The white rot fungi are exceptional in this regard, being able to produce cellulases at a C/N ratio of 2000/1.

Lignin content. So far as inherent decomposability goes, the chemical nature of organic residues seems to be more important than their C/N ratio. In general, hemicelluloses decompose more rapidly and are utilized as substrates by a wider range of microorganisms than cellulose, and lignin is more refractory than cellulose. As the lignin content of plant tissues increases, so too does the ability of the litter microflora to decompose them decline, and the capacity to use lignin as a substrate is not widespread among microorganisms (see sect. 4.1). According to Minderman (1968) it is the slowly decomposing litter components, such as lignin, which determine the shape of the long term decay curve once the more labile constituents have been oxidized.

Several authors have shown lignin content to be a good index of substrate quality for the prediction of litter decomposition rates. For various litter components – green needles, twigs, cones and bark – in old growth Douglas fir forests, Fogel and Cromack (1977) found that weight loss was better correlated with lignin concentration than with C/N ratio. The relationship between annual decomposition rate and lignin content of the various tissues was expressed as follows:

$$k = 1.126 - 0.275\ (\ln \%\ \text{lignin}). \qquad (4.6)$$

The correlation coefficient r for this relationship was 0.999, indicating that lignin concentration accounted for virtually all the variability in the decay constant. These findings support the contention (Bollen 1953) that structural complexity is a more important determinant of substrate decomposability than C/N ratio or nitrogen status *per se*. For tissues of the same kind and comparable states of maturity however, the C/N ratio may be a better index of decomposability than lignin concentration (Edmonds 1980).

Polyphenol content. To complete this discussion of substrate quality, brief reference should be made to the role of polyphenols and other water soluble extractives of litter. Polyphenols in conifer needles are thought to complex with proteins of the mesophyll cells just prior to leaf abscission. Such complexes are quite resistant to decay, resulting in mesophyll cells being a major recognizable constituent of mor humus in coniferous forests (Handley 1954). Polyphenols also complex with proteins of plant and microbial origin in the litter layer and, according to Williams and Gray (1974), there is good evidence that they are important in determining differential rates of litter decomposition. Their effect is influenced by the pH of the litter and soil nutrient status: there is a higher content and a wider variety of polyphenols on infertile soils, and complexes formed with water soluble extracts (tannins) are more stable at lower pH.

Climate

Decomposition proceeds more rapidly in tropical than temperate climates (Jenny *et al.* 1949; Olson 1963): decay constants range from about 0.1 in arctic and subalpine communities to 4.0 in tropical lowland rainforests. As pointed out by V. Meentemeyer (1978) however, there have been few critical studies of the role of environmental factors in decomposition. On a global continental scale, Meentemeyer found the annual decay rate to be strongly correlated with the actual evapotranspiration (AET), probably because AET is a measure of the energy and moisture concurrently available for decomposition processes in the litter layer.

Decomposition rates, based on litter bag studies, were compared along environmental gradients of temperature, moisture and altitude in the western coniferous forest biome project of the International Biological Program. Needle decay constants in old growth Douglas fir forest, about 450 years of age, decreased from 0.28 at cool, moist sites to 0.22 in warmer, drier habitats (Fogel & Cromack 1977). It was also shown that, although decay rates generally decrease with increasing altitude, decomposition occurs under a heavy snow pack in a subalpine *Abies amabilis* community (Edmonds 1980).

Interaction of climate and substrate quality

When the decay constant is plotted against time, the shape of the resulting curve suggests that decomposition proceeds as a series of exponential steps, despite the fact that one exponential function can generally be fitted to the data quite successfully. In other words, it seems probable that k values are not constant throughout time. This might be the expected response to extrinsic variables such as weather, for litter layers in many plant communities are normally subjected to alternating periods of wetting and drying, and to cycles of fluctuating temperature. Minderman (1968) implied that substrate quality has a similar effect, and suggested that the course of litter decomposition could best be represented as the summation of the decay curves of the several constituents, sugars, hemicelluloses, cellulose, lignin, waxes and polyphenols etc., in proportion to the amounts in which they occur.

V. Meentemeyer (1978) set out to examine the dual control of regional climate (subpolar to warm temperate) and substrate quality on litter decomposition, using lignin concentration as an index of substrate quality and AET as the most appropriate integrator of climatic variables. Multiple regression analysis revealed that 51 per cent of the variance in the annual percentage weight loss was accounted for by AET and a further 19 per cent by the AET/lignin ratio. On a global scale, climatic control should be even more dominant. On the other hand, within any region of uniform macroclimate, at sites of comparable topography and aspect and therefore relatively uniform in microclimate, one would expect substrate quality as measured by lignin status to govern the rate of litter decomposition. The relative control by lignin over decomposition rate is not uniform over different climatic regions however: the more favourable the temperature and moisture regimes the more rapid the decay for a given lignin content, but the higher the lignin content the more favourable the climatic conditions needed for litter breakdown. Meentemeyer visualizes AET as the climatic forcing function and lignin concentration as the substrate moderator of decomposability.

4.6 Population energy budgets

As indicated earlier (sect. 4.4), the traditional trophic model is inadequate for proper consideration of the energetics of the soil ecosystem. This matter has been discussed by R. G. Weigert and colleagues (Weigert *et al.* 1970). The problem is partly one of nomenclature, the traditional terminology (producer, consumer, decomposer) tending to obscure the primary trophic differentiation into autotrophs and heterotrophs. A further complication is the fact that the name 'decomposer' is applied generally to such a heterogeneous collection of soil organisms. An improvement over the simple trophic model, which shows a single energy pathway through the soil (the detritus food chain), is one which shows two parallel pathways, separated according to whether the energy (food) source is alive or dead (Fig. 4.5). This modified trophic model uses the classification of biotroph and saprotroph, introduced in Chapter 2. Saprotrophs might be regarded as the 'true' decomposers; biotrophs include both grazers and predators. The first order heterotrophs of the soil subsystem are mostly saprotrophs, though first order biotrophs may be present also, such as larvae and nematodes which feed on plant roots or amoebae that ingest soil algae. In addition, both higher order saprotrophs and biotrophs are a feature of the soil ecosystem.

The modified trophic model shown in Fig. 4.5 has the advantage

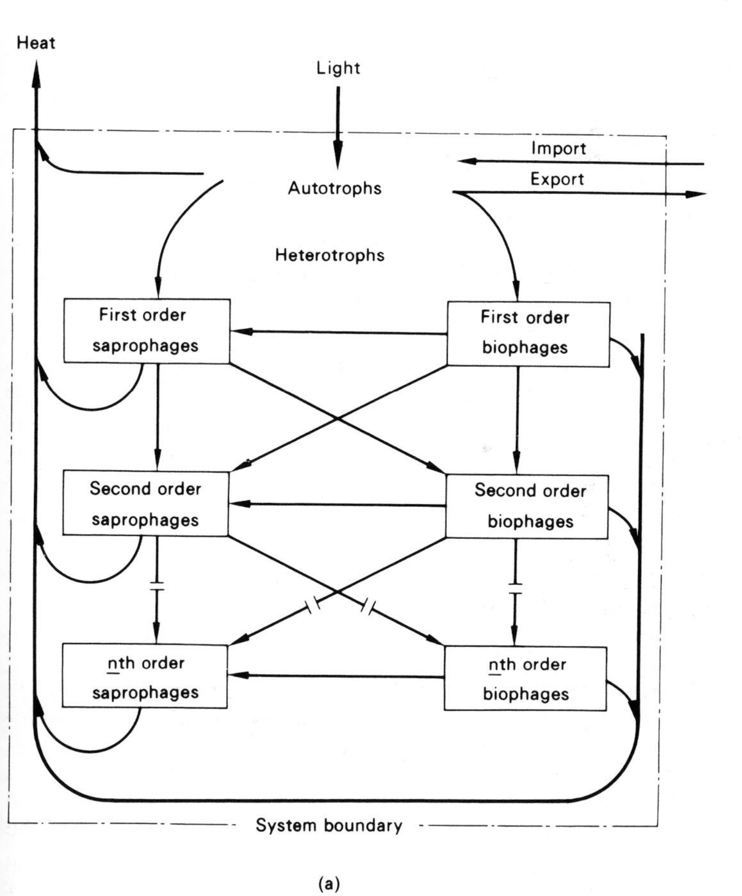

Fig. 4.5 (a) Modified trophic model for the soil-litter subsystem (From Weigert *et al.* 1970)

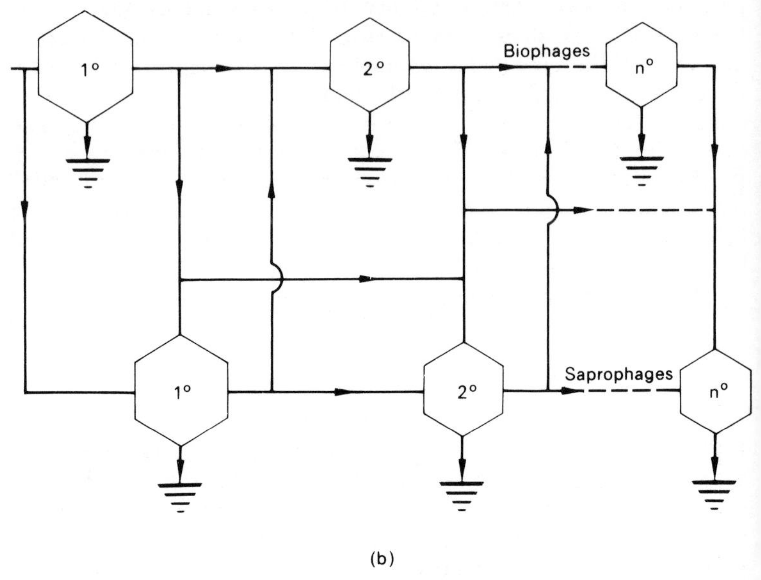

Fig. 4.5 (b) Energy network diagram showing details of the detritus food chain in the soil.

of permitting separate consideration of saprotrophs subsisting at different trophic levels. It is still limited, however, by the occurrence of a single species at more than one level, and by the fact that some animals apparently alternate between biotrophy and saprotrophy. Because of these limitations, it may be better to revert to a single species population model which, although it may not so readily distinguish between biotrophy and saprotrophy, at least will permit the contribution of a single species to energy flow to be evaluated. In any event, whatever model is used, the determination of the population energy budget is basic to any analysis of ecosystem energetics.

Following Weigert *et al.* (1970), a simplified population energy budget may be expressed by the following equation (neglecting work exchanges with the system surroundings, which are quantitatively negligible):

$$I = P + R + E \tag{4.7}$$

where I = ingestion or consumption, i.e. chemical energy content of ingested matter;

P = production, i.e. chemical energy content of materials synthesized;

R = respiratory heat loss, i.e. the net heat exchange between the population and its surroundings;
E = egestion, i.e. chemical energy content of egested matter.

The production component may be further subdivided into growth and yield, if necessary, and the energy equation then becomes:

$$I = G + Y + R + E \qquad (4.8)$$

where G = growth, i.e. chemical energy content of matter represented by the increase in biomass or standing crop;
Y = yield, i.e. chemical energy content or matter lost to a subsequent trophic level.

A simple flow diagram illustrating eq. (4.8) is given in Fig. 4.6. This model might serve as a basis for applying the techniques of systems analysis to a solution of the problem of energy flow in the soil ecosystem. This is by no means a simple task, however: it requires each important species component to be identified and its population energy budget determined; it also presupposes a knowledge of the food web of the system. There are formidable difficulties in the way of obtaining the requisite data, especially that pertaining to the Protista. Despite significant advances made in recent years, techniques for clearly distinguishing between active, inactive, and dead microbial cells in the soil have yet to be perfected. There also remains the problem of measuring soil respiration without interference from root respiration, a task made more

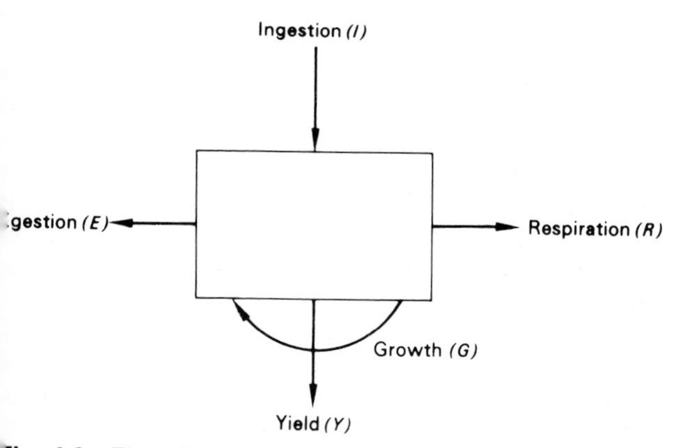

Fig. 4.6 Flow diagram for a balanced energy equation. See eq. (4.8). (From Weigert *et al.* 1970)

difficult by the fact that an important fraction of the microbiota is closely associated with living roots (see. Ch. 6).

Partitioning energy flow among the soil biota

The relative contributions of microbes and soil animals to soil metabolism is not known for certain. It is generally accepted that the microbiota (mainly fungi and bacteria) contribute far more to total soil metabolism than do the meso- or macrobiota. Something of the order of 10 per cent of the annual input of energy in litter is thought to be utilized for biosynthesis, growth and respiration by the soil fauna. The low biomass of soil animals relative to bacteria and fungi is indicative of their respective direct contributions to soil metabolism. The indirect involvement of the fauna, in stimulating microbial development (sect. 4.3), is of course much greater.

Role of soil fauna

For adult animals, if it is assumed that growth and yield are negligible and that respiratory heat loss is all directed towards maintenance, eq. (4.8) may be simplified thus:

$$I = (r \times B) + E \qquad (4.9)$$

where r = respiratory heat loss of an individual per unit of biomass;
B = biomass or standing crop.

An estimate of r, obtained in the laboratory, can be used to estimate $R(= r \times B)$, provided the biomass of the population is known, and provided the respiratory rate of the animals in question is the same under field and laboratory conditions. Such an estimate is unrealistic, however, because it ignores the normal age-class distribution which pertains in all naturally occurring populations. Because the respiratory rate of individuals varies with their state of maturity, a more accurate estimate could be derived by summation of the respiration of the various age (or size) classes represented in the sample. A further disadvantage of the method is that it is not applicable to the smaller members of the mesofauna, for which indirect estimates of respiration must be made, for example by calculating the regression of respiration on weight for a sample of the population.

Techniques are also available for making independent estimates of the other terms in eqs. (4.7) and (4.8). For example, the quantity of food ingested, I, and the amount of matter egested as faeces, E, may be determined by laboratory studies. Production, P, can be estimated by making successive measurements of biomass. This is comparatively simple and direct so far as the larger soil animals are concerned, but weighing of many of the individual smaller forms is difficult if not impossible, and for these animals, such as the smaller mites and collembola, indirect estimates of biomass must be made, for example from previously determined relationships between linear dimensions or surface area and weight.

An example of an energy budget for a soil animal population is provided by the data of M. D. Engelmann (1964), for cryptostigmatid mites in a grassland soil from Michigan, USA. The values (cal m^{-2} yr^{-1}) for the various parameters were as follows: I 10 248; P, 270; R, 1965; E, 7686. A further term which must be taken into account is mortality M, 430. Adding this term to the right-hand side of eq. (4.7) permits an independent estimate of I, as the sum of P, R, E and M, that is, 10 351 cal m^{-2} yr^{-1}. This compares quite closely with the experimentally determined value of 10 248 cal m^{-2} yr^{-1}.

The contribution of the various trophic groups of soil animals to total faunal respiration has not been estimated reliably for many soils. The significance of individual groups will vary with their biomass, which in turn varies with locality, so that the relative importance of any particular group will change from place to place. Some larger animals, isopods for example, may compensate for a low metabolic rate with high biomass, while conversely, the low biomass of groups such as collembola, nematodes and enchytraeids may be offset by their high respiratory activity. Groups such as crytostigmatid mites, which usually have a low biomass coupled with a relatively low rate of metabolism, probably make little direct contribution to organic matter decomposition in soils; the ecological significance of these animals lies in their ability to promote decomposition by bacteria and fungi. The lack of correlation between biomass and energy flow is clearly brought out in Table 4.3. Note that the coniferous (spruce) forest soils have a higher energy flow relative to biomass than do comparable broadleaved (beech) forest soils, the difference being due mainly to small decomposers and predators. Large decomposers are much more important than small decomposers in the energetics of broadleaved than coniferous forest soils; there are relatively very few of the former in the spruce soils. Note also that the difference between mull and mor, in both biomass and energy flow, is more pronounced in the broadleaved forest than in the coniferous forest.

Role of soil microbes

As pointed out previously, the activities of the chemoheterotrophic protists so dominate the processes of organic matter decomposition and energy flow in the soil subsystem that their contribution

Table 4.3 Contribution of trophic groups of soil animals to biomass and energy flow

Group[†]	Beech mor			Beech mull		
	Biomass ($g\ m^{-2}$)	Energy ($kcal\ m^{-2}\ yr^{-1}$)	flow (%)	Biomass ($g\ m^{-2}$)	Energy ($kcal\ m^{-2}\ yr^{-1}$)	flow (%)
Herbivores	4.7	33.6	20.4	9.3	106.5	27.3
Large decomposers	4.2	13.9	8.4	60.8	122.1	31.2
Small decomposers	3.8	76.0	46.1	2.4	93.5	23.9
Predators	2.9	41.5	25.1	1.5	68.7	17.6
Total	15.6	165.0	100.0	74.0	390.8	100.0

Source: Adapted from an analysis of nine soil habitats by Macfadyen (1963).
[†] Herbivores include nematodes, molluscs, bugs, beetles, flies, moths, etc.; large decomposers – earthworms, woodlice and millipedes; small decomposers – nematodes, enchytraeids, oribatid mites and collembola; predators – spiders, centipedes, beetles, etc.

to total soil metabolism exceeds that of the rest of the soil biota combined. Important as these decomposers are, there are as yet few reliable estimates of their contribution to soil respiration, nor is it possible to say for certain which of the two major microbial groups involved, namely bacteria and fungi, contributes the more.

Estimating microbial biomass

Great problems are involved in preparing energy budgets for soil microorganisms. In the first place, it is extremely difficult to determine microbial biomass, the usual method for bacteria being to estimate population density (numbers per gram), measure their average dimensions and calculate their volume, then multiply this figure by their presumed specific gravity. Errors may arise at every stage of this procedure – during enumeration, calculation of biovolume, and determination of specific gravity. The techniques available for these purposes, and their limitations, must therefore be appreciated.

Enumeration of soil microorganisms. Both direct and indirect methods may be used to estimate microbial population densities (Parkinson *et al.* 1971; Rosswall 1973). The most widely used technique for estimating the numbers of viable bacterial, actino-

Table 1.1 *continued*

Spruce mor			*Spruce mull*		
Biomass (g m^{-2})	*Energy* (kcal m^{-2} yr^{-1})	*flow* (%)	*Biomass* (g m^{-2})	*Energy* (kcal m^{-2} yr^{-1})	*flow* (%)
11.3	128.3	32.9	5.9	88.4	28.4
1.0	2.1	0.5	5.5	10.3	3.3
1.6	113.5	29.1	1.4	113.7	36.5
1.2	145.9	37.4	1.1	99.3	31.9
15.1	389.8	99.9	13.9	311.7	100.1

mycete or fungal propagules – cells, spores and hyphal fragments – in soil is the **dilution plate count**. This indirect method involves shaking up a known weight of soil in sterile water or saline solution, progressively diluting the resulting suspension, and dispersing aliquots from one or more of the higher dilutions with a cooled but still liquid agar nutrient medium in sterile petri dishes: if an appropriate dilution is chosen, discrete colonies of microorganisms will appear on the plates, and these may be counted and multiplied by the dilution factor to obtain an estimate of the number of viable microbes or propagules contained in the original soil sample. By varying the composition of the growth medium or the environmental conditions, selected groups may be encouraged at the expense of others. Since no single medium or environment is suitable for culturing all the physiologically diverse groups of microbes present, the plate count method has serious limitations as a basis for estimating soil biomass. Wherever absolute numbers of soil microorganisms are required, direct microscopic observation is preferable to dilution plating, even though it is much more time consuming. Direct counting techniques have been much improved in recent years by the use of epifluorescent microscopes and the availability of a wide range of fluorescent dyes (Babiuk & Paul 1970). This has made it easier to distinguish microbial cells, whether active or resting, from soil particles, and to separate live organisms from dead. The simplest means of direct observation involves the preparation of a **soil smear**. A suspension of soil in water is prepared, spread on a glass microscope slide, fixed with dilute hydrochloric acid and/or heat, then stained. This method was devised by the American bacteriologist H. J. Conn in 1918, who used it to demonstrate that fungi were normal inhabitants of soil, a matter of some dispute at that time. Because uneven drying causes organisms to concentrate on parts of the slide, and since the smear is of variable thickness, it is not possible to convert smear counts to soil population densities with any degree of accuracy. For quantitative purposes, the **agar film** technique, introduced by Jones and Mollison in 1948 (see Nicholas & Parkinson 1967) is preferred. A weighed soil sample is suspended in water, a measured quantity of dilute (1.5%) molten agar added, and a small amount of the resulting suspension pipetted onto the 0.1 mm platform of a haemocytometer slide. Here it is covered with a cover clip and allowed to solidify, producing an agar film which can be transferred to an ordinary microscope slide where it may be stained (if desired) and examined microscopically. Because the depth of the agar film and the area of the microscope field are known, the volume of suspension observed can be calculated, and population densities in the original soil sample estimated by applying the appropriate dilution factor.

Measurement of biovolume. The linear dimensions of samples of microorganisms may be measured on soil smears, and for this purpose bacterial rods are usually assumed to be cylinders with hemispherical ends. A major source or error in such biovolume measurements is the range of size encountered in bacteria and fungi growing under natural conditions; large samples are required to account for this variability. There is also the possibility that the process of fixing, drying and staining the smear may alter the dimensions of the organisms. This problem may be overcome to some extent by examining unstained preparations in agar films (Frankland 1974). However, even if the linear dimensions of microorganisms are not changed by the agar film technique, the apparent total biovolume can be increased by increasing the degree of particle dispersion during preparation of the soil suspension (Swift 1973). This is the result of hyphae enclosed in soil aggregates becoming exposed when these are disrupted; but too much homogenization can destroy some hyphae. Even the choice of magnification may influence the results: Bååth and Söderström (1980) measured twice the length of fungal hyphae at 1250× than at 800× or 500× magnification, though the reason for this is not readily apparent.

Another problem in determining biovolume is that the dimensions of microorganisms, and even their shape, have been observed to change with their growing conditions. For example, Clarholm and Rosswall (1980) found an increase in bacterial cell size in forest soil and peat following rain, and numerous authors have indicated that the volume of soil organisms *in situ* is less than that measured in the more favourable environment of laboratory cultures (van Veen & Paul 1979). Small coccoidal forms appear to dominate the bacterial flora of soils but many, if not all, of these are thought to represent starved rod-shaped forms, since they have been shown to turn into rods under better growing conditions.

Determination of specific gravity. On the basis of studies with pure cultures, carried out many years ago, the specific gravity of bacteria has usually been taken as 1.1 (wet weight basis) and, since 80 per cent of their live weight is assumed to be water, this is equivalent to a dry weight specific gravity of 0.22, according to van Veen and Paul (1979).[2] These authors found, by comparison, that the specific gravity determined by drying and weighing a known number of washed cells of a variety of soil microorganisms grown in shake cultures in the laboratory was as follows: bacteria (short rods), 0.6–0.8; yeast, 0.4–0.5; fungi (mycelial forms), 0.1–0.2. All the organisms tested generally had much higher specific gravities when grown under moisture stress (high osmotic tensions), by a factor of 1.5 or 2 in some instances. It is likely, therefore, that the specific gravity of soil microorganisms growing under natural conditions, where moisture stress is a frequent occurrence, is considerably higher than the value of 0.22 derived from laboratory cultures.

Calculation of biomass. The biovolume of the soil population is multiplied by the dry weight specific gravity to convert it to biomass. It is not yet possible to specify the proper conversion factor to apply though it is clear that the factor of 0.22 used by many workers is too low. For fungi, van Veen and Paul (1979) suggested an average figure of 0.33. They indicated that the conversion factor for bacteria may be even higher, but refrained from making any specific recommendation.

Indirect methods of measuring biomass. Soil biomass can be determined indirectly by chemical analysis for specific cell constituents. All such methods must be checked against estimates derived from direct microscopic examination, this being the accepted standard. One of the more useful techniques is the measurement of adenosine triphosphate (ATP) content but this suffers from the disadvantage that the carbon to ATP ratio is not constant; generally it is 250/1 but in actively growing cells the ratio is less than that found in dormant structures.

Another indirect method, first described by Jenkinson and Powlson in 1976, involves the monitoring of soil respiration during the incubation of soil after fumigation with chloroform ($CHCl_3$). Microbial cells lysed in this way are mineralized for a time at a fairly regular rate, and the flush of CO_2 which occurs during a standard incubation period represents a surprisingly constant fraction – about two-fifths – of the biomass carbon in the original population (Paul & Voroney 1980). Biomass is calculated from the expression $B = F/k$, where k is the proportion of fumigated microbes mineralized to CO_2, and $F = (X\text{-}x)$ where X is the quantity of CO_2 evolved in 10 days following incubation and x the amount produced by non-fumigated soil in the same period. Since sieving the soil during sample preparation often stimulates microbial respiration temporarily, the CO_2 flush may need to be calculated from $F = (X\text{-}y)$ where y is the CO_2 respired between 10 and 20 days in unfumigated soil. The constant k, that is the fraction of biomass carbon mineralized, is usually taken to have the value of 0.41 (Anderson & Domsch 1978).

Jenkinson and Ladd (1981) have compared soil biomass estimates from direct and indirect methods. Most values fall within the range 300–3000 μg C g^{-1} soil and there is good correlation between those derived from direct counts and those obtained by the ATP and chloroform lysis techniques. Assuming that carbon represents 50 per cent of the biomass, and that a square metre of mineral soil 15 cm deep weighs 180 kg, these values are equivalent to 108–1080 g m^{-2}. Between three-fifths and three-quarters of this is believed to be contributed by fungi (Parkinson 1973; Shields *et al.* 1973; Anderson & Domsch 1975), which is consistent with the advantages conferred by a mycelial growth habit in decomposing plant residues (see sect. 1.1). There is recent evidence however (Clarholm & Rosswall 1980) that the contribution of bacteria may have been somewhat underestimated. Despite the methodological problems involved, it is essential that soil ecologists obtain more reliable estimates of microbial biomass.

Microbial energy budgets

Solving eqs. (4.7) and (4.8) for microorganisms presents many difficulties. Consumption (I) may be determined experimentally as the rate of disappearance of substrate, and production (P) assessed concurrently, in a chemostat, which is an instrument designed to permit bacteria to grow indefinitely in a medium of constant composition, with independent control of growth rate and population density. While the growth rate of individual species populations can thus be measured accurately in the laboratory, this has yet to be achieved successfully in the soil. Estimates based on changes in biomass between successive samplings are subject to all the errors inherent in biomass determination (q.v.) and, in any event, microbial growth rates in the soil fluctuate rapidly in concert with changing environmental conditions (Clarholm & Rosswall 1980). Even in favourable circumstances they are much lower than in the laboratory; generation times for soil bacteria in nature are usually between 40 and 50 hours (Paul & Voroney 1980).

Egestion (E) is a concept not readily applicable to bacteria and fungi. The exudation of waste products by these microbes is not strictly comparable to egestion by animals, because they also excrete exoenzymes in addition to metabolic wastes, and so are able to 'digest' some of their substrates extracellularly, as well as disposing of their unwanted residues in this way. In any event, it seems unlikely that the rates of exudation of wastes could be satisfactorily determined under natural conditions, although the formation of some extracellular products can be measured in a chemostat.

The remaining term in the energy balance equations, R, is another which cannot yet be estimated for specific groups in the field. Equation (4.9) cannot be applied to microbial populations, because these organisms are so minute and have such a rapid rate of reproduction that it is impossible to measure the respiration of mature individuals. A further problem arises with bacteria because many are facultative anaerobes and therefore the efficiency of their substrate utilization varies with change in the oxygen supply. Since soil aeration is a continuously variable parameter, both spatially and temporally, it is doubtful whether measurements of CO_2 output from soils would provide an unequivocal estimate of the contribution of bacteria to energy flow, even if it were possible to suppress entirely the activities of the rest of the soil biota.

The generalized energy balanced equations of R. G. Weigert and colleagues, that is eqs. (4.7) and (4.8), have been expressed more explicitly by Erickson (1979) in the following form:

$$CH_mO_l + aNH_3 + bO_2 = y_cCH_pO_nN_g + zCH_rO_sN_t + cH_2O + dCO_2 \quad (4.10)$$

where CH_mO_l denotes the elemental composition of the substrate;

$CH_pO_nN_g$ is the elemental composition of the biomass;

$CH_rO_sN_t$ is the elemental composition of any extracellular products;

y_c is the fraction of substrate carbon converted to biomass;

z is the fraction converted to extracellular products;

d is the fraction respired as CO_2.

As eq. (4.10) illustrates, the energy released from organic substrates may be utilized for biosynthesis and growth (production of biomass), dissipated as respiratory heat, or incorporated into extracellular products. In addition, a small fraction of the energy consumed is used to replace unstable cell constituents and to protect cell integrity, and this is referred to as specific maintenance energy. It is very small in resting cells but may be greater in actively growing populations (Paul & Voroney 1980) though not directly related to the rate of growth. Thus Pirt (1965) found a specific maintenance requirement of 0.04 h^{-1} in liquid cultures but this value should be reduced by at least an order of magnitude if it is to be representative of the total soil microflora, in order to allow for the large number of inactive cells that are always present: several workers (Babiuk & Paul 1970; Gray & Williams 1971; Parkinson *et al.* 1978) used a figure of 0.001 h^{-1}, while others (Shields *et al.* 1973; Behara & Wagner 1974) used 0.002 or 0.003 h^{-1}. The higher values may encompass so-called 'cryptic growth' e.g. energy incorporated in uncounted biomass, as well as maintenance energy.

As Paul and Voroney (1980) have implied, the persistence of a large resting population of soil microorganisms, and its low maintenance energy requirement for continued viability, complicate the transposition of laboratory findings to the real world of the soil. Another factor which causes the growth and activity of microorganisms in nature to differ from those they exhibit in axenic culture is their interaction with soil colloids. It must be concluded that, at the present time, it is not practicable to prepare population energy budgets for soil fungi and bacteria, and hence it is not possible to solve the energy balance equations directly. Perhaps the best that can be achieved is, knowing the total soil metabolism, to subtract from this the experimentally determined value for the metabolism of the animal component. In this way, 'estimates' of 80–90 per cent as the contribution of microbial metabolism to the total have been obtained. Even these estimates are in error to the extent that total soil metabolism includes the respiration of plant roots.

4.7 Total system energy budget

To place the energetics of the decomposer subsystem in context, it is necessary to consider the metabolism of the total system. G. M. Woodwell and D. B. Botkin (1970) have approached this problem by analysing carbon flow in terrestrial ecosystems by means of a series of 'production equations'.

The increase in biomass, that is net primary production (NPP), of the autotrophic component of an ecosystem, namely the plant community, is given by:

$$NPP = GP - R_a \qquad (4.11)$$

where GP = gross production (total amount of photosynthate produced); and
R_a = photosynthate lost by respiration of the producer organisms, i.e. autotroph respiration.

The increase in biomass of the ecosystem as a whole (net community production, *NEP*) is, likewise,

$$NEP = GP - R_e \qquad (4.12)$$

where R_e is the total respiratory loss of all components of the system.

The total community respiration is the sum of the respiration of the heterotrophs (R_h) and the autotrophs, thus:

$$R_e = R_a + R_h. \qquad (4.13)$$

Heterotroph respiration is also made up of two major components, namely consumer respiration (R_c) and decomposer (soil) respiration (R_d), so that:

$$R_h = R_c + R_d. \qquad (4.14)$$

Substituting for GP from eq. (4.11) and R_e from eq. (4.13), we may rewrite eq. (4.12) as:

$$NEP = (NPP + R_a) - (R_a + R_h). \qquad (4.15)$$

that is

$$NEP = NPP - R_h$$

Equation (4.15) provides a means of estimating net community productivity. This parameter cannot be estimated from eq. (4.12) because there is no satisfactory method for measuring gross productivity directly (due to the impossibility of determining respiration losses by the producers during photosynthesis).

We may extend Woodwell and Botkin's treatment by further partitioning decomposer respiration:

$$R_d = R_z + R_m \qquad (4.16)$$

where R_z and R_m represent the respiration of the soil fauna and microflora, respectively.

Microbial respiration can likewise be broken down into its fungal (R_f) and bacterial (R_b) components, along with any minor additional contributions from unidentified elements of the microflora:

$$R_m = R_f + R_b + \ldots \quad (4.17)$$

These production equations are of paramount importance in studies of ecosystem function, for their solution permits the contribution of the various components to energy flow to be assessed. As already indicated, the preparation of an energy budget is an essential first step in any serious study of the relationship between structure and function in terrestrial ecosystems. R. H. Whittaker and G. M. Woodwell and their associates have attempted to analyse an oak–pine forest ecosystem at Brookhaven, NY, in this fashion (Whittaker & Woodwell 1969). Gas exchange techniques were used to provide several independent estimates of R_e, the average value (expressed as dry matter equivalent) being 2129 g m^{-2} yr^{-1}. This figure for total respiration of the ecosystem should theoretically comprise the sum of the contributions of the several components, such as leaves and twigs, branches and stems, ground cover and the soil surface. In practice, direct measurement is possible for only some of these components. For example, the respiration of leaves and twigs cannot be measured directly because there is no way of monitoring the respiration which occurs contemporaneously with photosynthesis. Furthermore, when measuring CO_2 evolution from soil, it is impossible to avoid confounding the respiration of roots (which is part of R_a) with that of the heterotrophs (R_h). The latter can be estimated indirectly from a knowledge of litter decay rates (see sect. 4.4) and the soil respiration data corrected accordingly to give a value for root respiration. When this is added to the respiration of stems and branches, an estimate of 1520 g m^{-1} for R_a is obtained.

There are no satisfactory data for net primary productivity of the plant community based on direct methods involving gas exchange, so an indirect estimate of 1195 g m^{-2} yr^{-1} derived from harvest techniques must be utilized, along with that for R_a, to solve for GP in eq. (4.11). Thus $GP = 1195 + 1520 = 2715$ g m^{-2} yr^{-1}. In turn, NEP can be calculated from GP and R_e using eq. (4.12): $NEP = 2715 - 2129 = 586$ g yr^{-1}.

From this oak–pine forest, therefore, we have a solution to all the production equations given above except eqs. (4.14), (4.16) and (4.17). Of the gross annual production of 2715 g m^{-2}, some 56 per cent is utilized for respiration by the producer organisms, leaving 1196 g m^{-2} as the net production of the plant community. Of this net primary production, a further 609 g or 51 per cent is consumed by heterotroph respiration. That an increment of

586 g m^{-2} remains indicates that accretion of organic matter is still ocurring in this ecosystem, in other words it has not yet attained a steady state. This is also illustrated by the fact that the ratio of gross primary production to total community respiration (GP/R_e) exceeds unity, since in ecological succession the P/R ratio approaches 1 as the ecosystem matures towards the steady state.

Even though biomass is still increasing, the fraction of net primary production which is respired by heterotrophs is substantial. The relative proportions utilized by consumers and decomposers can only be guessed, since eq. (4.14) remains unsolved. While some members of the soil biota, namely the biotrophic animals, might be classified as consumers, their contribution to heterotroph respiration is usually held to be slight, despite evidence that it may account for several per cent of net production in some forest ecosystems. Consumers other than soil biotrophs are usually more important. These include leaf-eating insects which may harvest a significant fraction of NPP, up to 30 per cent in some Australian woodlands (Springett 1978), or even more in European beech woods (Macfadyen 1970). The proportion is generally much less than this however, being only a few per cent in northern hemisphere forests according to Bray (1964), unless insect populations reach plague proportions and massive defoliation results. In the forest ecosystem under consideration, 98 per cent of NPP remains unharvested (Woodwell & Botkin 1970) and consequently passes directly into the decomposer subsystem.

Despite attempts to partition energy flow within terrestrial ecosystems, the relative significance of the various pathways remains unresolved. Woodlands and forests, because of their greater biomass, tend to utilize a larger fraction of gross production for producer respiration than do grasslands, leaving less to be passed on to other trophic levels. Beyond this, however, few generalizations are possible. It is not even certain that the grazing food chain figures more prominently in grasslands than in woodlands or forests. Ultimately, of course, most of the photosynthate produced in all terrestrial ecosystems finds its way into the detritus pathway where it provides the energy source for the soil biota.

4.8 Biogeochemistry of carbon

Geochemistry is the science that concerns the chemical composition of the Earth and the movement of elements between different parts of the Earth's crust and the hydrosphere and atmosphere. The exchanges of elements between the living and non-living parts of the biosphere (the region of the Earth which is inhabited by organisms) are referred to as biogeochemical cycles. At the ecosystem level, they are simply called nutrient cycles. Carbon, though not one of the more abundant elements on the Earth, nevertheless plays a central role in geochemistry, because carbon compounds are essential for every known form of life. The carbon cycle, as commonly understood in biology, consists of the photosynthetic reduction of CO_2 by green plants and its subsequent respiratory release to the atmosphere by plants and microorganisms and, to a lesser degree, by animals. From a geochemical viewpoint, however, this conception of the carbon cycle is a gross oversimplication (Mason 1966).

The most important part of the carbon cycle involves exchange between the air, the sea, and the terrestrial biosphere. Of these three reservoirs of carbon the greatest is the ocean, containing as it does about 90 per cent of the total. It is not a uniform reservoir, but consists of an upper mixed layer extending to a depth of about 75 m, that is, above the seasonal thermocline, a lower deep sea layer below about 1000 m, and an intermediate layer through which the cold surface water of higher latitudes interchanges with the deep sea layer by the process of convection. The mixed layer contains most of the marine life and may be considered to comprise two distinct reservoirs, namely warm surface water which is located in the region of the permanent thermocline, that is, below about 40° latitude, and cold surface water which makes up the remainder. There is a 'fast' carbon cycle involving the mixed layer of the ocean, the atmosphere, and the terrestrial biosphere (green plants, animals and soil organisms), in which the turnover time of carbon is measured in years or at the most, decades; exchange with the deep sea layer is slower, being measured in hundreds of years. Superimposed on this fast cycle is a much slower cycle, turning over about once every 100 000 years, which involves rock weathering and the dissolution and precipitation of carbonates in the ocean.

Accumulation of carbon in the lithosphere

Carbon becomes enriched in the lithosphere in a number of ways: by chemical erosion of rocks, as biogenic deposits or as humus. These will be discussed briefly in turn.

Chemical erosion

Carbonate may be formed from the silicate of primary rocks by the action of CO_2 in the presence of water, according to the reaction:

$$CO_2 + MSiO_3 \overset{H_2O}{\leftrightharpoons} MCO_3 + SiO_2$$

where M is a divalent metal. In the absence of other processes, this would tend to maintain a very low concentration of CO_2 (about 0.001 per cent) in the atmosphere. It is a common reaction in the biosphere, the CO_2 being supplied mainly by soil microorganisms; it therefore occurs most rapidly in soils high in readily decomposable organic matter.

Carbonate formed by this or any other process may be converted to bicarbonate by the action of water containing dissolved CO_2 as follows:

$$CO_2 + H_2O + CaCO_3 \rightarrow Ca(HCO_3)_2.$$

The bicarbonate thus formed is fully soluble and eventually finds its way to the ocean, where it is utilized for photosynthesis by phytoplankton. Approximately half of it will subsequently be deposited as exoskeletons in oceanic sediments and so returned to the lithosphere, while the remainder is respired to the atmosphere.

Biogenic deposits

Three kinds of biogenic deposits need to be considered. The most important of these, from the geochemical viewpoint, are calcareous sediments. The presence of great thicknesses of limestone ($CaCO_3$) in sedimentary deposits indicates that enormous quantities of CO_2 have left the atmosphere during geological times. Except in the case of coral, which is formed by coralline algae and coral polyps, nearly every instance of the deposition of CO_2 as $CaCO_3$ is a biological process involving microorganisms. At the present time, the chief agents of sedimentation in the deep sea are the foraminifera. Other kinds of deposit, namely argillaceous sediments such as shales, contain reduced forms of carbon; this carbon represents CO_2 which was originally fixed photosynthetically.

Coal and petroleum are two other biogenic deposits which, though negligible from the geochemical viewpoint, are further examples of the involvement of the soil biota in the carbon cycle. In the formation of coal, the role of soil organisms is restricted to the modification of organic matter during the accumulation of peat, before it is exposed to the mechanical forces which ultimately determine its physical nature. The precise role of microbes in the transformation of deep oceanic sediments into petroleum is a matter of debate, though much petroleum derives from the oils synthesized and stored by diatoms. There is good evidence that bacteria are involved in the early stages of conversion of these oils into petroleum, but the later stages seem to be largely if not wholly physicochemical.

Humus

The term humus (sect. 4.1) is used to encompass the organic fraction of the soil itself, as distinguished from the relatively unaltered organic matter of the surface litter, and as such humus is largely a product of the activities of soil organisms. It represents carbon which was fixed by the process of photosynthesis. The actual carbon content of soils depends on the rate of organic matter production by green plants and its rate of alteration by heterotrophic organisms once it enters the soil. The balance between the two processes usually results in a net gain of organic carbon in soils, so that humus becomes a more or less permanent constituent of the lithosphere. The humus content of different soils varies greatly with climatic and other factors. In general, temperate and cold humid climates favour the accumulation of humus, whereas hot tropical conditions favour the oxidative processes which return CO_2 to the atmosphere.

In temperate regions, annual grain crops return between 0.5 and 2 t C ha^{-1} to the soil each year as roots and stubble, while perennial pastures have an annual input of 2 t ha^{-1} or more, mainly as roots (Jenkinson 1981). Forests in comparable latitudes return 1 to 3 t ha^{-1} (100–300 g m^{-2} – see Fig. 4.4) as litter and, according to some workers (Edwards & Harris 1977; Cox *et al.* 1978), as least as much again as root detritus.

Turnover of soil carbon. Just how much of the organic carbon entering the soil each year is humified depends on the **turnover time**, which is defined as the time taken to replace a quantity of material equal to the amount of that substance in a given reservoir, that is, on the organic carbon content of the soil divided by the annual input of carbonaceous residues. In a steady state system, carbon flux is constant so that the turnover time in any one of its compartments is equal to the mass of carbon in that compartment. By assuming that soil which has supported relatively undisturbed vegetation for many centuries has reached the steady state condition, it has been shown that the turnover time of carbon is 30–40 years in grasslands (Greenland & Nye 1959) and 10–30 years in forests, depending on climate (Nihlgard 1970; John 1973). These values may be compared to 1–2 years for soils which have been cultivated and sown to wheat every year for one hundred years (Jenkinson & Rayner 1977). Calculations such as these are normally based on the carbon content of the upper 20 or 30 cm of the profile, which is where most of the biological activity is concentrated.

Turnover time may also be derived mathematically, using exponential decay models similar to that described for litter decomposition (sect. 4.4). The following treatment is based on Jenkinson (1981) who has discussed turnover models of soil organic matter in some detail.

If all the organic matter in the soil, designated as C, is considered to be equally decomposable, and r is the fraction of C which is decomposed each year, A the annual input of carbon in plant and animal residues, and f the fraction of A that is humified each year (i.e. converted to soil organic matter), then the rate of change of organic carbon in the soil is given by

$$dC/dt = fA - rC \qquad (4.18)$$

where t is the time in years.

Solving this,

$$C = \frac{fA}{r} + \left(C_0 - \frac{fA}{r}\right)e^{-rt} \qquad (4.19)$$

where C_0 is the initial organic carbon content of the soil.

At equilibrium, $dC/dt = 0$ and hence the equilibrium organic matter content of the soil is given by

$$C_E = fA/r. \qquad (4.20)$$

Substituting for fA/r in eq. (19), we have

$$C = C_E + (C_O - C_E)e^{-rt} \qquad (4.21)$$

The turnover time t_E for a soil in equilibrium is calculated as C_E divided by the amount of humified material added each year, that is,

$$t_E = C_E/fA = fA/rfA = 1/r. \qquad (4.22)$$

This model is an oversimplification because it is based on the false assumption that all fractions of soil organic matter decompose at the same rate, that is have the same r values. This is not so. For example, using radiocarbon dating techniques, Jenkinson (1973) showed that the ages of soil organic matter fractions in an unmanured field at Rothamsted, cropped to wheat continuously for 100 years, were as follows: fulvic acid, 420 years; humic acid, 750 years; and humin, 2395 years. Since the organic carbon content of the soil had remained unchanged at about 26 t ha^{-1} for 100 years, the soil may be assumed to be in steady state. If all humus fractions had the same r value, then no age fractionation would be evident. Furthermore, the soil organic matter as a whole had a radiocarbon age of 1450 years whereas its age, calculated from eqs. (4.20) and (4.22) and assuming $f = 1/3$, is estimated to be only 66 years.

More realistic models have been developed to overcome this problem. Such models are usually based on a number of organic matter fractions, all decomposing exponentially but at different rates. Thus Jenkinson and Rayner (1977) constructed a five-compartment model for the Rothamsted soil which predicted that it would contain the following:

0.1 t ha^{-1} C as readily decomposable plant material ($t_E = 0.2$ yr)
0.6 t ha^{-1} C as resistant plant material ($t_E = 3.3$ yr)
0.3 t ha^{-1} C as microbial biomass ($t_E = 2.4$ yr)
13.6 t ha^{-1} C as physically protected organic matter ($t_E = 71$ yr)
14.6 t ha^{-1} C as chemically stabilized organic matter ($t_E = 2900$ yr).

The total predicted carbon content is 29.2 t ha^{-1} (cf. actual value, 26 t ha^{-1}) and the predicted radiocarbon age of the whole soil is 1240 years (cf. 1450 years, measured).

Jenkinson and Rayner's model assumes that all organic matter fractions are continuously synthesized and decomposed, and predicts a uniform radiocarbon age throughout the profile. In practice it is found that radiocarbon age tends to increase with depth. O'Brien and Stout (1978) developed a model which explained this, by postulating that soil contains a small 'old' inert fraction, distributed uniformly throughout the profile, and a larger 'modern' fraction (less than 100 years old) which decreases exponentially with depth.

While simple models such as the single compartment model represented by eq. (4.21) fit experimental field data reasonably well over periods of several decades, models based on numerous compartments are needed to represent long term carbon turnover and the accumulation of humus under natural conditions.

Recalcitrant molecules

Microbiologists regard it as axiomatic that every naturally occurring organic compound is mineralized by one or more groups of microorganisms. If it were not so, then the carbon cycle would be interrupted, and huge accumulations of the resistant substances would occur in the biosphere. It is true that massive deposits of some organic materials, such as coal and petroleum, do exist, but the carbonaceous residues of organisms from which these biogenic deposits derive are susceptible to microbial decomposition, that is, are biodegradable, given appropriate environmental conditions.

There are however certain organic compounds, notably synthetic chemicals, which are either non-biodegradable or which decompose extremely slowly in all environments. M. Alexander (1965, 1979) refers to these substances as recalcitrant. They deserve close study because many of them have become pollutants of regional and even global significance. Among the most refractory synthetic molecules are the branched alkyl benzene sulphonate detergents and the chlorinated hydrocarbon insecticides. The latter are remarkably persistent in soil, some such as DDT being of especial concern because of the way in which they become more and more concentrated as they pass from one trophic level to the next, and because they become so widely distributed in nature as food chains ramify to form complex food webs.

Despite such potentially damaging consequences, the expectations of modern society cannot be satisfied without recourse to technologies dependent on synthetic chemicals. Herbicides and insecticides are essential for weed and insect pest control in agriculture, and many different organic polymers are used in the manufacture of clothing, packaging materials, and for other industrial purposes. These man-made compounds will be biodegradable only if they can be metabolized by the enzyme systems acquired by microbes during the course of evolution. S. Dagley pointed out in 1975 that biodegradation will take place only if two conditions are satisfied:

(a) microbial enzymes must be capable of using as substrates compounds having molecular structures similar to those found in nature;
(b) the artificial substrates must be able to induce or derepress the synthesis of the enzymes needed.

Synthetics are unlikely to be readily biodegradable if their chemical structure is one that is not encountered in nature. This can be illustrated by comparing the diphenyl methane insecticides, methoxychlor and DDT. The latter compound is very refractory, persisting in soil for at least 15 years (Alexander 1979). When the para chlorine atoms of DDT are replaced by methoxyl groups, the resulting molecule is methoxychlor. Since methoxy groups are common substituents of the benzene molecule in natural products (Dagley 1975), it is not surprising to find that methoxychlor is

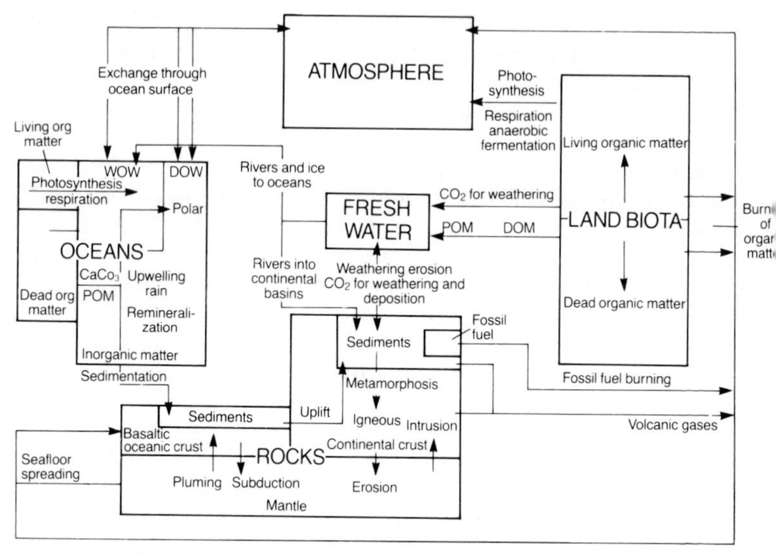

Fig. 4.7 Principal reservoirs and fluxes in the carbon cycle (WOW = warm ocean water, COW = cold ocean water; POM = particulate organic matter; DOM = dissolved organic matter). (From Bolin *et al.* 1981.)

much more readily metabolized by microorganisms than is DDT, and is therefore much less persistent.

The global carbon cycle

The global carbon cycle, as visualized by Bolin *et al.* (1981), is shown in Fig. 4.7. The approximate sizes of the major reservoirs participating in the fast cycle are as follows: ocean 35 000, humus 3000, land biota 800, atmosphere 700 $\times 10^{15}$ g C. These reservoirs are small compared to the amounts of carbon involved in the slow cycle. For example, sedimentary rocks contain more than 10 000 000 $\times 10^{15}$ g C of which only about 10 000 $\times 10^{15}$ g participates in the fast cycle, some 5000 $\times 10^{15}$ g being available for dissolution in the sea and another 5000 $\times 10^{15}$ g occurring as fossil fuels.

Twentieth-century increase in atmospheric carbon dioxide

About 99.5 per cent of the carbon in the Earth's atmosphere occurs as CO_2; the remainder is mainly methane, CH_4. The concentration of CO_2 in the atmosphere is very low, an average value for temperate latitudes being presently about 0.033 volume per cent, or 330 ppm. There are considerable variations in concentration however, variations which may be local and regional, diurnal and seasonal, or secular (long-term). The CO_2 concentration at high latitudes is less than that at low latitudes, because cold water dissolves more CO_2 than warm water: polar air therefore contains slightly less CO_2 than tropical air. At altitudes below about 300 metres, atmospheric CO_2 levels over forest and grassland rise to a maximum between the hours of 1.00 p.m. and 6.00 a.m. and fall to a minimum in the early afternoon, coinciding roughly with peak periods of respiration and photosynthesis, respectively. Also as a result of biological activity, the CO_2 concentration in air is greatest in spring and least in mid-summer, seasonal variation decreasing with increasing altitude; these fluctuations are more pronounced in the northern hemisphere.

The most important factor controlling the CO_2 content of the air is the CO_2 store, including the various ions and salts of carbonic acid, in the hydrosphere. As the partial pressure of CO_2 in air increases, more goes into solution in the ocean and the balance tends to be restored. Despite this buffering mechanism however, secular changes in the CO_2 content of the atmosphere have been observed. Based on a survey of available measurements in 1940, G. S. Callendar came to the conclusion that the amount of CO_2 in the atmosphere had increased during the previous half-century, and suggested that the increase was due to the burning of fossil fuel – coal and oil – by man. Subsequent careful analyses of a greater range of data (Callendar 1958) has shown that the CO_2 content of the Earth's atmosphere began to increase slowly towards the end of the 19th century, and thereafter much more rapidly. During the last 120 years, its concentration has risen from about 290 ppm (or less) to about 330 ppm. The increase would have been much greater but for the buffering capacity of the ocean and, according to some authorities (Goudriaan & Atjay 1981), increased immobilization in the terrestrial biomass.

Since coal and oil deposits are too ancient to contain any ^{14}C (half-life, 5570 years), the burning of fossil fuels should lower the $^{14}C/^{12}C$ ratio in the atmosphere. The observed decrease in this ratio however (determined by measurements on tree rings) is much less than anticipated, even when allowance is made for the transfer of much of the industrially produced CO_2 to the hydrosphere. Although other factors may cause a fall in the $^{14}C/^{12}C$ ratio (reduced production of ^{14}C by cosmic rays, for example), the most likely explanation of this apparent anomaly is the addition of 'young' CO_2 to the atmosphere during the last 100 years, that is CO_2 which contained ^{14}C as well as ^{12}C. This CO_2 is thought to come from increased respiration of soil microorganisms following on man's increased agricultural activities which accompanied the industrial revolution. It is argued that more intensive cultivation of farmlands already in use by the mid-nineteenth century, together with the cultivation of lands previously under forest or prairie, has contributed substantially to the rise in atmospheric CO_2. The burning of forests during their conversion to farms would supplement this cultivation effect, by releasing the accumulated CO_2 of centuries into the atmosphere. Accelerated clearing of tropical forests during recent decades has no doubt had a similar effect, and it has been suggested that deforestation is becoming an increasingly important factor in the build up of atmospheric CO_2 (Jenkinson 1981). This is likely to be true only where forests are replaced by agricultural crops or pastures: provided the cleared areas are reforested, either naturally or artificially, the higher net productivity of the regrowth stands will ensure that increasing amounts of the CO_2 released by felling and burning the mature forests are sequestered in the aggrading biomass.

Available evidence does not permit a firm conclusion to be drawn concerning the relative importance of the two main factors which have contributed to the recent secular increase in atmospheric CO_2. Whether the major cause is fossil fuel consumption or the extension and intensification of agriculture, microorganisms will have been largely responsible, either indirectly through their role in the formation of biogenic deposits (coal and oil, q.v.), or directly through their respiratory activities in the soil.

4.9 Notes

1. The main structural components of the bacterial cell wall are mureins, which are heteropolymers of two amino sugars and three or more amino acids. The amino sugars are N-acetylglucosamine and N-acetylmuramic acid, while lysine or the structurally related diaminopimelic acid is always present as one of the amino acids.
2. Suppose a cell of volume 1 μm^3 has a wet weight specific gravity (s.g.) of 1.1, and a moisture content of 80% (wt/wt), then:

volume of water in cell $= 1.1 \times 0.80 = 0.88\ \mu m^3$;
volume of cell contents $= 1 - 0.88 = 0.12\ \mu m^3$;
weight of cell contents $= 1.1 \times 0.20 = 0.22\ \mu g$;
s.g. of intact cell $= 0.22$ (dry weight basis);
s.g. of cell contents $= 0.22/0.12 = 1.83$.

5

Mineral cycling processes

By virtue of their effect on nutrient circulation in the biosphere, microbes have great significance in the biogeochemistry of the elements. Availability of nutrients markedly influences plant vigour and consequently, productivity. Microbial processes affecting nutrient availability thus have a vital role in ecosystem function. Microbes may however affect plants adversely as well as beneficially. For example, they can on occasion compete with plants for nutrients, but only if they have an adequate supply of energy. In other instances, plant nutrients are made available by microorganisms in the course of meeting their energy demands. Therefore, in order to determine how microorganisms affect the availability of plant nutrients, we must consider separately the various kinds of energy-yielding processes which were discussed in Chapter 2.

To recapitulate, the energy sources for life are two: the radiant energy of the sun, and the chemical bond energy contained in organic and inorganic compounds. Radiant (solar) energy is the basis of photosynthesis. Bacterial photosynthesis is a strictly anaerobic process, and this requirement for anaerobic conditions, coupled with the fact that they can utilize light of longer wavelengths than green plants, restricts the activities of photosynthetic bacteria to aquatic environments such as lakes and estuaries. They are therefore unlikely to play any part in promoting nutrient availability in soils and so will not be considered further. There are two other groups of photosynthetic microorganisms, the algae and the cyanobacteria or blue-green algae. Some cyanobacteria can fix atmospheric nitrogen, and in some ecosystems (e.g. in rice culture) they play a significant part in maintaining soil fertility. While photosynthesis by soil algae may be important in certain ecological situations, in general it is not the photosynthetic microorganisms, but those which obtain their energy from chemical processes which are of greatest significance in plant nutrition, and we will be concerned only with these. Three kinds of oxidation reduction reactions provide energy for microorganisms (Table 2.2) and these are termed, according to the nature of the final hydrogen acceptor, respiration, anaerobic respiration or fermentation. The fact that some microbes can respire aerobically so long as oxygen is present, but when the oxygen supply becomes limiting switch to anaerobic respiration or fermentation as a source of energy, does not detract from the value of examining the energy-yielding processes themselves, since it is these which are important in nutrient cycling rather than the microorganisms *per se*.

To provide a framework for this examination, we need first to outline the main facets of nutrient cycling in terrestrial ecosystems.

5.1 Mineral cycling in ecosystem function

Nutrient elements accumulate in ecosystems as a result of successional processes – primary and secondary – that result in increased biomass. The processes of community growth and nutrient accretion are thus closely interwoven, and so long as net primary production exceeds zero, the system's content of minerals will increase. As described in Chapter 3, the sequence of development is governed by the interaction of the independent state factors, namely climate, organisms, parent material and topography that are operative during the developmental time interval (Jenny 1961). Eventually, in the absence of disturbance, the system approximates a steady-state condition.

Several hypothetical strategies of ecosystem development have been advanced. Perhaps the best known is that of Odum (1969) who proposed that ecosystems conserve essential elements by evolving mechanisms which promote internal recycling, so that succession is seen as a process whereby plant communities accumulate sufficient nutrient capital to permit the rise of succeeding ones. Having reached the mature or steady state phase, the system is considered to exercise maximum control over its accumulated store of nutrients, and to have maximum ability to entrap fresh inputs. This hypothesis has been challenged in recent years (Vitousek & Reiners 1975; Vitousek 1977; Bormann & Likens 1979), and there is evidence to support an alternative hypothesis, namely that maximum regulation of nutrient inputs and outputs occurs during that stage of succession when net primary production, and hence the rate of nutrient accretion in biomass, is greatest. In contrast to Odum's view, this hypothesis envisages nutrient 'leakage' from the system, that is, the excess of outputs over inputs, as being greater at maturity than in mid-successional stages.

The model proposed by Bormann and Likens' (1979) typifies this viewpoint. While it relates specifically to secondary succession following a major disruption such as clearcutting of forests, it is applicable to primary succession also. Bormann and Likens recognized four main phases of ecosystem development after disturbance:

(a) a relatively short **reorganization** interval;
(b) an **aggradation** phase of major growth in biomass and mineral capital;
(c) a **transition** stage of varying duration when biomass and nutrient storage decline;
(d) a **steady state** phase during which the ecosystem stabilises in terms of biomass and nutrient storage.

This strategy of ecosystem development visualizes the aggradation phase as being the most stable biogeochemically. It culminates in an unstable transition phase as its essentially even-aged character is destroyed by random mortality of individuals or subpopulations. When the system eventually stabilizes again, at a lower level of biomass and nutrient storage, it consists of a structural and floristic mosaic of varying complexity, comprised of individual areas representative of all four developmental stages; this is designated the **shifting mosaic steady state**. The implications of such a model lie in the emphasis it places on natural agents of disturbance such as fire and wind, in determining structural, floristic and functional patterns in plant communities (cf. sect. 3.2). If the probability of autogenic succession proceeding uninterrupted to a steady state is less than previously supposed, then the probable effects of man-induced disturbance on mineral cycling may need to be re-evaluated. This suggestion is explored more fully by Richards and Charley (1983).

Pathways of nutrient cycling

Three interconnected mineral flow pathways affect the nutrition of terrestrial communities. Switzer and Nelson (1972) designated these pathways, or cycles, as geochemical, biogeochemical and biochemical.[1] The geochemical cycle links the external environment to the ecosystem by processes such as precipitation, rock weathering, leaching and groundwater flow. The biogeochemical cycle concerns the circulation of nutrients between soil and vegetation; it begins with the process of nutrient uptake by roots and ends with the release of minerals from decomposing litter and soil organic matter. The biochemical cycle involves the redistribution of nutrients within the plant biomass by mechanisms such as the withdrawal of nutrients from leaves during senescence, or from sapwood during its transition to heartwood. The relative significance of the several cycles varies from element to element, and with the stage of development of the system, and all three are linked in overall community nutrition. A conceptualized picture of mineral cycling pathways in a forest ecosystem is shown in Fig. 5.1.

We turn now to consider the role of the various energy-yielding processes of microorganisms on nutrient cycling in plant communities.

5.2 Aerobic processes and nutrient cycling

In the aerobic process of respiration, the hydrogen donor or substrate may be either organic or inorganic. Of the inorganic hydrogen donors which may be used, only reduced nitrogen compounds and reduced sulphur compounds are of any significance in the context of plant nutrition. We will return to these later, after first examining the role of respiration carried out with organic substrates.

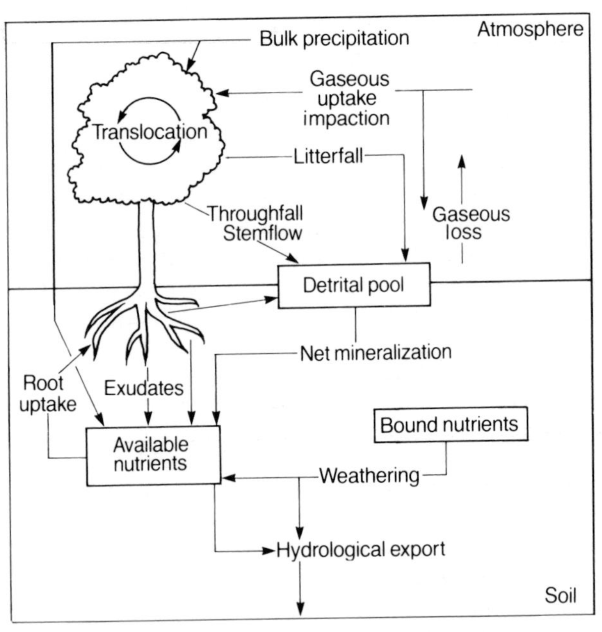

Fig. 5.1 Generalized picture of mineral cycling pathways in terrestrial ecosystems. (After Charley & Richards 1983.)

5.2.1 Respiration with organic substrates

A great variety of organic compounds can serve as hydrogen donors for microbial respiration. The effect of such energy-yielding oxidations on the supply of plant nutrients is entirely indirect, that is nutrients become available only as by-products of the microbial respiration of organic substrates. When inorganic ions are produced by the oxidation of organic compounds, the process is termed **mineralization**; when inorganic molecules are assimilated into microbial protoplasm, we speak of **immobilization**. It is a basic concept of soil microbiology that mineralization and immobilization of nutrients proceeds concurrently, so that there is a continual biological turnover, or **mineralization–immobilization cycle**, in the soil. The energy needed to keep this cycle running is that released during the oxidation of organic compounds, added to the soil as plant and animal residues or stored in the soil organic matter. The cycle operates continuously but more or less intensely, depending on the supply of readily decomposable substrates (Fig. 5.2).

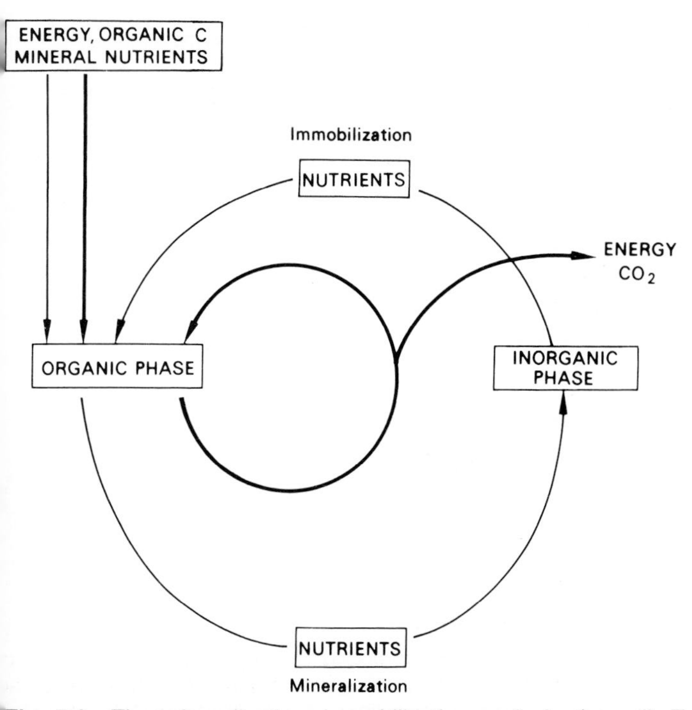

Fig. 5.2 The mineralization–immobilization cycle in the soil. Energy flow shown by broad lines, nutrient pathways by narrow lines.

Not all the energy released during oxidation is captured by the decomposing microbes, a considerable quantity being dissipated as heat. A large amount of carbon also escapes from the system as carbon dioxide, but the mineralized nutrients are not usually lost and are available again and again for microbial use. Unless the energy supply is renewed therefore, by the addition of fresh substrates in the form of plant debris, a stage will be reached when inorganic ions accumulate, and we can speak of **net mineralization** (Fig. 5.3). In these circumstances, nutrients will be readily available for plant uptake. Conversely, if readily decomposable substrates are added to the system, there tends to be a surplus of energy, the demands of the microflora for inorganic ions become greater than the mineralization outflow, and any source of inorganic ions in the soil will be drawn upon by microorganisms, resulting in **net immobilization** (Fig. 5.4). Under these conditions, the supply of nutrients available

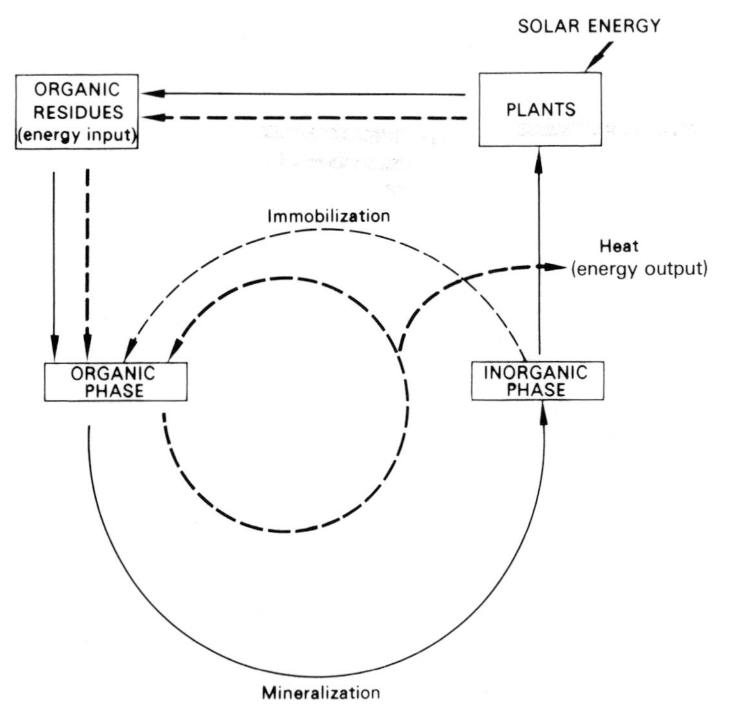

Fig. 5.3 The mineralization–immobilization cycle under conditions of energy deficit, resulting in net mineralization. Energy flow shown by broad lines, nutrient pathways by narrow lines. Broken lines indicate interrupted or intermittent flow.

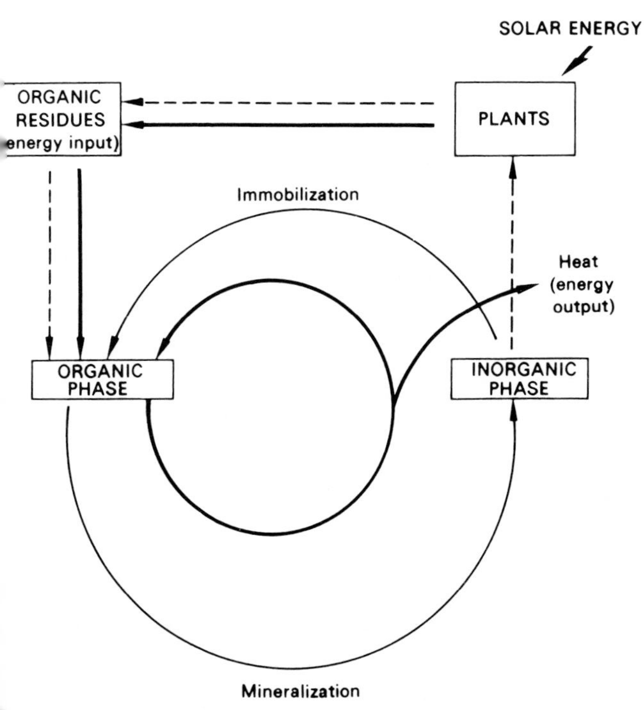

ig. 5.4 The mineralization–immobilization cycle under conditions of energy surplus, resulting in net immobilization. Energy flow shown by road lines, nutrient pathways by narrow lines. Broken lines indicate nterrupted or intermittent flow.

or plants becomes depleted. The three diagrams can be combined Fig. 5.5) using the energy network diagrams of H. T. Odum desribed in Chapter 1.

Implicit in the concept of a mineralization–immobilization cycle s the assumption that plants cannot compete successfully with nicrobes for inorganic nutrients, and only when there is net mineraization can they satisfy their nutrient requirements. This is most robably correct for nitrogen, the plant nutrient required in greatest mount by microorganisms, and may on occasions be true for sulphur nd phosphorus also. The demands of microbes for other nutrients re relatively modest, and plants are unlikely to suffer from deficiencies of them as a result of microbial competition. Not surprisingly herefore, research tends to be centred on the nitrogen turnover ycle. Even here, the addition of fresh substrates to soil will affect lant growth adversely only if these substrates are high in energy in

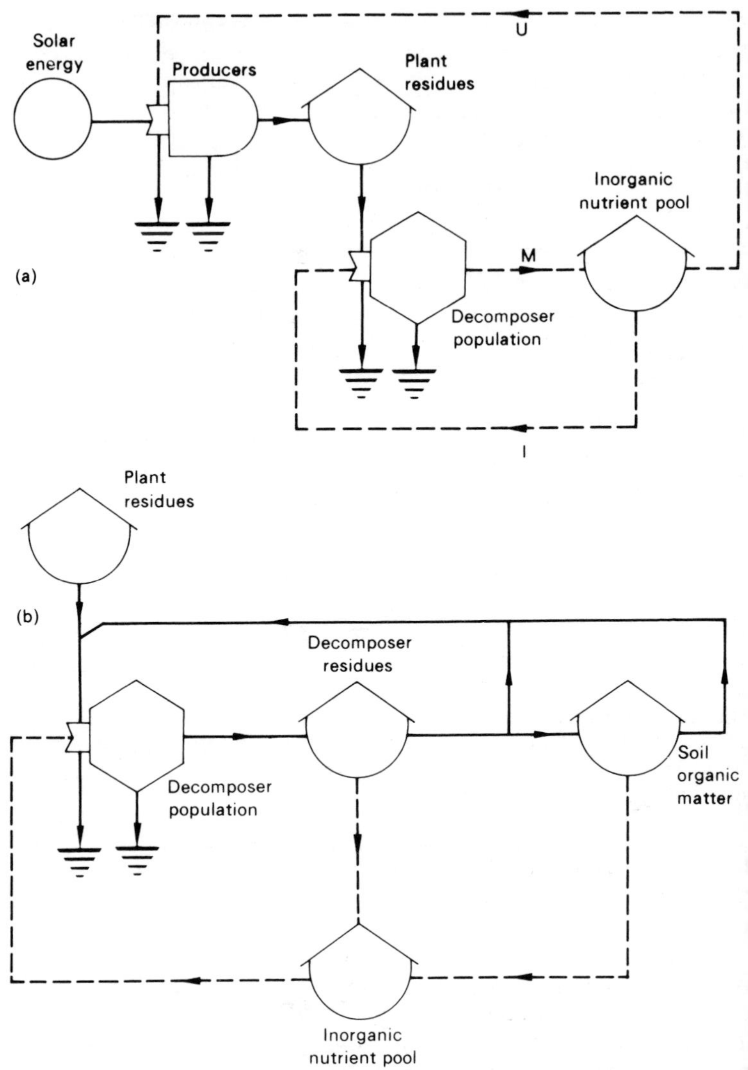

Fig. 5.5 Energy network diagram of the mineralization–immobilization cycle. (a) Generalized picture of the cycle in the soil–plant system. (b) Detail of the cycle in the soil. Major energy flows shown as solid lines, pathways of low energy compounds (inorganic nutrients) as broken lines. M, mineralization; I, immobilization; U, plant uptake.

relation to nitrogen. The primary limiting factor for growth of heterotrophic microbes is energy supply, and this in turn is a function of substrate quality, for the particular suite of organisms involved in the decomposition process.

Many kinds of chemoheterotrophs are involved in mineralization–immobilization cycles. Fungi are especially prominent, particularly in forest ecosystems where low soil pH often restricts the activities of bacteria. Ascomycetes and imperfect fungi, notably ***Penicillium, Fusarium, Aspergillus*** and ***Trichoderma***, and zygomycetes such as ***Mucor*** and ***Rhizopus***, are among the fungi most frequently isolated on soil dilution plates; population densities are usually in the range of 10^4–10^6 propagules (spores or hyphal fragments) per gram of soil. Other isolation techniques have shown that, in addition to these well known genera, various basidiomycetes are widespread and active in soils. As primary colonizers of fresh plant detritus, and secondary colonizers of the products of polysaccharide decomposition, the former group might be expected to have a larger role in organic matter turnover.

This is borne out by a study of mycelial production and depletion luring ten weeks' incubation of a sandy podzolic soil from subtropical Eastern Australia, where it was found that hyaline non-septate hyphae (phycomycetes) and pigmented septate hyphae (ascomycetes and imperfect fungi) together accounted for more than 90 per cent of the total mycelium colonizing nylon gauze buried in the soil, while clamp-bearing hyphae (basidiomycetes) represented less than 2 per cent of the total. The addition of glucose greatly stimulated the growth of the two major groups of fungi, resulting in net immobilization of added amino-nitrogen (Table 5.1). A closer analysis of mycelial production and depletion at intervals during the incubation period revealed that the activity of the phycomycetes was subject to greater fluctuations than that of the septate fungi.

Among bacteria, the Gram-positive rods *Arthrobacter* and *Corynebacterium*, and the aerobic sporeformer *Bacillus*, make up a large proportion of the microflora. The Gram-negative rods known as fluorescent pseudomonads (*Pseudomonas* spp.) are also active in organic matter decomposition; they have great biochemical versatility, being able to use a far wider range of organic compounds as carbon and energy sources than any other group of microorganisms. Bacterial populations in fertile soils fall between 10^6 and 10^9 cells per gram.

Mineralization and immobilization of nitrogen

Ammonia is a waste product of microbial metabolism, and any

Table 5.1 Influence of glucose on the production of fungal mycelium and the disposition of labelled amino-nitrogen in a sandy podzolic soil

	Category of hyphae (mm/mesh)					*Nitrogen fraction* ($\mu g\ g^{-1}\ ^{15}N$)	
Treatment	*Total colonization*	*Hyaline non-septate*	*Hyaline septate*	*Pigmented septate*	*Clamp-septate*	*NH^+-N*	*Organic-N*
Minus glucose	2.0	1.0	0.22	0.75	0.00	34.5	34.4
Plus glucose	5.1	1.9	0.26	2.8	0.12	7.8	56.9
Significance level†	***	**	ns	***	ns	***	***

Source: After Jones and Richards (1978).

† One, two or three asterisks indicate that the means differ at the 5%, 1% or 0.1% levels of probability, respectively; ns = not significant.

NH_4-N that accumulates in soil represents the quantity of substrate nitrogen in excess of microbial requirement. Since organic matter added to soil will normally supply both carbon and nitrogen, whether NH_4-N will accumulate depends in part on the relative proportions of carbon and nitrogen in the additive. This explains a well known phenomenon in agriculture, that incorporating the residues of one crop just before another is sown may result in a temporary nitrogen deficiency in the second crop. The duration of the period of net immobilization depends on the nature of the organic matter added, that is on its carbon and nitrogen status, and its incorporation in the soil must be properly timed in relation to the date of sowing if the new crop is not to suffer from nitrogen deficiency. As indicated in Chapter 4, this realization has led to much research on the effect of the ratio of readily available energy to readily available nitrogen in crop residues or organic manures, a rough approximation of which is given by the carbon–nitrogen (C/N) ratio of the additives. Agronomic experience has shown that when organic matter containing about 1.8 to 2.0 per cent N is added to soil there is neither net gain nor net loss of mineral nitrogen, whereas organic amendments containing less than 1.2 to 1.3 per cent deplete inorganic nitrogen reserves rapidly. In natural materials with a carbon content of about 40 per cent, these N levels correspond to C/N ratios of 20/1 and 30/1 respectively: adding organic matter with a C/N greater than 30/1 usually results in net immobilization for a period of weeks or even months, while incorporating material with a ratio less than 20/1 causes net mineralization almost immediately.

The concept of a critical C/N ratio of about 20/1 was developed for arable soils in the temperate regions of the northern hemisphere. In such soils, only materials having a C/N ratio of less than 20/1 are likely to provide available nitrogen for crop plants when incorporated in the soils at the beginning of the growing season. The concept cannot be applied uncritically to another situation, however. It is not only the C/N ratio of organic amendments that is important, but also their chemical nature. The nitrogen requirement for breakdown of lignified material is less than for more succulent tissue, seemingly because the less readily decomposable substrate promotes less microbial activity and hence there is less microbial protoplasm to be synthesized. Also there is evidence (sect. 4.5) that wood-decaying fungi are physiologically adapted to conserve nitrogen by a variety of mechanisms, and are thus able to promote rapid decomposition of lignified tissues despite their very high C/N ratios.

The influence of C/N ratio on the balance between mineralization and immobilization of nitrogen may also be modified by environmental factors. Less nitrogen is needed when plant residues decompose under anaerobic conditions than under aerobic, partly because decomposition is slower and partly because less cell material is synthesized per unit of substrate under anaerobic conditions (Jenkinson 1981). Decomposition rates are also reduced when soil temperatures are low, so that less nitrogen is required to decompose a given quantity of organic matter in winter than in summer.

Forms of organic nitrogen in soils

Only a very small portion of the organic nitrogen in soils appears to be protein, but this fraction, which probably represents the protein of living microbial cells, is very important as a source of mineral nitrogen for plant growth. Some of the ammonium released during the decomposition of proteins and amino sugars combines with quinones and polyphenols to form products of greater stability against microbial attack, and this reaction occurs also with a certain proportion of the ammonium added as fertilizer; it is often called ammonium fixation.[2]

Amino acids released during the process of decomposition can combine with quinones too, producing relatively resistant humic acid polymers. Although the chemical nature of about half the total nitrogen in soil remains obscure, 20–40 per cent of it is known to occur as bound amino acids and 5–10 per cent as combined amino sugars. That these compounds do not occur free in soils is evidenced by the fact that they provide excellent substrates for a wide variety of microorganisms in the laboratory. It seems that only a very small portion of the total soil organic matter participates in the mineralization–immobilization cycle at any one time: the nitrogen mineralized during the course of any one growing season may amount to no more than 3 per cent of the soil's organic N, according to Satchell (1974). The great bulk of organic matter that enters the soil, once partially decomposed, is more or less stabilized against further microbial attack. It is in this form that it is called humus, and as such it constitutes a passive organic phase existing contemporaneously with, but outside, the turnover cycle.

Urea hydrolysis

An important reaction which liberates mineral N is the hydrolysis of urea:

$$CO(NH_2)_2 + 2H_2O \rightarrow (NH_4)_2CO_3$$

The conversion of urea of ammonium carbonate brings about a rapid increase in soil reaction, and if the pH rises above 8, substantia amounts of ammonia may be evolved. This is because the equilibrium between ammonium ions in the soil solution and free ammonia shifts towards the production of more ammonia at high pH:

$$NH_4^+ + OH^- \rightleftharpoons H_2O + NH_3$$

The resulting combination of high pH and free ammonia makes the environment unsuitable for most other bacteria. When synthetic urea is used as a fertilizer, plants can be damaged as a result of urease activity in two separate but related ways: firstly, if the pH rises above 9.5, ammonia toxicity can result from the presence of as little as 1–2 ppm NH_4-N; secondly, nitrite toxicity can occur as a result of the accumulation of NO_2^- ions, because *Nitrobacter* spp., which oxidize NO_2^- to NO_3^-, are more sensitive to high pH than are *Nitrosomonas* spp., which oxidize NH_4^+ to NO_2^-.

Urea is a typical product of animal excretion and large amounts are constantly being added to soils and natural waters; smaller quantities derive from the decomposition of nucleic acids. Confining cattle to feedlots results in very heavy additions of urea, approximately 140 g being voided daily in the form of urine by a mature beast. Consequently, large amounts of ammonia escape to the atmosphere and this may lead, in turn, to local air- and water-pollution problems.

Urea hydrolysis is brought about by a number of heterotrophic soil bacteria but since the reaction liberates little energy, they cannot use it as a source of energy. Instead they satisfy their carbon and energy demands by oxidizing amino acids (most of them cannot oxidize carbohydrates). If urea is mixed with peptone and inoculated with soil, the organisms which predominate are spore formers, mainly *Bacillus pasteurii* and *Sporosarcina ureae*. These bacteria grow best in alkaline media, up to pH 11, and can hydrolyse urea in concentrations up to 10 per cent. Non-spore forming urea-hydrolysing bacteria also exist, but these can only operate at urea concentrations below 3 per cent; examples are *Micrococcus ureae, M. aureus*, and certain yeasts including *Rhodotorula* and *Cryptococcus*. Hydrolysis is due to urease, which is the trivial name for a group of enzymes known as urea amidohydrolases. Its production is not restricted to the urea bacteria, since many other bacteria and fungi, and a large number of higher plants, possess the enzyme and can therefore use urea as a source of nitrogen. While none of these other microorganisms can tolerate high concentrations of urea, they are nevertheless likely to be responsible for much of the urea decomposition which occurs in acid soils, since the urea bacteria are very sensitive to acidity.

The study of urease activity in soils is of increasing importance since, according to J. M. Bremner and R. L. Mulvaney (1978), urea is destined to become the most important fertilizer in world agriculture. In the absence of growing plants, the hydrolysis of fertilizer urea is controlled mainly by soil urease, which is remarkably stable under natural conditions. The protective mechanisms responsible for its longevity are not yet fully understood.

Heterotrophic nitrification

Oxidation of ammonium compounds, once thought to be the sole province of the chemoautotrophic nitrifying bacteria (sect. 5.2.2), is now known to be accomplished by a variety of chemoheterotrophs. In fact, heterotrophic nitrification, as it is called, is not an uncommon microbiological phenomenon, at least *in vitro*, though its biogeochemical significance may be slight. Unlike the autotrophs, heterotrophic nitrifiers can oxidize organic forms of nitrogen such as hydroxylamine, amino acids, peptones, oximes and certain aromatic compounds. The end products of these oxidations are many and varied, and tend to accumulate during the stationary phase of population development after exponential growth has ceased. The organisms concerned apparently derive no energy from the oxidations, although recent evidence cited by Verstraete (1981) indicates that *Methylococcus* spp. may do so. *Methylococcus* is representative of a widespread group of bacteria able to utilize methane (CH_4) and a few other one-carbon compounds as a sole source of energy and carbon; some can also 'co-metabolize' molecules containing more than one carbon atom, that is, they can oxidize such substances provided methane is being utilized concurrently.

Two groups of heterotrophic nitrifiers exist, one oxidizing ammonium and the other, various organic nitrogen compounds. Among the ammonium oxidizers are the actinomycete genera *Streptomyces, Nocardia* and *Micromonospora*, the bacteria *Bacillus* and *Pseudomonas*, and representatives of several fungal genera including *Aspergillus, Mortierella, Penicillium* and *Mucor*. In pure cultures, isolates of these organisms rarely produce more than 1 or 2 $\mu g\ g^{-1}$ NO_2-N, though amounts up to and exceeding 5 $\mu g\ g^{-1}$ have been recorded. Usually the process proceeds no further and nitrite disappears through assimilation, but some bacteria and fungi are credited with oxidizing ammonium to nitrate. The second group of heterotrophic nitrifiers produce mainly nitrate (though sometimes nitrite) from a variety of organic nitro-compounds such as the nitrophenols, and from oximes of several organic acids including pyruvic, oxaloacetic and α-ketoglutaric, and from peptones and amino compounds. Among the organisms reported to do so are the bacteria *Agrobacterium, Alcaligenes* and *Corynebacterium*, the actinomycete *Nocardia*, and various fungi including *Aspergillus flavus, A. ustus, Cladosporium herbarum* and *Penicillium citrinum*. The rate of production of NO_3-N by these organisms in pure culture is known to be less, by at least an order of magnitude, than in autotrophic nitrification, but is considerably higher than the amounts of NO_2-N formed by the nitrite producers. Many nitrate producers, notably *A. flavus*, form substantial amounts of nitrate when incubated in the laboratory in an appropriate medium. J. A. Duggin (1984) found that most of the heterotrophic nitrifiers isolated from strongly acid (pH 4.0) soils at Hubbard Brook, New Hampshire, were fungi and most of these formed nitrate: several produced more than 20 μg NO_3-N ml^{-1} during 21 days incubation in glucose-peptone broth, although the majority formed 3 μg NO_3-N ml^{-1} or less.

The evidence that fungal nitrification is ecologically significant is mainly circumstantial. For example, nitrate production takes place rapidly in some soils despite apparently low populations of autotrophic nitrifiers, or where the latter have been inhibited by bacteriostatic agents. In the competitive environment of the soil Wainwright (1981) suggests that the efficiency of heterotrophs might be enhanced, because ammonium or nitrite are not always readily available as substrates for the chemoautotrophs. While the consensus of opinion is that heterotrophs are unimportant in the formation of nitrate in soils generally, they may be of significance in unusual habitats, where their relative inefficiency might be compensated for by their large numbers or great biomass. In acid forest soils in particular, where only small populations of autotrophic nitrifiers exist, the contribution of heterotrophic nitrification to the mineralization process may be far from negligible. Indeed, Duggin (*loc. cit*) showed, by means of a series of selective inhibition experiments, that heterotrophic nitrification accounted for more than half the nitrate produced by incubating fresh soil samples in the laboratory. He further demonstrated that autotrophic nitrification was totally inhibited at a soil pH of 3.5 whereas heterotrophic nitrifiers continued to function in these very acid conditions. In contrast, at pH 6.5 autotrophic nitrification increased dramatically but heterotroph activity ceased between pH 4.5 and 6.5. The optimum temperature for heterotrophs was 30 °C and for autotrophs, 20–25 °C. Based on these observations, he suggests that the two processes have overlapping or complementary environmental requirements and can proceed concurrently in the same soil.

Mineralization and immobilization of sulphur

Transformations of sulphur in soils resemble in many ways the microbial conversions of nitrogen. Soil organic matter constitutes an important pool of reserve sulphur in ecosystems, but few details are known of the chemical nature of the sulphur compounds themselves. About half of the soil organic sulphur can be reduced to hydrogen sulphide by hydriodic acid (HI) and is therefore assumed to be not directly bonded to carbon atoms; it probably occurs mainly as sulphate esters which are predominantly products of microbial metabolism (Russell 1973). About two-fifths of the remainder is classified as 'carbon-bonded' and a substantial amount of this is known to be in the form of the amino acids cystine, cysteine and methionine. The residual fraction is of unknown composition but is exceptionally stable, and therefore probably of little significance as a potential source of sulphur for plants (Briederbeck 1978). The HI-reducible fraction can be readily hydrolysed to inorganic sulphate by acid or alkali, and consequently is considered to be the most labile form of organic S in soil. By comparison, C-bonded sulphur is relatively stable.

Sulphur is taken up by plants mainly as sulphate (SO_4^{2-}) and is reduced to sulphydryl (–SH) within plant tissues. Mineralization of plant debris is therefore an essential part of the soil sulphur cycle. As in the mineralization of organic nitrogen, the extent of mineral sulphur formation is governed to some degree by the sulphur content and the C/S ratio of the organic matter. The amounts of mineral S produced on incubation of soil samples in the laboratory are about an order of magnitude less than those of mineral N, nevertheless S mineralization proceeds actively enough to permit rapid recycling of the more labile components. For example, Till (1980) found that 15 per cent of the sulphur in ^{35}S-labelled clover residues was re-incorporated into grass within two months of application, while even greater proportions of S were recycled from dead clover roots mixed with the soil.

Heterotrophic sulphur oxidation

M. Wainwright (1978) listed a variety of heterotrophic bacteria, actinomycetes and fungi capable of oxidizing reduced inorganic sulphur compounds, or elemental sulphur, *in vitro*. These organisms presumably respire aerobically, using sugars in the growth medium as energy sources. The process should therefore be seen as quite distinct from the formation of sulphate as an end product of the mineralization of soil organic sulphur. It is somewhat analogous to heterotrophic nitrification but whether it is of any significance in soils *in situ* is unknown.

Mineralization and immobilization of phosphorus

As with nitrogen and sulphur, there is a large reservoir of organic phosphorus in soils which is unavailable to plants. While much of the phosphate taken up by plants is provided by the weathering of primary minerals, the microbial oxidation of organic substrates is an important supplementary source of inorganic phosphate. The mineralization process is especially important in virgin soils, both the total amount of phosphorus mobilized and the percentage of total organic phosphorus mineralized being greater in virgin soils than in their cultivated counterparts.

Soil organic matter contains a range of inositol phosphates (sect. 2.3) from mono- to hexa-, but the main forms present are the penta- and hexaphosphates; furthermore, inositol is not only present as the ubiquitous myoinositol but occurs in other isomeric forms also. These mixed inositol polyphosphates are loosely known as 'soil phytate' or 'soil phytin', although strictly speaking phytic acid is myoinositol hexaphosphate and phytin or phytate its calcium/magnesium salt (Cosgrove 1977). The amount of soil phytate varies from soil to soil, generally being 15–30 per cent of the total organic phosphorus though the range may be as wide as 10–60 per cent. Smaller quantities of organic P (2–5%) exist as nucleotides; nucleic acids in living microorganisms could account for this. Still smaller amounts (1–2% or less) are found as phospholipids. Traces of ribitol and glycerol phosphates, which probably derive from the techoic acids of bacterial cell walls, have also been recognised, but much of the organic phosphorus in soil remains to be identified.

Phospholipids, nucleic acids and sugar phosphates undergo rapid enzymatic degradation to inorganic phosphate, but the mineralization of inositol polyphosphates is relatively slow. Nevertheless these soil and litter phytates are potentially an important source of PO_4^{3-} for plants, since phytase activity is common among soil and rhizosphere microorganisms (Greaves & Webley 1965) and mycorrhizal fungi (Moser & Haselwandter 1983). This might be especially important in forest soils which seem to have a greater proportion of their organic P present as inositol hexaphosphates than grassland soils (Caldwell & Black 1958). Some mor humus soils have very low inorganic phosphate levels, and phytates are probably the most common sources of P for plant growth therein. Whether or not root-infecting fungi or the root surface microflora in general stimulate phytase activity in the root environment, it should be kept in mind that increased mineralization of organic P in the rhizosphere may not necessarily lead to increased PO_4^{3-} uptake by plants. Indeed, as Barber and Loughman (1970) have shown, at low phospate levels rhizosphere microbes accumulate PO_4^{3-} from organic sources at the expense of plant roots. This is a situation of net immobilization with respect to phosphorus: as with nitrogen, inorganic P derived from organic sources will be available for plant use only when there is net mineralization. The mineralization–immobilization cycle of phosphorus is analogous to the turnover cycle of nitrogen but plants rarely suffer from phosphorus deficiency as a result of microbial competition. Nevertheless such competitive effects have been recorded: Benians and Barber (1974) found that less P was taken up by plants grown in non-sterile basaltic loam soil than under aseptic conditions, and that these microbial effects could be obviated by adding phosphate.

Incomplete oxidations and biological weathering

The major end product of the decomposition of organic matter under aerobic conditions is carbon dioxide, and in this event the oxidation of organic compounds is regarded as complete. Certain microbes, however, carry out incomplete oxidations, so that carbon compounds other than CO_2 may accumulate. The best known example is the oxidation of ethyl alcohol to acetic acid by the acetic acid bacteria, a reaction which is utilized in the commercial production of vinegar. Other bacteria produce organic acids, for example many bacteria of the genus *Pseudomonas* oxidize glucose to gluconic acid and subsequently to 2-ketogluconic acid via the Entner-Doudoroff pathway:

$$\begin{array}{ccccc} & & & & \mathrm{COOH} \\ & & & & | \\ \mathrm{CHO} & & \mathrm{COOH} & & \mathrm{CO} \\ | & & | & & | \\ (\mathrm{CHOH})_4 & \rightarrow & (\mathrm{CHOH})_4 & \rightarrow & (\mathrm{CHOH})_3 \\ | & & | & & | \\ \mathrm{CH_2OH} & & \mathrm{CH_2OH} & & \mathrm{CH_2OH} \end{array}$$

Another product of the incomplete oxidation of sugars is oxalic acid (HOOC-COOH), which is the simplest of the dibasic organic acids. It is especially characteristic of fungi, where it accumulates as the calcium salt often in the form of a crystalline precipitate on the surfaces of hyphal strands and rhizomorphs.

Primary minerals in nature are subject to the combined influence of physical and chemical weathering. Organisms contribute to both kinds of weathering, but especially to the latter. Physical forces result in the comminution of rocks thereby increasing the surface area of exposed silicate minerals to the action of chemical agents. Direct evidence of the changes in mineral structure which occur in the presence of plant roots and microorganisms is provided by X-ray diffraction analyses which reveal, for example, that potassium is released from the crystal lattice of biotite, altering it to vermiculite

(Boyle & Voigt 1973; Malquori *et al.* 1975). This removal of metallic ions from primary minerals by living organisms has long been assumed to be the result of metabolic acid production, and Boyle *et al.* (1967) showed that the mineral alterations produced by cultures of *Aspergillus niger* were similar to those caused by organic acids such as oxalic and citric. However, while it is widely accepted that carbonic acid and organic acids of microbial or plant origin act as agents of solubilization and as sources of hydrogen for ion exchange (see Ch. 6), there is considerable doubt as to their significance in the weathering process, especially since carbonic acid is weak and highly labile, and many of the organic acids credited as weathering agents, with the possible exception of oxalic acid, occur only at low concentrations. Apart from acidity *per se*, however, the chemical degradation of rocks and minerals could be mediated by chelation. Most of the metallic cations found in silicate minerals, such as felspar and mica, readily chelate with appropriate organic molecules, for example hydroxy acids, to form stable ring structures. Soil organic matter contains numerous chelating agents. Some of these, such as the tricarboxylic acid cycle (TCA) intermediate, malic acid, are products of intermediary metabolism, while others, like ketogluconic acid, are end products of incomplete oxidations. Still others, including humic and fulvic acids, are formed during the microbial conversion of plant residues to humus.

Chelation might also be important in the weathering of phosphate minerals. This is especially relevant because all the phosphorus needed by organisms for metabolic purposes derives ultimately from the mineral fluorapatite, which consists mainly of insoluble calcium phosphate, $3Ca_3(PO_4)_2.CaF_2$, which weathers to hydroxapatite, $3Ca_3(PO_4)_2.Ca(OH)_2$ in slightly acid to alkaline soils. In more acid soils, phosphate is associated with iron or aluminium as insoluble hydrated phosphates (e.g. strengite, $FePO_4.2H_2O$ and variscite, $AlPO_4.2H_2O$) or bound to their oxides (Fe_2O_3,Al_2O_3) or fixed to the crystal lattices of clays. The phosphorus in both primary and secondary minerals can be released into the soil solution as orthophosphate ($H_2PO_4^-$) by the chelation of Ca^{2+}, Fe^{3+} or Al^{3+}; in this form it is readily absorbed and metabolized by plant roots and microorganisms. Alternatively, the phosphate itself may be held in stable organic complexes which, being soluble, are potentially available for uptake by organisms. According to Halstead and McKercher (1975) however, it is uncertain whether this organic phosphate must be mineralized first before utilization by green plants.

Speculation on the role of microorganisms in the weathering process arises as a result of the presence of bacteria and fungi in the weathered crusts of rocks. The microorganisms isolated from this harsh environment have been tested in pure cultures for their ability to dissolve silicates such as felspars and micas. When presented with finely ground mineral particles suspended in an agar medium containing glucose or some other simple carbon and energy source, many common fungi and bacteria may be shown to possess this attribute. D. M. Webley and colleagues (Duff & Webley 1959; Duff *et al.* 1963) reported that the most active silicate-dissolving fungi were species of *Botrytis, Mucor, Penicillium* and *Aspergillus*, while the most prominent bacteria were the fluorescent pseudomonads capable of producing 2-ketogluconic acid. The latter also had the capacity to release phosphate from insoluble phosphates, including apatite. On the basis of chromatographic evidence, these workers assumed that the increased acidity of the medium, which invariably accompanies the solubilization of phosphatic minerals, was due to the chelation of calcium by the 2-ketogluconate anion.

In 1978, A. Moghimi and co-workers (Moghimi *et al.* 1978) confirmed the ability of 2-ketogluconic acid to release phosphate from di-and tricalcium phosphate and apatite rock but queried the mechanism by which this was achieved, because the chelating ability of sugar acids is generally low, especially in acid conditions. Moghimi and Tate (1978) therefore set out specifically to answer the question of whether dissolution is due to chelation or protonation. They found that 2-ketogluconic acid had a very low pK_a value which placed it among the strongest of the monobasic carboxylic acids while its negligible calcium stability constant showed it to be a very poor chelator over the pH range 2.4 to 6.4. They concluded that 2-ketogluconic acid dissolves hydroxyapatite by acting as a ready source of protons (H^+ ions) rather than as a chelating agent. It is probable that oxalic acid, which is another very strong, low molecular weight organic acid that has been credited by Cromack *et al.* (1977) with being an effective chelator of Ca^{2+}, Fe^{3+}, and Al^{3+}, also acts primarily as a source of hydrogen ions. The marked ability of certain fungi to mobilize and accumulate Ca^{2+} and associated ions from otherwise insoluble sources has been attributed by Cromack and his colleagues (Graustein *et al.* 1977; Cromack *et al.* 1977) to the copious production of oxalic acid.

It is one thing however to demonstrate a mechanism for the breakdown of finely ground phosphate and silicate minerals in the laboratory, but quite another to determine the significance of such processes in nature. In the field, three factors are operative which are normally not taken into account in the laboratory; other microorganisms are present, reducing the concentration of solubilizing acids by competing for available substrates with the organisms that produce them, or by utilizing the acids themselves as substrates; minerals are not pulverized and are therefore more difficult to dissolve; and an energy source or carbon substrate is scarce or lacking. The latter constitutes the major obstacle to accepting the significance of biological weathering since (with the exception of algae in lichens) all the microorganisms so far implicated are heterotrophs. In view of their dependence on an organic carbon source, the most likely habitat in nature where they could act as primary agents for rock weathering is the rhizosphere. Various sugars including glucose are exuded by plant roots and these could conceivably be converted by the rhizosphere microflora into acids and used to solubilize primary minerals. This will be discussed further in Chapter 6.

5.2.2 Respiration with inorganic substrates

The bacteria which use inorganic hydrogen donors as substrates for aerobic respiration are able to utilize CO_2 as a source of carbon for cell synthesis. One group of these chemoautotrophic bacteria is of particular significance to plant nutrition and mineral cycling, namely the nitrifying bacteria. Another group of some importance is the sulphur-oxidizing bacteria.

Nitrification

The mineralization of organic nitrogen can be differentiated into the processes of ammonification and nitrification (Fig. 5.6). Ammonification concerns the production of ammonium by heterotrophic soil microorganisms, and has been dealt with previously

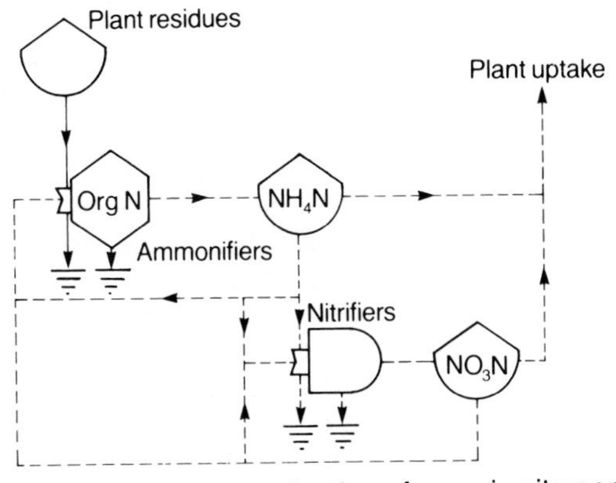

Fig. 5.6 The mineralization of organic nitrogen by soil microorganisms, showing the sequential stages of ammonification (production of NH_4^+) and nitrification (production of NO_3^-).

(sect. 5.2.1). Nitrification refers to the oxidation of ammonium to nitrite and nitrate by autotrophic bacteria. T. Schoesling and A. Muntz were the first to prove that nitrification was a biological process, by demonstrating in 1877 that it could be stopped by sterilizing agents such as chloroform. The nitrifying bacteria fall into two physiological groups. The first, typified by *Nitrosomonas*, oxidizes ammonium to nitrite:

$$2NH^+ + 3O_2 \rightarrow 2NO_2^- + 4H^+ + 2H_2O$$

The second, represented by *Nitrobacter*, oxidises nitrite to nitrate:

$$2NO_2^- + O_2 \rightarrow 2NO_3^-$$

Because of the difficulty of isolating them, few details are known of the distribution and frequency of occurrence of individual species of nitrifiers. E. L. Schmidt (1978) listed five genera of ammonium oxidizers and three of nitrite oxidizers. All have long been considered to be obligate chemoautotrophs, but *Nitrobacter* is capable of utilizing acetate as a sole source of carbon and energy so that, strictly speaking, the term 'facultative autotroph' is more appropriate. Switching to a heterotrophic mode of nutrition does not increase the growth rate of *Nitrobacter*, but presumably it has some survival value in environments where nitrite is limiting.

The energy yields for both stages of the nitrification process are low, and the nitrifiers grow very slowly; even under optimum conditions, their mean generation time is about 10 hours. The growth rate of heterotrophic nitrifiers (sect. 5.2.1) is much faster. Nevertheless, the autotrophs are thought to be responsible for the bulk of the biological oxidation of ammonium in most soils.

The activity of the nitrifying bacteria is markedly influenced by certain environmental conditions, chief among which is soil pH. In acid environments their activity is greatly reduced, even in the presence of an adequate amount of substrate. In pure culture, their optimum pH is between 7.5 and 8.0. In soils, nitrate production falls off rapidly below pH 6.0 and generally is negligible below pH 5.0. Nitrification can however occur at a soil pH of 4.0 and nitrifying bacteria have been detected in even more acid soils. Another important environmental variable is the soil oxygen supply; both *Nitrosomonas* and *Nitrobacter* are obligate aerobes, hence adequate soil aeration is essential to their function. Population densities are very low, and in acid soils may be only of the order of 100 cells per gram. They rarely exceed 10^5 per gram even in neutral and alkaline soils, unless ammonium fertilizers have been added.

Under most conditions in nature, the oxidation of ammonium proceeds as far as nitrate, so that nitrite does not normally accumulate in soils; this is fortunate because nitrite is toxic to plants and microbes. Where it does occur it is usually the result of the combined effects of alkalinity and high ammonium levels, as described previously. In arable soils, nitrate is considered to be the principal form in which nitrogen is assimilated by plants, hence from the crop agronomist's viewpoint the nitrifying bacteria play a key role in the soil nitrogen cycle and in the maintenance of soil fertility. In natural grassland and forest soils, the nitrifiers may not be so important (Fig. 5.7). Arable soils of the temperate regions, unless they are very acid, contain a fairly constant but very low level of ammonium, and a variable but higher amount of nitrate. The NO_3-N concentration commonly ranges from 2–20 ppm $\mu g\ g^{-1}$), and in favourable circumstances rises to over 50 ppm. It varies throughout the year according to agronomic practice, tending to accumulate in fallow soils and to disappear under cropping. As E. W. Russell (1973) has pointed out, however, soil in fallow will accumulate nitrate only if four conditions are satisfied. First, sufficient readily decomposable organic matter must be present to provide ammonium as an energy source for the nitrifiers. Second, the soil must be kept free of weeds otherwise these may utilize the nitrate as rapidly as it is produced. Third, the soil must be kept moist but not necessarily continuously. Fourth, there must not be too much rain or else nitrate will be lost by leaching.

It is generally assumed that nitrification plays little or no part in the nitrogen economy of permanent grassland and forest soils, and that grasses and trees absorb nitrogen almost entirely as ammonium. While direct experimental proof of this assumption is lacking, there is much circumstantial evidence to support it. According to F. E. Chase and colleagues (Chase *et al.* 1967) there are large areas of the Earth's land surface 'in which, because of soil or climatic conditions, coupled with dense perennial plant cover, nitrate does not occur and plants must subsist on an ammonium diet'.

Grassland soils of the cool temperate zone in the northern hemisphere maintain a low but fairly constant level of NH_4-N, about 3–9 ppm throughout the year, but they contain very low or negligible amounts of nitrate nitrogen (1–2 ppm) which distinguishes them from soils under crops or fallow (Russell 1973). A depression of nitrification might be expected in old, permanent grassland soils since these are frequently acid, and contain few nitrifying bacteria. In addition, there is some evidence which points to a direct inhibition of nitrifiers by grass root exudates. In many instances however, the absence of nitrification under pasture cannot be explained, and many grassland soils nitrify well if incubated in the laboratory; furthermore, it is common experience that nitrification will

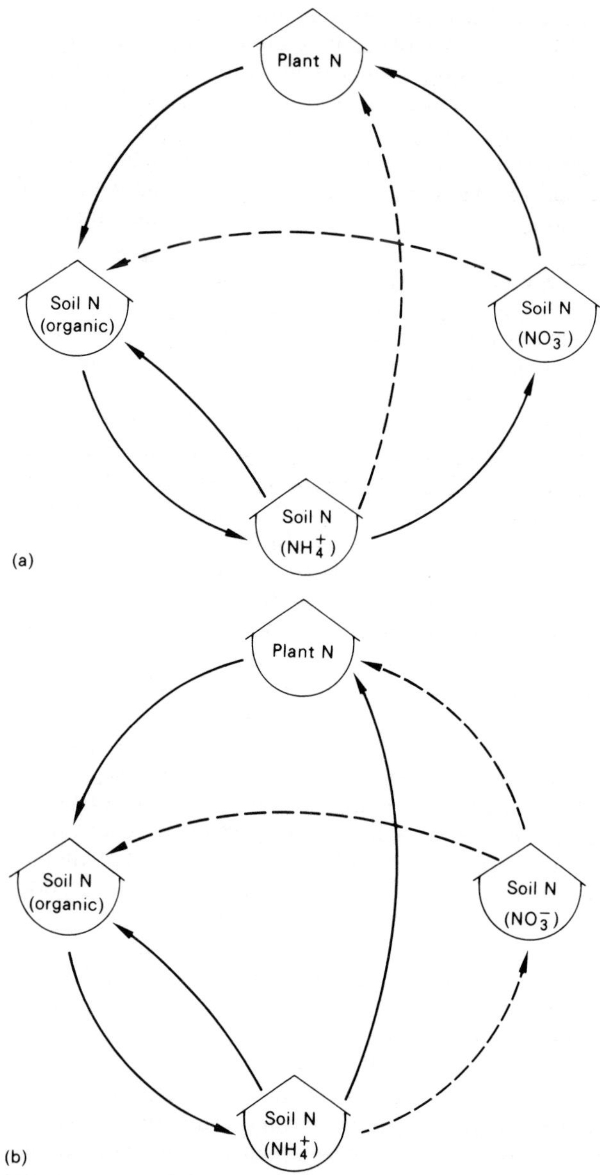

Fig. 5.7 The soil nitrogen cycle. (a) The 'nitrate model', as typified by arable soils. (b) The 'ammonium model', as found in many permanent grassland and forest soils. Major pathways shown as solid lines, minor pathways as broken lines.

proceed rapidly in the field if old pastures are ploughed out. Not all grassland soils behave in this fashion, however. In Southern Australia, with a warm, intermittently dry summer and a cold, wet winter, J. R. Simpson (1954) found that soils from an old *Phalaris*–subterranean clover pasture accumulate nitrate during summer and autumn (cf. fallow soils), the levels sometimes reaching 40 ppm NO_3-N during the summer and falling to about 2 ppm in winter and early spring; NH_4-N tends to remain at about 4–5 ppm throughout the year.

The levels and forms of mineral nitrogen in forest soils depend on their pH and the type of humus. Nitrification is generally much more active in mull than in mor soils. Many forest soils are moderately to strongly acid, and in consequence the activity of nitrifying bacteria in these soils is likely to be inhibited. The level of NH_4-N in lateritic podzolic soils from dry sclerophyll eucalypt forest in South East Queensland varies between 3 and 8 ppm, while NO_3-N rarely occurs in detectable amounts. The native microflora in these soils shows a marked preference for ammonium and seemingly has limited ability to utilize nitrate. Table 5.2 shows how nitrate remains largely unincorporated during incubation even though the bacterial population rises with the addition of lime, whereas ammonium is rapidly immobilized by the microflora and subsequently mineralized.

Whether these observations apply in general to forest soils is not known. If forest soil microbes are indeed limited in their capacity to utilize nitrate then what nitrate is produced under natural conditions must be either taken up by plants or else lost to the system by denitrification or leaching. Alternatively, the low levels of nitrate commonly found in undisturbed forest ecosystems might be explained by postulating that nitrification plays a minor role in such ecosystems. This is difficult to accept in view of the known presence of nitrifying bacteria in many forest soils and the necessity of providing nitrate as an electron acceptor in the denitrification process (without which the nitrogen cycle is incomplete). In any event, there are certainly some forest soils which have considerable potential for nitrification when the tree cover is disturbed or when they are limed, even though they may contain little or no nitrate in the virgin condition. Others do contain significant amounts of nitrate naturally, especially in the tropics where the phenomenon may be related to an alternating cycle of wetting and drying imposed by prevailing climatic regimes. For reasons that are not entirely clear, there is rapid oxidation of organic matter in a soil that has been remoistened after drying, the flush of decomposition lasting several days. This could well explain the high, fluctuating

Table 5.2 Effect of lime on microbial populations and on mineralization and immobilization of $^{15}NH_4$-N or $^{15}NO_3$-N during incubation of a sandy podzolic soil

			Bacteria per gram			Organic N		Inorganic N	
Treatment	*Soil pH*	*Fungi per gram* ($\times 10^5$)	*Ammonifiers* ($\times 10^6$)	*Nitrifiers*	*Denitrifiers* ($\times 10^6$)	*Non-hydrolysable* (%)†	*Hydrolysable* (%)†	NH_4^+ (%)†	NO_3^- (%)†
$K^{15}NO_3$ minus lime	4.7	2.91	2.94	ca. 60	0.07	2.4	2.6	0.4	94.5
$K^{15}NO_3$ plus lime	7.4	1.91	11.18	ca. 80	0.53	1.9	2.5	0.4	95.1
Significance level‡	***	***	***	ns	***	ns	ns	ns	ns
$(^{15}NH_4)_2SO_4$ minus lime	4.6	3.43	3.88	ca. 40	0.24	6.0	22.4	64.6	6.9
$(^{15}NH_4)_2SO_4$ plus lime	7.3	2.55	18.81	ca. 1400	1.36	14.6	27.7	42.4	15.2
Significance level‡	***	***	***	***	***	**	***	***	***

Source: From Jones, J. M. (1968) PhD thesis, University of New England, Australia

† Percentage of total labelled nitrogen.

‡ One, two or three asterisks indicate that means differ at the 5%, 1% or 0.1% levels of probability, respectively; ns = not significant.

nitrate levels (5–25 ppm N) and the low but fairly constant ammonium level (4 ppm N) found by D. J. Greenland (1958) in soil from tropical high forests in Ghana. The stimulating effect of alternating dry and moist phases on mineral N production was first recorded by A.N. Lebedjantzev in 1924. Both physicochemical (Paul & Tu 1965; van Schreven 1964) and biological (Birch 1960) interpretations have been proposed to account for the phenomenon.

Nitrification and plant succession

In 1972 and 1973, E. L. Rice and S.K. Pancholy published the results of studies comparing the levels of mineral nitrogen in soils from three stages of secondary succession in prairie and forest communities from Oklahoma, in the United States. They found ammonium concentrations increased from low in the early stage to high in the late or climax stage, while nitrate levels and the numbers of autotrophic nitrifiers decreased throughout the succession. Based on these findings, they advanced the hypothesis that soils under climax vegetation were low in nitrate because of the allelopathic influence of climax plant species on the nitrifying bacteria, and showed that tannins were able to inhibit nitrification in soil suspensions. Furthermore, they demonstrated that tannins in leaf litter accumulated in the soils of intermediate and climax communities, to levels in excess of those necessary to inhibit nitrifiers in the laboratory. In 1974, Rice and Pancholy presented further evidence in support of their hypothesis, by finding that many aromatic compounds other than tannins, for example phenolic acids and phenolic glycosides, are produced by plants prominent in mid and late succession, and showing that these substances were also strongly inhibitory to nitrification at low concentrations (10^{-6}–10^{-8} M).

Many other workers have found some evidence of inhibition of nitrification in climax ecosystems. In grassland communities, where relatively little surface litter accumulates, inhibition is thought to be associated with living roots (Moore & Waid 1971). In 1974, however, B. S. Purchase (1974a) attempted to evaluate the claim that grass root exudates inhibited nitrification, and found no evidence to substantiate it. Instead, he concluded that the inhibition was more apparent than real, and the lack of nitrification could be explained by assuming it resulted from competition for ammonium by the heterotrophic soil microflora. The idea that nitrifiers could be denied access to ammonium by heterotrophs had been proposed by S. L. Jansson as early as 1958. In another study, Purchase (1974b) showed that phosphorus could become a limiting nutrient for nitrifiers in phosphorus deficient soils, while Jones and Richards (1977) and Johnson and Edwards (1979) gave further evidence of the importance of heterotrophic competition for ammonium in restricting nitrification in forest soils. Moreover, a number of workers (e.g. Johnson & Edwards 1979; Montes & Christensen 1979) have been unable to detect nutrification inhibitors in leaf, litter or root extracts of late successional species.

The hypothesis that nitrification is inhibited by allelopathic substances generated by plants characteristic of late successional communities leads to speculation that it represents an adaptation to conserve nitrogen (by minimizing leaching losses) and energy (by reducing the energetic costs of N uptake and utilization by plants). As such, it is in keeping with the strategy of ecosystem development envisaged by E. P. Odum (1969; sect. 5.1). However, while some workers (e.g. Jordan *et al.* 1979; Richards *et al.* 1985) find little or no nitrification in apparently mature, undisturbed ecosystems, as predicted by Rice and Pancholy's hypothesis, other studies have shown that it proceeds actively in some old growth forests (Runge 1974; Adams & Attiwill 1982; Richards *et al.* 1985). Furthermore, there is clear evidence, from either incubation studies in the field and laboratory (Montes & Christensen 1979; Lamb 1980) or from an examination of streamwater nitrate levels (Vitousek & Reiners 1975; Vitousek 1977), that nitrification sometimes increases throughout succession. The evidence is thus conflicting but certainly, as M. Runge (1983) pointed out in summarizing European experience, the generalization that inhibition of nitrification is a characteristic of climax ecosystems 'does not seem permissible in consideration of the almost exclusive NO_3^--production in the soils of numerous forest ecosystems'. There are some indications that the soils of highly productive communities are more likely to nitrify actively right throughout succession than those of less productive ones (Lamb 1980; Adams & Attiwill 1982), and this suggestion accords with the consensus reached by a number of American workers (Vitousek *et al.* 1979) that nitrate losses to groundwater and streamwater in disturbed ecosystems are probably greatest on soils regarded as fertile in a regional context, that is on high quality sites.

Regulation of mineralization processes. The nitrate inhibition hypothesis is of more than theoretical interest, because it is central to the problem of predicting nutrient losses from forested catchments should these be cleared for agriculture or harvested for wood production. Practical interest in the latter has heightened in recent years with the wider use of more intensive logging techniques, notably clearcutting. Nitrate losses are particularly significant in this context, not only because they may lead to downstream eutrophication, but also because the hydrogen ions released during nitrification and the high mobility of the NO_3^- ion itself, promote the mobilization and loss of soil cations (Vitousek *et al.* 1979). There are several mechanisms which might govern the formation and consequent loss of nitrate by leaching, apart from the possibility of allelopathic inhibitors. The most likely means of regulating the activity of nitrifiers involve the removal of their substrate, that is, ammonium, either by plant uptake or immobilization by the heterotrophic ammonifiers. There is now good evidence that autotrophic nitrifying bacteria may suffer competition for ammonium from the heterotrophic bacteria and fungi which require it as a nutrient. This is well illustrated by the results of J. M. Jones and B.N. Richards (1977), who studied the effect of lime and macerated pine needles on the transformation of ^{15}N-labelled $(NH_4)_2SO_4$ in an infertile podzolic soil. Both treatments enhanced the immobilization of $^{15}NH_4$-N, as evidenced by increased amounts of ^{15}N found as organic forms (Table 5.3). The immobilization caused by

Table 5.3 The influence of ground lime and macerated pine needles on the utilization of $(^{15}NH_4)_2SO_4$ by the microflora of a sandy podzolic soil

Microbial group or N fraction	*Minus lime*	*Plus lime*	*P*†	*Minus needles*	*Plus needles*	*P*
Bacteria ($g^{-1} \times 10^6$)	6.3	18.4	***	8.8	15.9	***
Fungi ($g^{-1} \times 10^5$)	3.4	2.6	***	2.4	3.5	***
Organic ^{15}N ($\mu g\ g^{-1}$)						
Non-hydrolysable	5.4	13.3	***	8.3	10.4	*
Hydrolysable	20.0	25.2	***	14.9	30.3	***
Mineral ^{15}N ($\mu g\ g^{-1}$)						
NH_4^+	57.7	38.5	***	52.1	44.2	***
NO_3^-	6.2	13.8	***	14.4	5.6	***

Source: From Jones and Richards (1977).
The table shows the 'main effects' of the two amendments, i.e. the effect of lime on soil samples half of which had been amended with needles and half of which had not. Likewise, the effect of pine needles represents an average of samples with and without lime.
† Probability level. One, two or three asterisks indicate that the means differ significantly at the 5%, 1% or 0.1% levels, respectively.

Table 5.4 The interacting effects of lime and pine needles on the ammonification and nitrification of $(^{15}NH_4)_2SO_4$ by the soil microflora

Treatment	*Bacteria* g^{-1} $\times 10^6$	*Fungi* g^{-1} $\times 10^6$	$^{15}NO_3$-*N* $\mu g\ g^{-1}$	$^{15}NH_4$-*N* $\mu g\ g^{-1}$
Nil	5.2	2.7	7.9	63.9
Lime	12.4	2.1	20.9	40.2
Pine needles	7.3	4.1	4.6	51.6
Lime and needles	24.5	3.0	6.7	36.8
LSD for $P < 0.001^{\dagger}$	4.1	ns	9.3	7.4

Source: From Jones and Richards (1977).
† Least significant difference for a probability of 0.1%; ns indicates that the interaction was not significant at the 5% level of probability.

adding lime could be attributed to a larger population of heterotrophic bacteria whereas increased growth of fungi as well as bacteria accounted for the immobilization due to pine needles. Both amendments reduced the amount of $^{15}NH_4$-N present, but lime raised the level of $^{15}NO_3$-N while pine needles lowered it. From an examination of the interaction between the two factors (Table 5.4), it became apparent that their contrasting effects on nitrate levels were due to different organisms and different processes. Amending the soil with lime ameliorated the acid environment, as shown previously in Table 5.2. This favoured the development of bacteria, including nitrifiers, thus promoting the utilization of $^{15}NH_4$-N as a substrate (energy source) for nitrification and leading to an increase in the concentration of $^{15}NO_3$-N (Table 5.4). When pine needles were added as well as lime, they served as a substrate for heterotrophic bacteria, enabling the population of these to enlarge even more and thereby greatly increasing their demand for available nutrients, including $^{15}NH_4$-N. In these circumstances, the autotrophic nitrifiers were deprived of substrate and no further nitrification took place.

A corollary of these findings is that nitrification will not occur at substantial rates unless there is net mineralization. This statement serves as a basic premise in predicting how mineralization processes are regulated during ecosystem development. As a first approximation, W. A. Reiners (1981) has assumed that the nitrification rate follows the rate of NH_4-N production, and has developed a model to illustrate this (Fig. 5.8). The model indicates that little NH_4-N is available for nitrification early in succession because nitrogen is accumulating in the biomass and detritus, as a result of

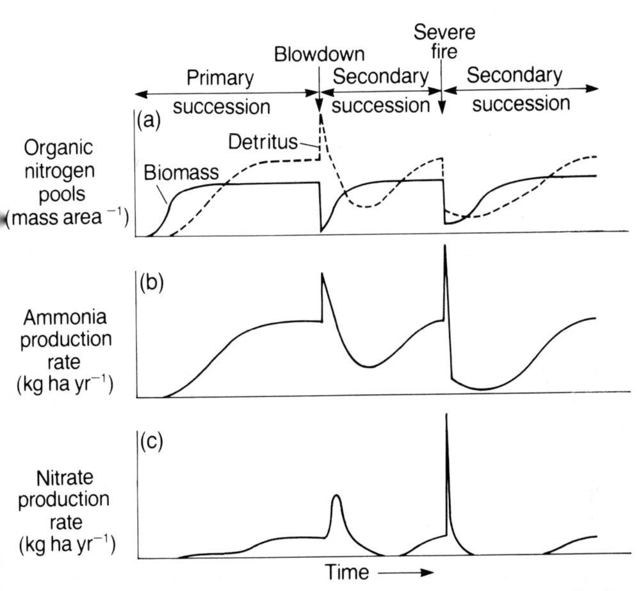

Fig. 5.8 A model for changes in organic and mineral nitrogen during primary and secondary succession in a terrestrial ecosystem. (a) Accumulation of organic N in biomass and detritus. (b) Ammonification rate with respect to changes in pool sizes of organic N. (c) Nitrification rate relative to ammonium production. (From Reiners 1981 – for explanation, see text.)

plant uptake and microbial immobilization. The hypothetical NO_3-N production curve shows little or no nitrification until biomass N begins to stabilize, that is, at the stage when plant uptake is reduced. The model assumes that nitrogen will continue to accumulate in organic detritus after it ceases to build up in the biomass. This will be true only if the detrital nitrogen pool is considered to comprise both the organic matter incorporated in the mineral soil and the unincorporated litter lying on the soil surface. It could explain why nitrate production occurs in some apparently mature or climax communities, as the model indicates.

Sulphur oxidation

The colourless sulphur bacteria use inorganic sulphur compounds, or elemental sulphur, as substrates for aerobic respiration. During this process, hydrogen sulphide is oxidized to sulphur, and sulphur and thiosulphate are oxidized to sulphate:

$$2H_2S + O_2 \rightarrow 2S + 2H_2O$$
$$2S + 2H_2O + 3O_2 \rightarrow 2SO_4^{2-} + 4H^+$$
$$S_2O_3^{2-} + H_2O + 2O_2 \rightarrow 2SO_4^{2-} + 2H^+$$

There are two groups of autotrophic sulphur-oxidizers. One, the filamentous sulphur bacteria, are closely related to the cyanobacteria. They occur in specialized habitats such as hot springs and on the surface of estuarine mud, and need not concern us here. The other group comprises gram-negative rods with polar flagella (if motile), and are assigned to the genus *Thiobacillus*[3].

The thiobacilli are primarily responsible for the oxidation of reduced forms of inorganic sulphur in soils. They are not present in large numbers unless sulphur compounds are added deliberately. Under normal conditions, most soils contain less than 200 thiobacilli per gram, though higher populations, of the order of 1000 cells per gram, occur when appropriate substrates are present. At the low population densities normally holding, they can hardly play a major role in sulphate formation. It seems more likely that sulphate production in most soils is achieved by other means, for example as a by-product of the decomposition of sulphur-containing proteins, which is brought about by a variety of heterotrophic bacteria and fungi that are more abundant and widespread than the thiobacilli (see sect. 5.2.1). Of more than 200 strains of sulphur-oxidizing microbes isolated by Vitolins and Swaby (1967) from Australian soils, 61 per cent were heterotrophic bacteria (31% *Arthrobacter* spp., 11% *Bacillus* spp.) and 22 per cent were facultative *Thiobacillus* spp., while only 13 per cent were autotrophic thiobacilli. Nevertheless, in soils where microbial transformations of sulphur are confined to the activities of heterotrophs, the rate of sulphur oxidation may be inadequate for satisfactory plant nutrition. Swaby and Fedel (1973, 1977) found that almost half the soils they examined from temperate Australia oxidised sulphur slowly or not at all, and contained very few thiobacilli; lack of sulphur-oxidizing capacity was even more widespread in tropical soils although the activity of thiobacilli remained moderately high.

Enrichment of soils in thiobacilli, by adding an appropriate substrate such as finely ground sulphur, can be used to ameliorate plant nutrient deficiencies due to excessive alkalinity. Manganese deficiency is especially responsive to this treatment. Inoculation may also be necessary if the soil is altogether lacking in thiobacilli. Increased growth of thiobacilli following the application of sulphur is also credited with controlling the disease potato scab, the causal agent (*Streptomyces scabies*) being sensitive to soil acidity.

Not all of the sulphate in soils is the product of biological oxidations *in situ*. A substantial proportion of soil sulphate derives from rain; its origin will be discussed later when anaerobic processes are considered (sect. 5.3). It should also be kept in mind that in well aerated soils sulphides, sulphur and thiosulphate can be slowly oxidized by purely chemical processes. Such processes are however insignificant in comparison with biological oxidations of reduced sulphur, provided soil temperature and moisture regimes are favourable for microbial activity.

5.2.3 Climatic control of nitrogen mineralization

Temperature and moisture are two environmental factors which exert strong influence on mineralization processes. Except at extreme temperatures the availability of water generally determines the intensity of mineralization. The process is inhibited, however, at high soil moisture contents by the lack of aeration: optimal mineral N production occurs when about 10–20 per cent of the pore space is filled with air, that is at soil water potentials of approx. −0.1 to −0.5 cm (Runge 1983).

Ammonium production occurs over the whole range of temperatures at which microorganisms are active, namely 0 to 70 °C. Since ammonification is a composite process made up of the integrated activities of many different bacteria and fungi, each with its distinctive temperature requirements, no universally optimal temperature range can be established. Nitrification is more strongly dependent on temperature than ammonification, in the sense that its optimal range is narrower. The autotrophic nitrifiers, *Nitrosomonas* and *Nitrobacter*, grow in liquid culture at temperatures between 5 and 40 °C, but limited nitrate production may occur in soils outside this range. D. D. Focht and W. Verstraete (1977) believe that much of the nitrate produced in soils above 40 °C is due to heterotrophic nitrification. The optimum temperature for nitrate formation in soils of the temperate region is about 25 to 28 °C but the figure may be considerably higher in tropical and arid zone soils.

Spatial and temporal variability in nitrogen mineralization

Since the availability of substrates for heterotrophic microorganisms varies in space and time (Ch. 4), it follows that there will be considerable spatial and temporal variability in ammonification and, consequently, nitrification. Vertical gradients of nitrogen mineralization are well known in field and meadow soils, with mineral N production decreasing more or less steadily with depth, while forest soils may show more contrast still, depending on humus type: mull soils generally react similarly to arable soils, but in mor soils mineralization is concentrated in the F, H and A_1 horizons (Runge

1983). In soils of semi-arid rangeland, most mineral N production occurs in the upper few centimetres of the profile (Charley 1977). In the opinion of M. Runge, the vertical gradation in mineral N production is ecologically significant, because plant species vary in their capacity to adapt root system geometry to the zones of maximum net mineralization.

Horizontal patterns in nitrogen mineralization have been less well studied than vertical gradients. They are however clearly evident in rangeland communities characterised by perennial shrubs (Charley & West 1977) and in some forests dominated by long-lived, large trees (Charley & Richards 1983). Horizontal variation of a different kind was described by K. E. Lee and T. G. Wood (1971) in some tropical and subtropical communities, where much of the organic matter which would be subject to mineralization in soils of temperate regions is sequestered in termite mounds.

Since the rate of litter deposition, or that of any other process whereby organic residues are added to the soil, is not constant, and because diurnal and seasonal fluctuations in soil temperature and moisture occur, one would expect considerable temporal variability in nitrogen mineralization. In communities whose productivity depends on low and unreliable rainfall, such as the rangelands of semi-arid regions, substrates accumulate during dry periods and a pulse of mineralization takes place following rain. J. L. Charley pointed out in 1977, that falls of rain too small to elicit a plant response may result in the build-up of a pool of mineral N, which is available to be drawn upon by plants as soon as more substantial rains occur. Similar temporal patterns of mineralization were found by Greenland (1958) in monsoon climates which experience a marked seasonal pattern of precipitation (sect. 5.2.2). According to Richards *et al.* (1985), less pronounced temporal variation in mineral N production exists in soils of more equable climates.

5.2.4 The soil–plant nitrogen cycle

S. L. Jansson (1958) has proposed that two soil nitrogen cycles must be considered, namely an external cycle, involving uptake and assimilation of nitrogen by plants, and an internal cycle in which various forms of nitrogen are cycled between the organic and inorganic phases by microbial processes (Fig. 5.9). The external cycle is governed and determined by the internal cycle, in which the key form of nitrogen is ammonium. This is subject to continuous consumption and renewal by the microflora, whereas nitrate

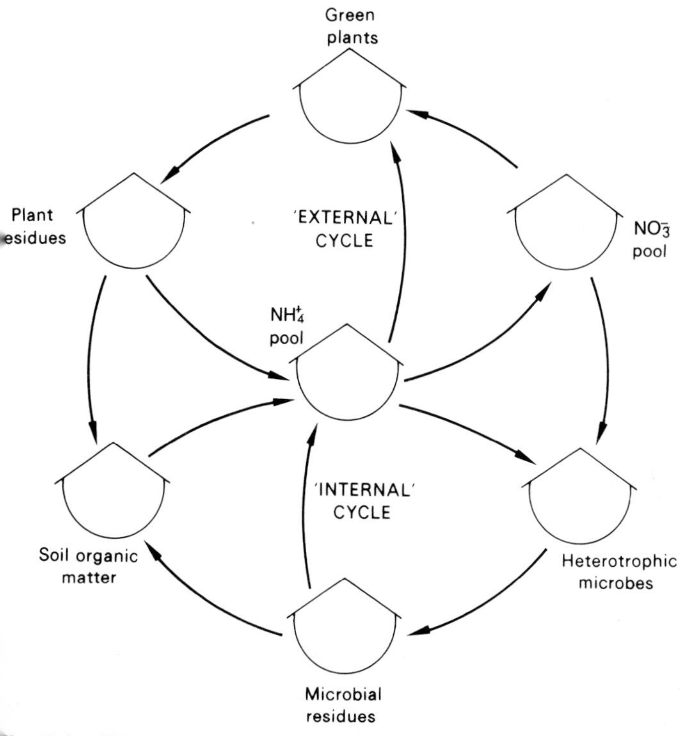

ig. 5.9 The flow of nitrogen through the soil–plant system, based on . L. Jansson's (1958) concept of ammonium as the key form of soil itrogen.

onstitutes a more or less transitory pool of surplus inorganic itrogen not needed in the internal cycle. The workings of the nternal cycle are depicted in Fig. 5.10. If there is insufficient itrogen in decomposing plant detritus (5.10b) to meet the nitrogen equirements of the decomposer population, then these are met at he expense of ammonium and nitrate reserves in the soil. If crop esidues contain sufficient nitrogen (5.10a) to satisfy the needs of he decomposer microbes, then the ammonium and nitrate pools vill be replenished.

Jansson's view of the soil nitrogen cycle envisages nitrifiers being veak competitors for NH_4-N compared with heterotrophic micro-rganisms, and green plants weak competitors for NH_4-N in omparison with nitrifiers. His emphasis on the overriding import-nce of ammonium is compatible with the hypothesis that nitrifica-on is not essential for the nutrition of vascular plants in many

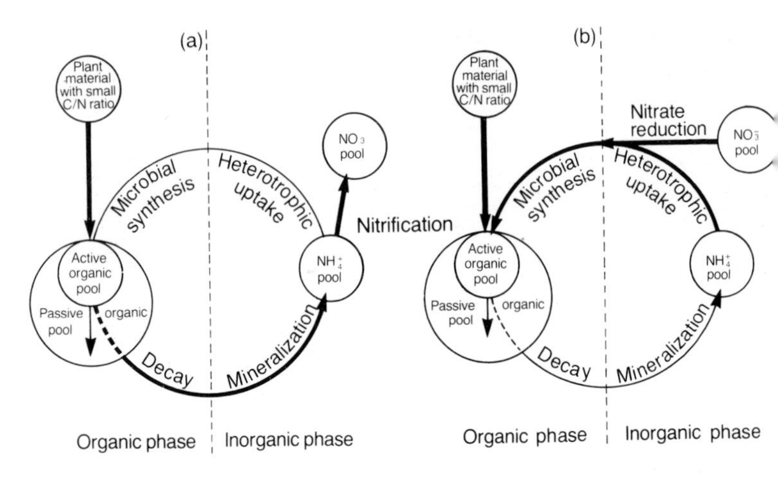

Fig. 5.10 The internal nitrogen cycle in soil containing actively decomposing plant residues (From Jenkinson, 1981; adapted from Jansson 1958.)

grassland and forest ecosystems. The agronomic significance of nitrate is not to be doubted, but it is likely that its major functional role in natural ecosystems is not to provide nitrogen for the producers, but rather to serve as an electron acceptor in denitrification. The biogeochemistry of nitrogen is discussed in Chapter 8.

5.3 Anaerobic processes and nutrient cycling

Two kinds of anaerobic energy-yielding processes are known among chemoheterotrophic microbes, namely anaerobic respiration where the hydrogen acceptor is inorganic, and fermentation in which an organic molecule serves as the hydrogen acceptor. Some anaerobic processes affect the supply of plant nutrients directly others indirectly. In well aerated soils, direct competition between anaerobes and plant roots for available nutrients seems unlikely Since the energy-yielding reactions of anaerobes are independent of oxygen however, they are able to maintain ion uptake when conditions are no longer suitable for active ion accumulation by roots. Therefore in badly aerated and infertile soils, it is conceivable that anaerobes could compete with plants for nutrients.

5.3.1 Anaerobic respiration

Of the three possible hydrogen acceptors known to participate in this process, only two – nitrate and sulphate – have any significance for plant nutrition. Nitrate reduction is a process of very great geochemical significance, being the principal means by which nitrogen escapes from the biosphere to the atmosphere. This aspect of the process will be taken up in Chapter 8. In the present context – that of plant nutrition – nitrate reduction represents a loss of available nitrogen. Sulphate reduction plays an analogous role in the sulphate cycle: not only does it reduce the availability of sulphur to plants, but it also represents the major pathway for transferring sulphur from the biosphere to the atmosphere. The analogy between the two cycles is not so close as might appear however, as we shall see.

Nitrate reduction: denitrification

Organisms may reduce nitrate for two purposes: to assimilate the reduced product (ammonium) into cellular materials or to provide metabolic energy when oxygen is not available as the hydrogen acceptor. The latter is sometimes known as dissimilatory nitrate reduction, and when its products are gaseous nitrogen or nitrous oxide, the process is termed **denitrification** (Fenchel & Blackburn 1979).

The organisms capable of utilizing NO_3^- as an electron acceptor in energy-yielding oxidations are all bacteria. They have a wide distribution over the surface of the Earth, both in terrestrial and aquatic habitats and in the air. Soil is a particularly efficient denitrifying system. Ammonium ions derived from the mineralization of organic matter first appear near the soil surface, and by virtue of their positive charge are adsorbed on the negatively charged soil colloids and so prevented from moving down through the soil profile. At or near the surface, oxygen levels are high and conditions favour nitrification. The nitrate ions formed, being negatively charged, are no longer held by the colloids but are leached to lower levels where the oxygen concentration is lower. Provided suitable organic substrates are present, the nitrate can serve as a hydrogen acceptor for denitrifiers, the end products being nitrous oxide or molecular nitrogen, which diffuse out of the soil and are lost to the atmosphere (Fig. 5.11). Whenever anaerobic conditions prevail in soil, for example following waterlogging, denitrification becomes the dominant process of the nitrogen cycle. It not only occurs in waterlogged soils however, but is a normal reaction in all soils, since pockets of anaerobiosis exist temporarily even under conditions of good aeration.

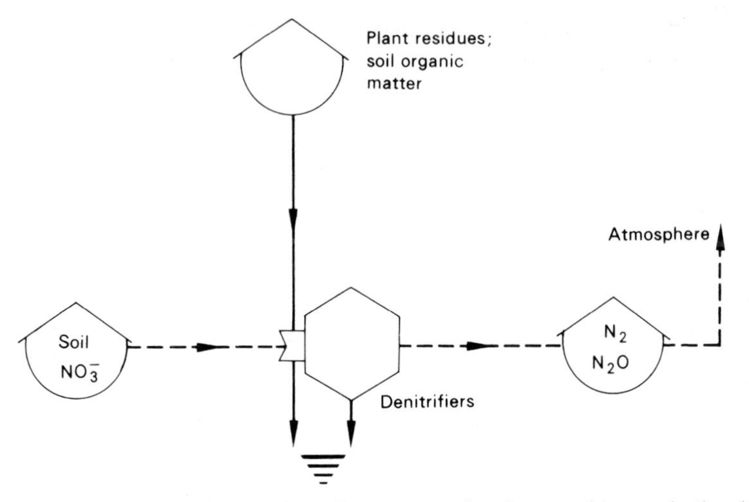

Fig. 5.11 Denitrification in soil, an example of anaerobic respiration in which nitrate ions are used as electron acceptors for the oxidation of reduced carbon compounds.

The significance of soil aeration. The interrelationships between soil water and the soil atmosphere were described in Chapter 1. These are such that aerobic respiration dominates organic matter decomposition processes wherever there is a supply of oxygen at the active microsite. Assuming that other factors (available substrate, soil temperature and moisture) do not change, aerobic respiration continues at a constant rate until the respiring microbes have reduced the oxygen concentration by a factor of about 0.01, after which the respiration rate declines rapidly and ultimately ceases (Rowell 1981). This can happen quickly within soil peds that contain only pores with a diameter of less than about 50 μm since these will remain full of water at field capacity, when the larger pores have drained under the action of gravity. Since oxygen diffuses very slowly through water, peds do not have to be very large before anaerobic centres develop, so even a well drained soil cannot be considered uniformly aerobic.

When the soil profile becomes partially waterlogged, oxygen diffusion is greatly restricted, and in a flooded soil only a very thin layer at the surface, perhaps no more than 1 cm thick, remains aerobic. D. L. Rowell (1981) has shown diagrammatically how the drainage characteristics of a soil and its pore size relative to the dimensions of microorganisms, determine the availability of oxygen and hence the likelihood that anaerobic processes, such as denitrification, will replace aerobic metabolism (Fig. 5.12).

Microbiology of denitrification. Only a few genera of bacteria can denitrify, but those which do are widespread. *Pseudomonas* and *Achromobacter*, and to a lesser extent the spore-forming *Denitrobacillus*, are among the principal genera of soil bacteria involved. Many of these same bacteria are active in proteolysis, ammonification and other organic matter decomposition processes, and begin to reduce nitrate only when conditions for aerobic respiration no longer exist.[4] Although most denitrifiers are heterotrophs, at least two autotrophic denitrifying bacteria are recognized, namely *Thiobacillus denitrificans* which oxidizes sulphur, and *Micrococcus denitrificans* which oxidizes hydrogen. Fungi, being for the most part obligate aerobes, are not involved in denitrification.

Denitrification is generally considered to take place in several stages, nitrate first being reduced to nitrite, then nitrite to molecular nitrogen via nitrous oxide:

$$NO_3^- + H_2 \rightarrow NO_2^- + H_2O$$
$$2NO_2^- + 3H_2 \rightarrow N_2O + 3H_2O$$
$$N_2O + H_2 \rightarrow N_2 + H_2O$$

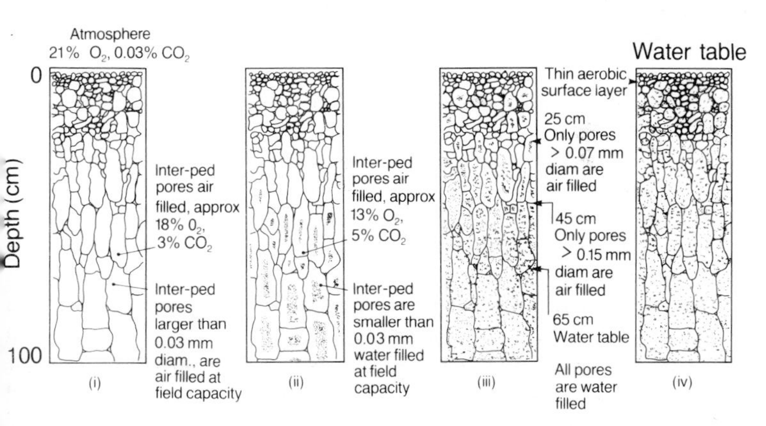

Fig. 5.12 The significance of soil aeration in determining the balance between aerobic and anaerobic processes. (a) The distribution of aerobic and anaerobic zones in soils under different drainage regimes – a model for static conditions. (i) Well drained at field capacity – fully aerobic; (ii) imperfectly drained at field capacity – anaerobic microsites; (iii) poorly drained at field capacity – high water table or impermeable subsoil; (iv) waterlogged soil. The dotted regions are anaerobic.

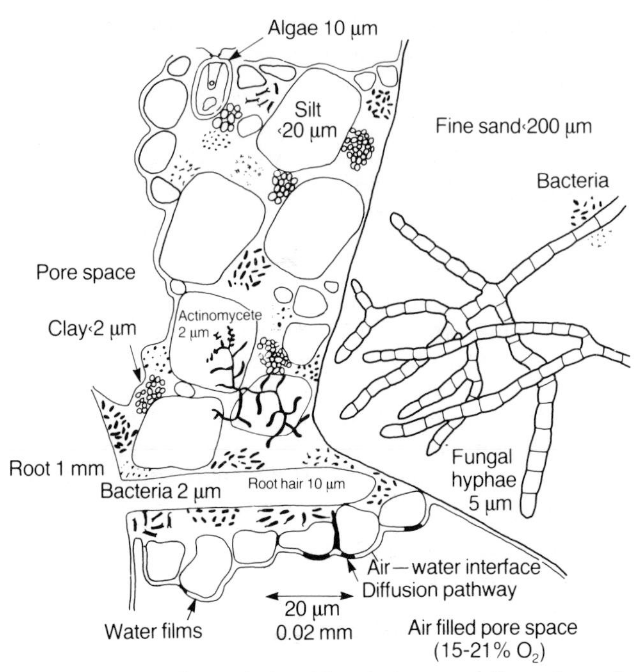

Fig. 5.12 (b) The relationships between active sites of respiration, mineral particles, and water films in a soil crumb. (From Rowell 1981.)

When the concentration of nitrate is high, some denitrifying bacteria tend to accumulate nitrous oxide (N_2O), which escapes to the atmosphere as such, or is further reduced to N_2 when the nitrate concentration falls. The reason for this is that the reduction of N_2O to N_2 is delayed by even low concentrations of NO_3^- while high concentrations of NO_3^- almost completely inhibit this process (Blackmer & Bremner 1978). Other factors that help determine the relative production of N_2O and N_2 during denitrification are soil pH, aeration and temperature (Focht & Verstraete 1977), and the identity of the denitrifying bacteria. The effect of soil reaction is complex, the optimum pH varying with nitrate concentration and the organisms involved. Active denitrification, by removing nitrate, tends to cause a rise in the pH of the system, hence the buffering capacity of the soil exerts a strong influence on the amount of N_2O evolved.

Although most of the nitrous oxide produced in soil is the result of denitrification, there are two other mechanisms by which it might be formed. One is by the chemical decomposition of nitrite in acid soils, especially following the addition of ammonium fertilizers which promote the activity of nitrifying bacteria. The second mechanism is through the microbial oxidation of ammonia: at low soil moisture levels, Freney *et al.* (1979) contend that this process is an important means of maintaining the N_2O concentration in the atmosphere.

In conclusion, it should be recognized that the phenomenon of denitrification has been studied intensively only in man-made ecosystems, which is to be expected in view of its agronomic significance. Little is known of the rates of denitrification under natural, undisturbed vegetation.

Sulphate reduction

Like nitrate reduction, sulphate reduction may be either assimilative or dissimilative. Sulphate assimilation in bacteria, fungi, algae and higher plants takes place when it is incorporated in cellular materials after reduction to thiols then cysteine. In contrast to this, dissimilative sulphate reduction occurs when sulphate (SO_4^{2-}) acts as the electron acceptor in anaerobic respiration, and in the process is reduced to sulphide (S^{2-}):

$$SO_4^{2-} + 4H_2 \rightarrow S^{2-} + 4H_2O$$

Sulphite (SO_3^{2-}) and thiosulphate ($S_2O_3^{2-}$) are alternative electron acceptors. The microbial reduction of sulphate is a step in the soil sulphur cycle analogous to denitrification in the nitrogen cycle. Unlike denitrification however, which is carried out by a number of facultative anaerobes, sulphate reduction is restricted to relatively few species of bacteria, the most widespread of which are *Desulfovibrio desulfuricans* and *Desulfomaculum aceto-oxidans*. These organisms influence soil fertility by reducing the amount of SO_4^{2-}, which is the principal form in which sulphur is absorbed by plants. Their activity in soils, like that of denitrifiers, is greatly enhanced by waterlogging, which provides the necessary reducing conditions. In flooded soils, as in rice paddies, the sulphide concentration may exceed 30 ppm ($\mu g\ g^{-1}$) within a few weeks of inundation.

Sulphate reduction is most characteristic of estuarine muds where it is one of the dominating microbiological processes. Some 100 million tonnes of H_2S is emitted into the Earth's atmosphere each year as a result, according to D. J. D. Nicholas (1980). In an indirect way, the anaerobic respiration of sulphate in coastal estuaries has more profound influence on plant nutrition than have the activities of sulphate reducers in the soil itself. The ocean is the great reservoir of sulphate and is the major source of sulphur for terrestrial plants and animals, which require it as a component of proteins and certain vitamins. Sulphate moves from the ocean to the land as a component of rainwater, and the greater part of the sulphate found in rainwater apparently derives from the spontaneous oxidation of H_2S which emanates from the shallow waters of the littoral zone. On the surface of marine sediments, sulphate-reducing bacteria convert dissolved sulphate to H_2S, in which form it escapes to the atmosphere where it is oxidized spontaneously to sulphate, taken up in water and finally deposited on the land in rain. The sulphur cycle is illustrated diagrammatically in Fig. 5.13.

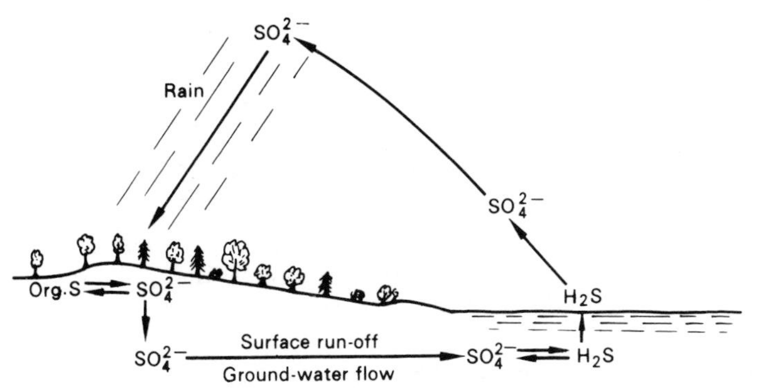

Fig. 5.13 A simplified outline of the sulphur cycle. Sulphate reduction, a form of anaerobic respiration, produces H_2S in the littoral zone which diffuses into the atmosphere where it is oxidised spontaneously to sulphate, dissolved in water and returned to the land in rain. Sulphate absorbed by plants is incorporated in organic compounds which are subsequently mineralized (usually aerobically) to produce sulphate again which returns to the sea via ground water and surface streams.

5.3.2 Fermentation

In fermentation, organic molecules act as both hydrogen donors and hydrogen acceptors. Being a strictly anaerobic process fermentation makes a minor contribution to organic matter decomposition in well aerated soils. In such soils aerobes, especially fungi, are the major decomposers of plant residues. There are fermentative bacteria capable of decomposing cellulose, but they are more abundant than fungi only in anaerobic habitats, such as waterlogged peats. Nevertheless since most soils, especially those of heavy texture, contain micro-habitats where anaerobic conditions exist from time to time, fermentation no doubt plays some part in mineralization–immobilization cycles. Sugars are the most readily and widely used substrates for fermentation, but a variety of other substances, including various organic acids, amino acids, purines and pyrimidines can be fermented by some bacteria. The different kinds of fermentation are classified according to their major end products.

The simplest possible fermentation consists of the splitting of a simple six-carbon sugar into two molecules of the three-carbon lactic acid. It is not only typical of the so-called lactic acid bacteria, but is also characteristic of animal tissues as well as certain protozoa and fungi. In the lactic acid fermentation, pyruvic acid, produced by glycolysis of a monosaccharide, acts as the final hydrogen acceptor. Two kinds of lactic acid bacteria are known:

(a) homofermentative forms, in which lactic acid is virtually the sole product of glucose fermentation; and
(b) heterofermentative forms, which produce carbon dioxide, ethyl alcohol and sometimes acetic acid, in addition to lactic acid.

The heterofermentative forms are characteristically plant inhabitants. They are thought to exist normally on plant surfaces, growing at the expense of exudations. They also flourish in decaying plant tissues, where they soon dominate the bacterial flora because their acid production inhibits the growth of most other bacteria. They are not considered a part of the normal soil microflora.

The butyric acid fermentation is typical of *Clostridium*, a genus of anaerobic spore formers. It is also found in *Bacillus macerans* and some anaerobic protozoa. The process involves the conversion of sugars and related compounds to butyric and acetic acids, carbon dioxide and hydrogen, plus a variety of alcohols. Many butyric acid bacteria are nitrogen-fixers, for example *C. pasteurianum*. The chief substrates for these fermentations are soluble sugars, starch and pectin. Pectin-decomposing clostridia are the principal agents in the retting (i.e. softening) of flax and other fibres. Another group of clostridia ferment amino acids, with the production of fatty acids, ammonia and carbon dioxide.

Many of the bacteria which carry out fermentations are, like the denitrifiers, facultative anaerobes. The end products of fermentation include many organic acids which may assist in the solubilization of primary minerals. The carbon dioxide produced may also contribute to rock weathering, through the formation of carbonic acid. It is doubtful however whether these effects are any more pronounced than the effects of organic acids and carbon dioxide produced by microorganisms which respire aerobically.

5.4 The role of soil fauna in nutrient cycling

The significance of invertebrate–microbe interactions in the process of organic matter decomposition was indicated in Chapter 4. Clearly there are implications for mineral cycling but while this has become more widely appreciated in recent years (Reichle 1977; Kitchell *et al.* 1979), our understanding of these interactions has not progressed very far.

J. F. Kitchell and his collaborators have suggested that consumers affect nutrient cycling processes either by transporting elements across subsystem boundaries or by altering mineralization rates through changing the particle size distribution of microbial substrates. Mobile soil animals have considerable potential for redistributing substrates and concomitantly, nutrients, and consumption in one locality followed by death or defaecation in another may change substrate quality and nutrient availability. Earthworm activity may increase the level of nitrate and exchangeable cations, for example, while frass from canopy-dwelling herbivores may contribute significantly to nutrient throughput via litter fall.

In addition to this physical transfer of nutrient elements, soil animals may effect indirect chemical transformations by changing surface/volume relationships of microbial substrates. Although direct chemical alteration due to invertebrate consumption may be very slight, even small changes could be magnified in the sequence of steps involved in humification. As indicated in sect. 4.3, D. A. Crossley and M. Witkamp reported in 1966 that the action of invertebrates doubled the loss of radioactive caesium (an analogue of potassium) during litter breakdown, even though the litter fauna utilized only 1 per cent of the energy expended by the decomposer population. Repeated utilization of substrates by the microflora and coprophagous fauna might enhance mineralization rates still further. The consumption and mixing of mineral soil and litter by millipedes, earthworms and springtails probably raises the mineral content of faeces, increasing the likelihood of net mineralization during subsequent microbial decomposition. According to Witkamp and Ausmus (1976), the channelling or tunnelling of wood borers and termites in fallen tree trunks hastens microbial succession (see sect. 4.3) and increases the nitrogen content of decomposing wood by translocation through fungal hyphae and by bacterial N_2 fixation in faeces. Net mineralization of nitrogen could be enhanced as a result, although M. J. Swift (1977b) has pointed out that fungal decay of branch wood is characterized by net immobilization of this element.

The condition under which soil fauna–microflora interactions favour net mineralization therefore remain to be elucidated. Such interactions do however have an important, albeit hypothetical, role in regulating mineral cycling processes. Along with mechanisms such as the absorption of nutrients by plant roots and the adsorption of cations on soil colloids, the continual interchange of nutrients among the various decomposer groups would tend to reduce the leakage of chemical elements from the soil–plant system and so help to maintain primary production. This potential regulatory role is well illustrated by the results of Malone and Reichle (1973), who studied the loss of ^{134}Cs from fescue litter in the field. Under natural conditions, when both fauna and microbes had access to the litter, they found that decomposition proceeded linearly for ten months, by which time about 50 per cent of the labelled caesium remained. When the soil fauna was excluded however, there was rapid net mineralization, resulting in 40 per cent of the label being lost in the first month; thereafter the mineralization outflow continued at a slower but regular rate for the next nine months when only 45 per cent of the initial amount of ^{134}Cs remained. The temporary immobilization of this nutrient due to the fauna–microflora interaction thus prevented its accumulation in inorganic form where it may have been subject to leaching below the root zone.

5.5 Notes

1. To a geochemist, all three cycles would be considered to be biogeochemical, but it is useful to distinguish among them here.
2. 'Fixed' ammonium is also found in the crystal lattices of secondary clay minerals.
3. In addition to the chemoautotrophs that oxidize H_2S, there are the photoautotrophic green and purple sulphur bacteria, which use H_2S as a source of reducing power for photosynthesis. They occur commonly in brackish waters polluted with organic matter, where H_2S is formed in large quantities by the anaerobic decomposition of organic matter and by sulphate reduction; the purple sulphur bacteria are found also in warm sulphur springs. Like the chemotrophic filamentous sulphur bacteria, the photosynthetic sulphur bacteria are of no ecological significance in terrestrial ecosystems, except perhaps in flooded soils.
4. There is now evidence that many strains of *Rhizobium* (sect. 8.1) can also denitrify. Unlike denitrification in general, rhizobial denitrification is stimulated by low concentrations (5% v/v) of oxygen (O'Hara *et al.* 1983).

6

Microbiology of the rhizosphere

The topics dealt with in Chapters 4 and 5, concerning the microbial contributions to energy transformations and nutrient turnover in soils, involve rather loose associations between plants and microorganisms. These plant–microbe relationships exist because the heterotrophic soil microflora depends primarily upon plant detritus for its energy supply, and because the growth of plants is determined in large measure by the activity of both heterotrophic and autotrophic soil microbes. The present chapter is concerned with a much closer plant–microbe interaction, that which occurs in the immediate vicinity of plant roots. The soil in this region is a highly favourable habitat for microorganisms, and has a characteristic microflora which is quite distinct from the general soil population. This unique environment which is under the influence of plant roots is called the **rhizosphere**. Within this zone, interactions between plants and microorganisms can greatly affect crop production and soil fertility and hence, at the ecosystem level, energy flow and nutrient cycling.

The concept of the rhizosphere was introduced in 1904 by L. Hiltner. Its extent is variable, there being no sharp boundary between it and the neighbouring soil. The roots of plants are surrounded by a mucilaginous layer varying in composition from a relatively simple oligosaccharide to a complex pectic acid polymer permeated by loose cellulose microfibrils. Electron micrographs give a graphic picture of this boundary zone between roots and soil, indicating that the space between the cell walls and mineral soil particles is filled with gelatinous material known as 'mucigel' (Fig. 6.1). The influence of plant roots extends beyond this mucilaginous zone, which may be 10–20 μm thick, into the surrounding soil especially if the soil is not a fertile one. The extent of the rhizosphere will be discussed later, but first it is necessary to consider its physical nature.

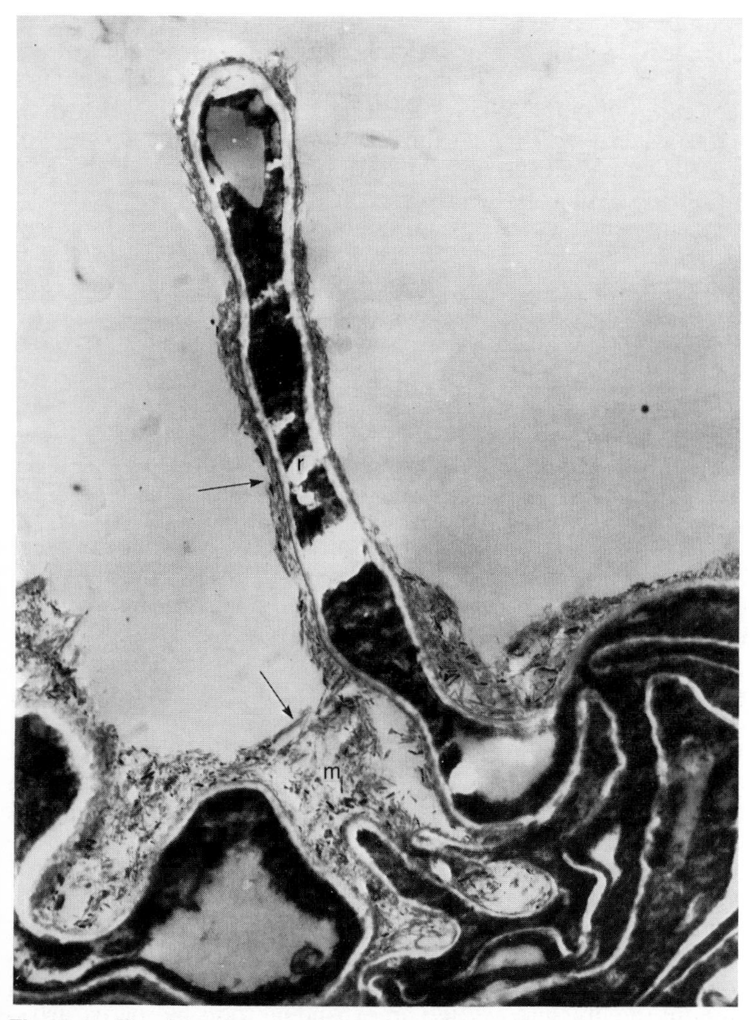

Fig. 6.1 The boundary layer between plant roots and soil. Electron micrograph of root hair (r) of pea, showing the zone of mucigel (m) outside the epidermis. The arrows indicate soil particles embedded in the mucigel. (From Greaves & Darbyshire 1972.)

6.1 Fine structure of the rhizosphere

Definitive study of the rhizosphere begins with an examination of the root surface itself. In the zone of root extension, the primary walls of epidermal cells are made up of cellulose microfibrils embedded in a thick, mucilaginous matrix of pectins and hemicelluloses. The root tip as it moves through the soil is protected by a root cap, the outer cells of which have mucilaginous walls which act as a lubricant. Behind the root cap is a meristematic region, 0.2–0.4 mm long, succeeded by a zone of elongation extending a further 1–10 mm. Behind this again is a region of maturation or root hair zone, up to several centimetres long, where vascular tissues are differentiated. When extension growth ceases, a secondary wall is laid down within the primary wall. This inner wall comprises closely packed lamellae of crossed, cellulose microfibrils, and consequently has a dense texture quite distinct from the open, mucilaginous nature of the outer wall. During the phase of extension growth, epidermal cells and their associated root hairs release mucilages into the surrounding soil. Mucilaginous materials, in the form of slime layers or capsules (see sect. 1.7.3), are also exuded or secreted by microorganisms. At the root–soil interface of older roots, plant and microbial exudates and secretions become mixed to produce the substance which Jenny and Grossenbacher (1963) termed mucigel. As indicated earlier, this takes the form of a fibrous matrix occupying most of the space between the root surface and the clay or organic soil particles.

R. C. Foster has made a particular study of the ultrastructure and histochemistry of the rhizosphere of roots fixed *in situ* so as not to disturb the soil fabric (Foster & Rovira 1976, 1978; Foster 1981, 1982). This approach has greatly clarified the concept of rhizosphere, because the techniques used enable a distinction to be made between polysaccharides produced by soil microbes and those that arise from epidermal root cells. Furthermore, the findings derive from observations made on the roots of mature plants growing in natural soils where they are subject to mechanical abrasion and microbial lysis. Foster's interpretation of the observations of various workers on the ultrastructure of the rhizosphere is illustrated diagrammatically in Fig. 6.2. In a sterile environment and without the pressure exerted on expanding roots by soil particles, epidermal cells would have a structure similar to that shown at A in this figure. Microbial lysis and root pressure in the natural environment of the soil results first in the loss of the cuticle, as at stage B. Further lysis removes the outer parts of the primary cell wall in which event the mucilage or gel will consist mainly of bacterial capsules (stage C). Eventually, when the primary wall has been completely lysed and the inner microfibrillar secondary wall subjected to bacterial attack, the epidermal root cell wall appears as shown at D.

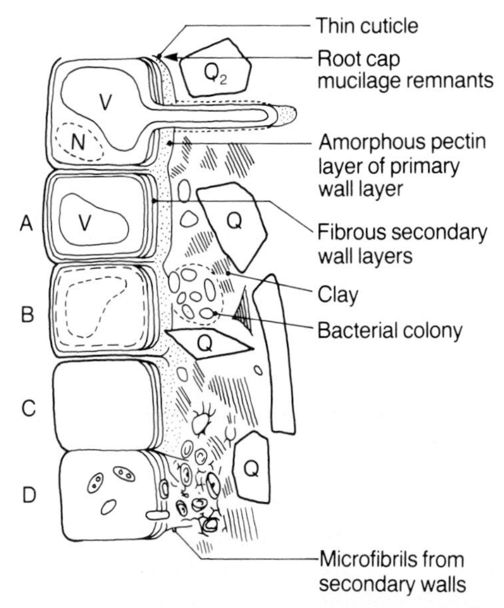

Fig. 6.2 Diagramatic representation of the ultrastructure of the rhizosphere, showing stages in degradation of epidermal cells of roots. (From Foster 1982 – for explanation, see text). V, vacuole, N, nucleus, Q, quartz particle.

Range of root influence: extent of the rhizosphere

Although a rhizosphere effect can be discerned at a distance of several millimetres from the root surface in sandy soils (Papavizas & Davey 1961), and although the cylinder of rhizosphere soil which adheres to roots removed from soil is at least a millimetre thick, the major zone of stimulation is measured in micrometres. There is also evidence of zonation within the rhizosphere (Table 6.1). That part of the rhizosphere immediately adjacent to the root surface is termed the **rhizoplane**. The thickness of the rhizoplane is, according to Foster and Rovira (1978), defined by the mean diameter of rhizosphere microorganisms, namely about 1 μm. These authors further differentiate between an 'inner rhizosphere', extending up to 10 μm from the root surface and including the rhizoplane, and an 'outer rhizosphere' located 10–20 μm from the root. The inner rhizosphere is characterized by small, discrete colonies (5–10 μm in diameter) of closely packed bacteria.

Table 6.1 clearly shows how microorganisms increase in abundance as the root surface is approached. There is also a greater variety of organisms in the inner rhizosphere than in the outer.

Table 6.1 Numbers of microorganisms in the rhizosphere of subterranean clover, as determined by transmission electron microscopy of ultrathin sections of undisturbed soil cores

Zone and distance from root (μm)		*Number of microbial types*[†]	*Frequency in soil* (cells $cm^{-3} \times 10^9$)
Rhizoplane	0–1	11	120
Inner rhizosphere	0–5	12	96
	5–10	5	41
Outer rhizosphere	10–15	2	34
	15–20	2	13

Source: After Foster and Rovira (1978).
[†] Organisms which can be distinguished cytologically.

Microbial density in the rhizosphere as a whole is nearly 5×10^{10} cm^{-3} which is several orders of magnitude greater than estimates derived by conventional plating techniques.

As is evident from Fig. 6.2, bacteria may invade the empty cells of the epidermis and cortex behind the root hair zone; indeed the cortex of normal, healthy roots may be inhabited by a wide range of microorganisms (Rovira *et al.* 1983). This specialized micro-environment has been termed the **endorhizosphere** by Balandreau and Knowles (1978).

6.2 Origin and nature of organic substrates in the rhizosphere

Stimulation of microorganisms in the rhizosphere is assumed to be caused by substances coming from roots, partly as a result of exudation and secretion and partly due to autolysis of moribund and dead root cells. Different plant species and also different parts of the root system of the same plant, may have distinctive rhizosphere microfloras. Rovira (1969a, 1973) used ^{14}C-labelling techniques to show that the major zone of release of organic substrates (mainly insoluble polysaccharides and sloughed-off root cap cells) from the seminal roots of wheat was the region of root elongation.

The causes of these differences, which may be both qualitative and quantitative, are not known for certain, though some of the effects can be attributed to differences in rooting habit, tissue composition, and in the products of exudation and secretion. The age of the plant also affects the composition of the rhizosphere flora. A rhizosphere effect is detectable in seedlings only a few days old, and this must primarily be due to the stimulatory properties of root exudates or secretions. As the plant matures, dead and sloughed-off root tissues add appreciably to the rhizosphere effect. With the approach of senescence, the rhizosphere population gradually declines, and eventually becomes indistinguishable from the normal soil microflora. Since the existence of an assured energy source makes the rhizosphere a habitat more or less independent of fluctuations in substrate availability, which is the major limiting factor for the heterotrophic soil microflora generally, it might be expected that the rhizosphere population would not be appreciably altered by the addition of crop residues or animal manures. However, while the direct effects of soil amendments on microbial growth in the rhizosphere are probably slight, their indirect effects, through their influence on plant growth, may be considerable.

Initially, microbiologists studying the rhizosphere effect assumed that most of the organic substances released by roots were exuded or leaked, that is they escaped passively from roots independently of metabolic processes. It is now thought that some compounds are actively secreted as a result of the expenditure of metabolic energy[1]. Furthermore, it is now known that the rhizospheres of plants growing in soil contain substrates which differ in kind and amount from those released into the axenic nutrient solutions on which most of the earlier investigators relied. In order to clarify the situation, Rovira *et al.* (1979) have proposed the following terminology (Fig. 6.3):

1. **Exudates**. Compounds of low molecular weight (e.g. monosaccharides and amino acids) which leak from all cells into the soil either directly or via the intercellular spaces. Exudation is not metabolically mediated.
2. **Secretions**. These include both low and high molecular weight compounds released by metabolic processes. Mucilages are representative of the more complex substances secreted.
3. **Plant mucilages**. These are of four kinds:
   (a) mucilage originating in the root cap and secreted by the golgi apparatus;
   (b) hydrolysates of primary cell wall polysaccharides which are found between the epidermis and sloughed root cap cells;
   (c) mucilage secreted by epidermal cells and root hairs which only have primary walls[2];
   (d) mucilage derived from bacterial breakdown of the outer, multilamellate walls of dead epidermal cells.

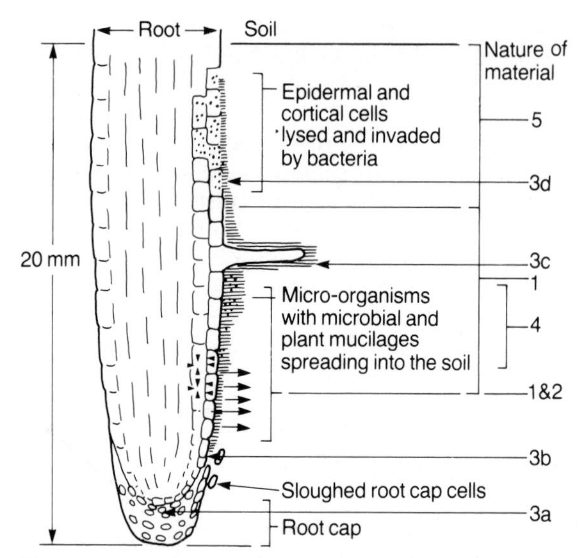

Fig. 6.3 Diagram of a root showing the nature and origin of organic materials in the rhizosphere. The numbers refer to the various classes of material described in the text. (From Rovira *et al.* 1979.)

4. **Mucigel**. This is the gelatinous material on the surface of roots growing in soils, and comprises a heterogeneous mixture of natural and modified plant mucilages, bacterial cells and their metabolic products such as capsules and slime layers, and colloidal mineral and organic soil particles[3].
5. **Lysates**. Compounds released by autolysis of older epidermal cells when the outer protoplasmic membrane (plasmalemma) ruptures. In the course of time, the entire cell walls are decomposed by microorganisms, the cell contents colonized, and the products of microbial degradation are released into the rhizosphere.

The nature of root exudates

Root exudates include carbohydrates, amino acids and other organic acids, vitamins, nucleic acid derivatives, and various miscellaneous compounds (Rovira 1965, 1969b). In studying root exudates, careful attention to aseptic technique is essential, since it has been shown that non-sterile roots exude greater quantities of amino acids than roots grown in axenic culture. For example, Vancura *et al.* (1977) reported that, following inoculation of wheat or maize with *Pseudomonas putida*, the amounts of exudate released into the ambient nutrient solution doubled. Whether or not microorganisms actually stimulate exudation by roots, there is no doubt that the microflora itself contributes significantly to the production of exudates in the rhizosphere. The relative contribution of the two components – roots and microbes – is difficult to assess in nature. Growth factors in particular may be contributed primarily by microorganisms, in other words bacteria with less specialized growth factor requirements may synthesize and release metabolites needed by more specialized nutritional types.

Among the carbohydrates exuded, at least ten sugars, including an oligosaccharide, have been identified in the exudates of a wide range of plant species. The hexoses glucose and fructose are generally the most abundant, and others which have been detected include the pentoses xylose, ribose and arabinose, and the disaccharides sucrose and maltose. The wide range of sugars recorded in root exudates suggests that exudation of sugars is a general phenomenon which probably has little or no effect in determining the actual composition of the rhizosphere microflora associated with any particular species of plant.

In contrast to their exudation of carbohydrates, plants appear to differ greatly in the amounts and kinds of amino acids they exude (Rovira 1956). The most abundant amino acids exuded by 10-and 21-day-old pea plants are homoserine, threonine and glutamine and by oats, lysine, serine and glycine; a total of twenty-two amino compounds has been found in pea root exudates compared to only fourteen in oat root exudates. In addition the total amount of exudate from peas is many times that produced by oats. Smith (1976) also found considerable interspecific variation in the nature and amount of organic compounds released from the unsuberized tips of lignified tree roots. Such qualitative and quantitative differences between the exudates of different plant species grown under the identical conditions are likely to have far-reaching effects on the composition and density of the rhizosphere population.

Various other organic acids, including several members of the tricarboxylic acid cycle, occur in root exudates of several plant species. These organic acids could not only provide readily available substrates for many microorganisms, but in addition they might have important secondary effects in the rhizosphere soil such as alteration of pH and chelation of metal ions (see sect. 5.2.1).

The most common vitamins found in root exudates are biotin and thiamine. The levels recorded are generally low and probably insufficient to meet the needs of all the vitamin-requiring microbes in the rhizosphere. Many of these microbes, perhaps the majority, depend on the exudates of less fastidious microorganisms rather than on root exudates. The nucleic acid constituents, adenine, guanine, uridine dan cytidine, are also released from the roots of plants. Exoenzymes such as phosphatase and invertase have been found associated with roots, too. And finally, a great variety of miscellaneous compounds is found in root exudates including phenol derivatives and other substances toxic to microorganisms. Thus not all the effects of plant root exudates are necessarily beneficial for the rhizosphere microflora (see sect. 6.4).

Most investigations into the nature of root exudates have been conducted on seedlings grown in nutrient solutions or in artificial, inert substrates such as glass beads. W. H. Smith devised a technique (Smith 1970) for collecting exudates from the roots of mature trees in the field. This involves growing the unsuberized tips of woody roots in specially designed culture vessels while still attached to their parent trees, and subsequently collecting exudates in test-tubes of distilled water. By this means, Smith (1976) was able to show that the organic fraction of mature tree root exudates was qualitatively similar to that reported by Bowen and Theodorou (1973) for forest tree seedlings, that is they contained principally simple sugars, organic acids and amino acids or amides. Quantitatively, organic acids were more important than carbohydrates and these in turn were exuded in greater amounts than amino compounds. Very large quantities of inorganic ions, relative to organic constitutents, were also found in the root tip exudates, and Smith inferred that such losses might make a significant contribution to mineral cycling. To what extent these findings apply to undisturbed roots in the natural environment of the soil remains unknown.

Factors affecting exudation and secretion

As indicated previously, the age of plants influences the kinds and amounts of organic materials emanating from roots. Apart from the observations already referred to, the results of many other workers (Vancura *et al.* 1977; Martin 1978) confirm the view that both the quantity and quality of exudates and/or secretions change significantly during ontogeny.

The conditions under which plants are grown also affect exudation (Hale & Moore 1979). Factors that alter the rates of production and translocation of photosynthate have an indirect effect on the release of organic substances from roots. Defoliation of sugar maples causes changes in the amounts of sugar, amino compounds and organic acids exuded by these trees (Smith 1972). Exposing pine seedlings to water stress in the laboratory reduces the incorporation and translocation of $^{14}CO_2$ and results in diminished exudation (Reid 1974; Reid & Mexal 1977). In contrast to this, an increase in the secretion of mucilage in the rhizosphere of wheat may result from increasing soil water stress (Martin 1977). In addition, if field soils are allowed to dry out to the stage where plants wilt, then re-wetted, there is a rapid release of amino acids from the roots as the plants regain turgor (Vancura 1964).

Microorganisms themselves can affect the pattern of exudation by changing the permeability of root cells, by altering root metabolism, and by modifying some of the materials released from roots. Furthermore, the root surface is not uniformly populated and this may change the exudate pattern from place to place along the root. Infection by soil-borne root pathogens generally stimulates root exudation (Mitchell 1976). Exudates from nodulated legume roots differ from those of non-nodulated roots. Similarly, the metabolites released by mycorrhizal and non-mycorrhizal root-systems are not identical.

Roots growing through soil displace their own volume and so increase soil bulk density in their immediate vicinity. As bulk density increases, pore size decreases and water per unit volume of soil increases, as does its tendency to move (that is its unsaturated conductivity). According to Baker and Cook (1974), this should facilitate the diffusion of exudates away from roots.

6.3 Composition and growth of the rhizosphere population

It has long been the practice to express the influence of roots on the microflora by means of the R/S ratio, which is the ratio of microbial numbers per unit weight of rhizosphere soil (R) to the numbers in a unit weight of adjacent non-rhizosphere soil (S). The R/S ratio, which is usually determined by dilution plate counts, is typically about 10–50/1 for bacteria, 5–10/1 for fungi, and less for other microorganisms.

A consideration of R/S ratios does little to increase our understanding of population dynamics in the rhizosphere. A better guide is provided by more recent work which aims at measuring growth rates on or around roots. G. D. Bowen and A. D. Rovira have both stressed the need to use generation times of bacteria as a basis for studying colonization of the rhizosphere, because this provides a more valid comparison of growth on a surface (the root) with that in a volume (the soil) (Bowen 1980). Using this approach, Bowen and Rovira (1976) reported that pseudomonads had a generation time of 5.2 h on roots of *Pinus radiata* compared to 77 h in comparable unplanted soil, while the corresponding values for bacilli were 39 h and more than 100 h, respectively. This not only shows the marked stimulatory effect of plant roots but also indicates a greater 'rhizosphere effect' for *Pseudomonas* than for *Bacillus*.

Such qualitative effects are common with bacteria, and can readily be demonstrated by viable plate counts. Generally, Gram-negative bacteria are favoured over Gram-positive, and non-spore formers over spore formers. Short Gram-negative rods respond most, and invariably make up a greater percentage of the rhizosphere microflora than of the normal soil population. The short, Gram-negative rods that are greatly stimulated in the rhizosphere fall mainly into the three genera, namely *Pseudomonas, Achromobacter* and, to a lesser extent, *Agrobacterium*. Anaerobic bacteria may also increase in numbers, and this is attributed to reduced oxygen supply resulting from root and microbial respiration. It should be kept in mind, however, that oxygen diffusing from aerial parts can apparently satisfy the requirements of roots and, possibly, those of rhizoplane and inner rhizosphere microorganisms as well. This would tend to minimize the effects of temporary anaerobiosis in the rhizosphere.

The high bacterial density in the rhizosphere undoubtedly results in a high degree of microbial competition and the selection pressures arising from this tend to favour rapidly growing and biochemically versatile organisms over slower growing and less versatile strains. This would suggest that the rhizosphere microflora has a greater ability to effect rapid biochemical changes than the general soil population. As emphasized by G. D. Bowen, however (Bowen 1980), natural selection will not necessarily lead to the dominance of the most productive strains of microorganisms in the rhizosphere, since evolution tends to select for survival rather than productivity. Where a high growth rate enhances survival, as in some zymogenous organisms, high productivity and ecological dominance in the rhizosphere may go hand in hand.

The nutrient requirements of the bacterial component of the rhizosphere biota have been studied by means of an empirical scheme for classifying soil bacteria, devised in 1943 by A. G. Lochhead and F. E. Chase. This classification recognizes seven nutritional groups of bacteria, distinguished according to their ability to grow on media of varying degrees of complexity. Some bacterial isolates can grow on a simple medium containing only glucose and mineral salts whereas others require preformed amino acids or B-vitamins, or unidentified growth factors contained in yeast extract or soil extract. Lochhead and Chase's nutritional

Table 6.2 Incidence of nutritional groups of bacteria in the rhizosphere of oats

	Percentage of whole population		*Total count per gram*	
Requirements for maximum growth	*Control soil*	*Rhizosphere*	*Control soil* (X 10^6)	*Rhizosphere* ($\times 10^6$)
No special growth factors[†]	13	33	32	690
Amino acids	2	11	5	230
B vitamins	9	11	22	230
Amino acids + B vitamins	2	1	5	21
Yeast extract	27	19	67	400
Soil extract	23	18	57	380
Yeast extract + soil extract	24	7	60	150

Source: After Wallace and Lochhead (1949).

[†] Capable of growing in a simple medium containing sugar and inorganic salts only.

classification reveals a consistent, preferential enhancement of organisms that either require no special growth factors or else require amino acids only (Table 6.2). At the same time, the proportion of bacteria with complex nutritional requirements declines, even though their actual numbers increase. The selection for bacteria whose growth is enhanced by amino acids is presumably the result of the high level of amino acids in the rhizosphere. These amino acids are derived from plant and microbial exudates and secretions, and the decomposition products of dead roots and microorganisms. In general, the rhizosphere bacteria are less fastidious than those which live beyond the region of root influence.

In contrast to their effect on bacteria, roots do not have such a marked influence on the total number of fungi. However, specific genera are stimulated, in other words the rhizosphere effect for fungi is more qualitative than quantitative. The imperfect fungi *Fusarium* and *Cylindrocarpon* are among the more prominent rhizosphere inhabitants but many other genera are represented, especially the zygomycetes *Mucor* and *Rhizopus*. An extreme example of this tendency is the virtual dominance of a single fungal species on the surface of mycorrhizal roots of the ectotrophic kind. This special case of the rhizosphere effect will be considered in Chapter 7. The spores of many common soil fungi, which normally lie dormant, will often germinate in the rhizosphere.

Microbial colonization of roots

Despite the high population densities in the rhizosphere, the surfaces of roots are no longer believed to be wholly covered with a mantle of microorganisms[4]. Rather, it is now thought that only 5–10 per cent of young root surfaces is colonized (Bowen & Theodorou 1973; Rovira *et al.* 1974), though the proportion may be higher on older roots. Furthermore, microbes are not randomly distributed but occur preferentially at cell junctions, which are not only the major sites of exudation but also serve as migration routes (Bowen & Rovira 1976). Fungi are able to spread from these points of colonization by hyphal extension, translocating nutrients across regions where exudates are scarce. Migration of bacteria is less well understood, but it is possible they are able to move along the meniscus of water between a fungal hypha and the root surface; at least this is thought to be a more likely migration route than surface water films (cf. sect. 3.5, under 'Commensalism').

It may be that migration of microorganisms along roots is not a common phenomenon in nature. This view certainly accords with observations that root surfaces are only partially colonized, that bacteria tend to occur as discrete microcolonies embedded in mucilage, and that roots frequently grow faster than fungal colonies can spread. If migration along roots is indeed inconsequential, then colonization of the rhizoplane must take place mainly from propagules germinating in the rhizosphere or nearby soil.

Other factors affecting the proliferation of microbes in the rhizosphere

Although organic materials originating from roots – exudates, sloughings, etc. – appear to be the primary determinants of the rhizosphere effect, there are additional factors which may be important in making the soil adjacent to plant roots a favourable habitat for some microorganisms. For example, the assimilation of nutrients by plants may lower the concentration available for microbial use, favouring those groups which can compete more successfully or which have very low nutrient requirements. Again, root respiration may alter the pH and the availability of some nutrients in rhizosphere soil, and by utilizing oxygen favour the growth of microaerophiles and anaerobes. Furthermore, soil physical properties can be altered by plant growth, as when root penetration improves soil structure, leading to increased aeration and a consequent stimulation of the oxidative processes of aerobic microbes.

Foliar application of chemicals, such as growth regulators, herbicides and fungicides, can also alter the composition of the rhizosphere population. In this instance, however, the effect seems to be an indirect one, associated with a change in the nature of root exudates (Hale & Moore 1979).

6.4 Microbial interactions in the rhizosphere

Interest in this topic stems from the possibility of manipulating the rhizosphere population to enhance the growth of economically important plants, either by encouraging the proliferation of strains that increase nutrient availability or by suppressing disease-producing organisms. Since the rhizosphere is a zone where a general stimulation of microbial growth occurs, one would expect the effect of microbial interactions (described in Ch. 3) to be heightened, despite the fact that the root surface is incompletely colonized. The possibility of interactions occurring comes about, as Bowen (1980) has indicated, because microorganisms tend to occupy the same microhabitats, namely the cell junctions.

Most of the evidence for the existence of microbial interactions in the rhizosphere is nevertheless circumstantial. *Bdellovibrio*-like organisms have been found in the rhizospheres of wheat and subterranean clover (Foster & Rovira 1978) which suggests that parasitism has a role to play in this environment. Many common saprophytic fungal inhabitants of the rhizosphere are known to produce antibiotics (see sect. 3.5) and it is conceivable that they might antagonize other organisms, including pathogenic root-infecting fungi. Some ectomycorrhizal fungi are likewise recognized antibiotic producers, and reference has already been made (sect. 3.6) to the potential significance of this in describing the inhibition of *Phytophthora cinnamomi* by the mycorrhizal fungus *Leucopaxillus cerealis var. piceina*. Certain Californian soils are known to suppress vascular wilts caused by fungi of the genus *Fusarium*, and Baker and Cook (1974) have pointed out that bacteria multiply more rapidly in these soils than in soils conducive to such disease. They infer that root exudates of plants growing in suppressive soils favour the development of bacteria which are antagonistic to *Fusarium*.

Positive interactions are likely to occur in the rhizosphere also, and some examples have already been described. A commensalistic relationship is inherent in the proposition that fungal hyphae provide a transport route for motile bacteria colonizing the root surface. Preferential stimulation of amino-acid requiring bacteria (Table 6.2) is further evidence that commensalism features strongly in rhizosphere biology although, given the nature of root exduates, there is no reason to assume that amino-acid exuding bacteria need be present. Whether mutualism occurs between micro-organisms in the rhizosphere, as distinct from mutualistic relationships between plants and microbes, remains to be demonstrated.

Experimental evidence for the existence of both positive and negative microbial interactions was provided by Bowen and Theodorou (1979). They studied the effect of bacterial inoculation on the colonization of roots by mycorrhizal fungi, and found that some bacteria stimulated fungal growth while others depressed it. Their results are given in Table 6.3, which also illustrates large differences in colonization ability among fungi, even in the absence of competitors or antagonists. Further investigations of this kind are needed to increase our understanding of microbial interactions in the rhizosphere, and to serve as a basis for the biological control of soil-borne plant pathogens.

Table 6.3 Rhizoplane interactions between bacteria and mycorrhizal fungi

Fungus	*Bacterium*				
	Experiment 1			*Experiment 2*	
	Nil	*Pseudomonas*	*Bacillus No. 1*	*Nil*	*Bacillus No. 2*
Rhizopogon luteolus	27.6	8.3	23.7	30.7	34.1
Suillus luteus	16.2	3.2	8.7	19.1	26.2
Corticium bicolor	17.2	3.2	15.3	15.8	27.1
LSD for $p < 0.05$		5.4		6.2	
$p < 0.01$		7.1		8.2	

Source: After Bowen And Theodorou (1979); values shown are lengths of root colonized by fungi (mm) after four weeks

6.5 The rhizosphere in relation to nutrient cycling

The activities of the rhizosphere microflora affect the circulation of nutrients through their influence on plant nutrition. The effect of microbial processes on nutrient availability in the rhizosphere thus provides a basis for considering the regulatory role of this root–microbe interaction. Before discussing it, however, it is pertinent to review briefly the state of knowledge concerning the uptake of nutrients from the soil. More detailed treatment may be found in the works of Barley (1970), Epstein (1972), Russell (1973), and Nye and Tinker (1977).

Nutrient absorption takes place in an aqueous environment. Provided there is water available for plants to use, there will always be a water film on particle surfaces. This is because at soil water suctions corresponding to the wilting point, that is approx. 15 bars or 1.5 MPa, the soil atmosphere has a relative humidity of 98.6 per cent and a layer of water many molecules thick condenses on mineral surfaces (Greenland 1979).

Soil as a source of mineral nutrients

Salts exist in soil in a number of different states:

(a) as water-soluble salts dissolved in the soil solution;
(b) as sparingly soluble or insoluble substances containing exchangeable ions;
(c) as insoluble substances from which ions are not readily obtained by exchange reactions.

The composition of the soil solution varies with soil type, and for all ions except phosphate its concentration increases as soil moisture content decreases. The level of phosphate is relatively independent of soil moisture content, presumably because the soil solution is nearly always saturated with rather insoluble phosphates.

With few exceptions, soil water may be regarded as a dilute solution of electrolytes. For the metallic cations, concentrations range from less than 1 $\mu g\ ml^{-1}$ in soils derived from acid igneous or sedimentary rocks to over 10 $\mu g\ ml^{-1}$ in soils developed on basic igneous rocks (Charley & Richards 1983). Phosphate levels generally range from less than 10^{-8} M in very poor tropical soils to more than 10^{-4} M in fertile soils of the temperate zone, but in most soils of reasonable phosphate status the value is around 10^{-5} M or 0.03 μg P m^{1-} litre (Russell 1973).

An equilibrium is maintained between the exchangeable ions and the soil solution, and in this way exchangeable ions become available for plant growth, that is by replacing ions in the solution phase as these are utilized by plants. Thus the concentration of ions in the soil solution is buffered by ions adsorbed on soil surfaces. Although both anions and cations can exist in exchangeable form, exchangeable anions are present in limited supply only. This is because the soil colloids possess a net negative charge, resulting in a general tendency for anions to remain in solution. Indeed it is the concentration of unadsorbed anions (Cl^-, NO_3^-, HCO_3^- and SO_4^{2-}) which primarily determines the strength of the soil solution.

The exchangeable cations are of greater importance in plant nutrition, and these comprise principally Ca^{2+}, Mg^{2+}, K^+ and Na^+. In acid soils, H^+ ions make up a substantial proportion of the exchangeable cations, and this leads to infertility, since the nutrient cations are displaced into the soil solution where they are subject to leaching. The relative proportions of different cations on the exchange complex is one of the factors that determine how many of each move into the soil solution to balance the dissolved anions, and so maintain electrical neutrality.

The greater part of the mineral salts content of soil is present as relatively insoluble substances from which ions do not readily exchange. For example, montmorillonite contains much potassium in non-exchangeable form and apatite consists of insoluble calcium phosphates. Such substances are slowly transformed by the process of weathering, which leads gradually to increased solubility and exchangeability of ions. The rate of weathering varies greatly for different minerals and is affected by pH, soil moisture content and temperature. Contact between plant roots and soil increases the rate of weathering, both by mechanical effects which tend to break up larger soil particles, and by the chemical effects of carbonic acid and organic acids produced by the roots.

Phosphorus is a nutrient which deserves special consideration because of its central role in biological processes. Soil parent material is the only significant source of this element, and it occurs in rocks as apatite, mainly calcium fluorapatite, $3Ca_3(PO_4)_2.CaF_2$ (Ch. 5). This weathers to several secondary minerals including hydroxyapatite, $3Ca_3(PO_4)_2.Ca(OH)_2$. Most mineral phosphate, whether primary or secondary, is insoluble, being bound to iron and aluminium in acid soils ($pH < 6$) or to calcium in neutral and alkaline soils. A certain amount of insoluble soil phosphate is also present in organic form, much of it as phytates (see Ch. 5). In arable soils, especially those of temperate regions, organic phosphates are of minor importance in plant nutrition, but in strongly leached tropical soils and those formed on ancient peneplains, they constitute the main reserve of plant-available phosphorus.

Insoluble soil phosphate equilibrates with the dissolved fraction only very slowly whereas the exchange between adsorbed and soluble phosphate ions is very rapid. If the exchangeable pool is large relative to the amount of phosphate in solution, then the soil solution will be quickly replenished as soon as it is depleted by plant uptake. If however the soil's phosphate adsorption capacity is small, then the concentration of phosphate in the soil solution will decline with the growth of crops. Any factor which speeds the dissolution of mineral phosphates, or the mineralization of organic phosphates, will help replenish the soluble pool and might consequently make a significant contribution to the biogeochemical cycle of phosphorus.

Mechanisms of ion uptake

Certain aspects of nutrient uptake by cells were discussed in Chapter 2 in relation to the nutrition of microorganisms. The entry of ions into plant roots is accomplished by both passive and active mechanisms, and the relative significance of these varies with soil conditions and plant species. The part of the root in which passive uptake processes such as diffusion and ion exchange operate consists of the cell wall and intercellular spaces. This 'apparent free space', as it is termed, lies outside the outer cytoplasmic membrane or plasmalemma. It is made up of the so-called 'outer space', which is freely accessible to diffusion, and the 'Donnan free space' which represents that fraction of the tissue available for ion exchange reactions. The latter take place on cell wall surfaces which, due to the presence of dissociated carboxyl groups, carry a negative charge and consequently are able to adsorb cations.

The apparent free space of roots therefore comprises those tissues which can be traversed by ions without their having to pass through a living cell membrane. It extends from the epidermis to the innermost layer of cortical cells, the endodermis, but does not include the tissues of the stele. The cortex is the tissue of greatest importance of nutrient uptake. Its cells are elongated and become highly vacuolated during extension growth; intercellular spaces are conspicuous (Fig. 6.4). The inner boundary of the cortex, namely the endodermis, is comprised of cells whose radial walls are impregnated with suberin to form the so-called Casparian strip, and it is this which prevents further movement of water and solutes via the cell walls.

By means of diffusion and exchange reactions, nutrient ions

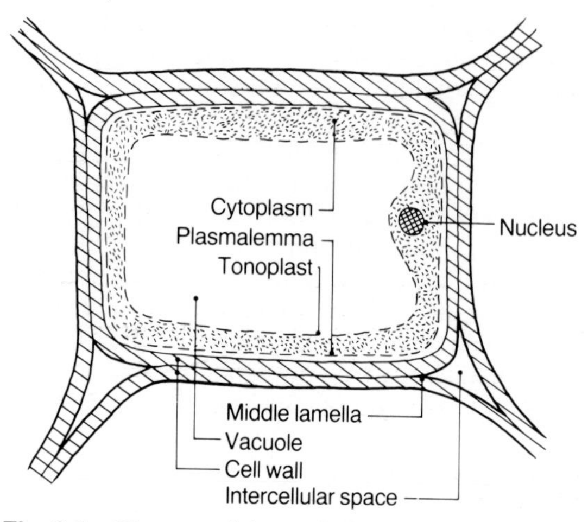

Fig. 6.4 Diagram of the main features of a cell typical of the root cortex, the tissue most concerned in nutrient uptake. (From Nye & Tinker 1977.)

from the external medium migrate into the apparent free space of roots. This process may be hastened by the mass flow of ions in water moving towards roots under the influence of the transpirational pull of leaves. Once the ions reach the plasmalemma, however, the semipermeable and selective properties of this membrane impede their further progress by passive means. Only by active processes can ions be transported across cell membranes and accumulated in vacuoles at concentrations many times that found in the soil solution. Ions which reach the endodermis by passive transport through the cortical free space must be actively absorbed before they can pass through this barrier into the stele. Like other ions which have been actively transported, they move from cell to cell via cytoplasmic connections known as plasmodesmata. As indicated in Chapter 2, active transport is thought to involve carriers, that is compounds which form complexes with ions and so render them more 'soluble' in the membrane.

Absorption of ions from soil

In any given environment, the uptake of a solute by a root is, in general, determined by the concentration of this solute in the soil solution and the extent to which this is buffered by the soil. Early ideas about ion uptake from soils were based on the premise that CO_2 evolved by root respiration dissolved in water to form carbonic acid, H_2CO_3, which dissociates to provide H^+ ions to exchange with cations adsorbed on soil colloids. Since roots also have cation exchange properties (Drake *et al.* 1951), the possibility arises of direct exchange between adsorption sites on soil and roots. H. Jenny and R. Overstreet (1939a, b) elaborated this concept of 'contact exchange', suggesting also that cations could migrate along clay surfaces and pass from one clay particle to another by this means.

In the sense that ions can only be transferred from soil to root through a water film, contact exchange does not differ fundamentally from uptake from solution. In any event, it cannot by itself account for all the cations taken up by roots, and furthermore, as indicated by Nye and Tinker (1977), it assumes (like the respiratory CO_2 hypothesis) that plant roots make the soil more acid. In fact, as these authors point out, plants generally take up more anions than cations and consequently liberate more bicarbonate (HCO_3^-) ions than H^+-ions in order to balance the electro-chemical potential of the soil solution: the soil in the immediate vicinity of roots thus becomes more alkaline (except where NO_3^- is not the major source of nitrogen – see further below).

Root characteristics relevant for nutrient uptake

The dimensions of the rootsytem and its absorbing roots are significant factors in the uptake of nutrients from soils moist enough to support plant growth. The primary roots of most plants have diameters greater than 1 mm; indeed the roots of temperate cereals usually exceed 2 mm diameter while the young roots of most trees and herbs are larger still (Russell 1973). These dimensions are about two orders of magnitude larger than the equivalent diameter of soil pores drained at field capacity. Water tension at field capacity varies with the nature of the soil and its temperature, ranging from 5 Pa in some temperate soils during winter to 35 Pa in subtropical soils in summer. At these suctions, pores larger than about 60 μm and 10 μm equivalent diameter, respectively, will be empty of water. As E. W. Russell (1973) has pointed out, this means that soils can be well drained yet contain no pores wide enough for roots to enter, unless they compress the soil.

Root hairs are finer than roots, being typically 5–15 μm in diameter, and consequently are able to enter some pores small enough to hold water at field capacity. Although the root hair zone is generally considered to be relatively short, quite old roots may bear hairs, their frequency and durability varying with species and depending also on a variety of environmental factors (Nye & Tinker 1977). While root hairs obviously increase the plant's capacity to tap a greater volume of the soil solution, it should be recalled (sect. 1.4) that storage pores can be as small as 0.5 μm. How then do roots maintain contact with the soil solution as the soil dries below field capacity?

In that part of the root where most absorption of water and nutrients takes place, a layer of mucigel (sect. 6.2) is interposed between the root proper and soil particles. This mucigel, which forms a continum from soil to root, protects the root against desiccation through its ability to become adsorbed on fine clay particles. As the soil dries and the root shrinks, a layer of soil containing many water-filled pores closes in around the root, maintaining a moisture film on its surface. According to Russell (1973), this would allow the root to maintain liquid contact with the soil as the suction rises appreciably above, say 50 Pa. Furthermore, since this mucilaginous layer has properties of a true gel, ions would diffuse through it in much the same way as they move through water (Greenland 1979), though it might restrict the process of mass flow.

Another factor affecting nutrient influx across the soil–root interface is the area of active root surface relative to the demands of the shoot for biosynthesis and growth. In annual plants, influx of nutrients is greatest during the first few weeks after germination and declines with age, although the uptake of some nutrients continues by mass flow even after the roots have been suberized. Plants are capable of compensating for nutrient deficiencies by increasing the root surface absorbing area relative to shoot demand, for example by increasing root hair length and density. The formation of mycorrhizas (Ch. 7) may also increase the effective absorptive surface of plant roots. The efficacy of mycorrhizas in this respect is further enhanced by the ability of their subtending fungal hyphae, which generally have diameters less than 5 μm, to provide access to soil pores too small to be penetrated by root hairs.

In the final analysis, it is the properties of the rootsystem as a whole which are likely to be of greatest ecological significance, and in this regard it is necessary to distinguish between species growing in natural and man-made ecosystems. The distinction was nicely made by P. H. Nye and P. B. Tinker in 1977:

> In the exploitation of the soil by plant roots there is a marked difference between a root system developing from seed in a fertilized soil, and progressively exploiting deeper layers of the soil – characteristic of an annual crop – and the established root system of perennial plants, such as occurs in many natural communities, e.g. a humid woodland. Nutrient levels in the soil solution in the latter are usually much lower; the overall rate of uptake will depend on rates of mineralization of humus, release by weathering, and leaching from living vegetation and litter by rain; and finally, competition between roots is intense.

Solute transport from soil to roots

As indicated by Nye and Tinker (1977), the essential transport processes occurring in the soil near roots are as follows:

(a) diffusion and mass flow towards the root in the solution phase;
(b) very slow diffusion towards the root in the exchange phase;
(c) rapid interchange of ions between exchange and solution phases;
(d) very slow release of ions from the solid phase into the soil solution.

If a solute is absorbed at a faster rate than water, as happens with phosphate and potassium, then its concentration in the soil solution at the root surface will fall. This change is of course buffered by the release of ions adsorbed on colloidal surfaces surrounding soil pores. If, on the other hand, water is absorbed faster than the solute, the latter accumulates at the root surface and will then tend to diffuse away from the root. This may happen with sulphate, for example.

The mobility of ions is determined principally by mass flow and diffusion. These two processes are the main ones governing the rate at which solutes move towards uptake sites on the root surface, the amounts transported by 'contact exchange' being negligible by comparison. In a moist loam an ion in solution may move 1 cm in 10 days whereas a strongly adsorbed ion would move about 1 mm; in a dry soil the movement may be an order of magnitude less.

Root influences affecting absorption processes

Roots may modify the rhizosphere soil physically, chemically and microbiologically. Compaction and reorientation are two possible physical effects: roots tend to follow pores and channels and usually displace some soil as they extend, and may reorient clay particles parallel to their surfaces.

Chemical effects. Roots may influence the diffusion of solutes by any or all of the following processes:

(a) release of H^+ or HCO_3^- ions;
(b) evolution of CO_2 by respiration;
(c) changing the concentration of other ions or solutes;
(d) excretion of organic substances.

As indicated earlier, plants generally absorb more anions than cations, though there are some important exceptions. Those which do must excrete HCO_3^- ions rather than H^+ ions, in order to maintain electrical neutrality across the root surface; the rhizosphere of such plants therefore becomes more alkaline (Riley & Barber 1969). This situation certainly applies when NO_3^- is the major source of nitrogen, as it is in most agricultural ecosystems. However, if nitrogen is supplied as NH_4^+, which appears to be the more general case for many of the world's natural plant communities (see Ch. 5), then the pH of the rhizosphere will fall. One consequence of this is to raise the concentration of PO_4^{3-} in the soil solution, and so increase its uptake from neutral soils (Miller *et al.* 1970; Riley & Barber 1971). Likewise, symbiotic nitrogen-fixing plants (Ch. 8) take up nitrogen as the neutral molecule N_2, and such plants therefore take up more cations than anions, excreting H^+ ions as a result.

The effects of root exudates are difficult to separate from those of microbial exudates in the rhizosphere. In any event, the role of exudates is speculative and based on the premise that certain solutes, especially polycarboxylic acids such as citric and oxalic, are likely to displace PO_4^{3-} from adsorption sites. Alternatively, organic acids might form soluble chelates with metallic cations including some trace elements, and so increase the concentration of these nutrients in the soil solution.

Microbiological effects. The ways in which rhizosphere microorganisms might affect the supply of plant nutrients have been stated by Nye and Tinker (1977) to be as follows:

(a) by a change in the morphological or physiological properties of roots or rootsystems;
(b) through alteration of the phase equilibria of nutrients in that they are more readily transported to root surfaces and/or absorbed, e.g. change in pH or redox potential, or complex formation;
(c) through a change in the chemical composition of soil, with similar results, e.g. mineralization of organic matter, weathering of minerals;
(d) by symbiotic processes involving direct transfer of nutrients from microbial symbiont to host;
(e) by the blocking of root surfaces, or competition for nutrients.

It is known that microorganisms can alter the morphology of roots and root hairs, probably by producing growth regulating substances such as cytokinins, and this may conceivably affect nutrient uptake. The possibility that bacteria might physically block access to absorption sites seems remote, in view of the small fraction of the root surface that is normally colonized (sect. 6.3), but there is evidence (Barber 1968) that microorganisms can compete with plants for nutrients at low solution concentrations. The most profound microbial effects on nutrient supply involve mutualistic associations between plants and soil microbes, and these are discussed separately in Chapters 7 and 8. Apart from these well known symbioses, the major likely influences of the rhizosphere microflora on solute transport from soil to root are those listed under (b) and (c) above. Particular effects are now discussed in the context of nutrient availability.

Availability of nutrients to plants

The availability of nutrient ions to plants depends on their mobility and proximity to absorbing surfaces. Dissociated ions in the soil solution, and the majority of exchangeable ions adsorbed on colloidal surfaces, are generally thought to be readily available to plant roots. Non-exchangeable ions in primary minerals and organic matter are usually believed to be not readily available although it is known that certain clay minerals such as illite may supply sufficient non-exchangeable potassium to support natural vegetation indefinitely or even some crops. The mechanism which initiates this slow release of potassium from the solid phase is the lowering of the K^+ ion concentration in the soil solution by plant uptake. This effect is most marked near root surfaces and, according to Nye and Tinker (1977), should stimulate the release of potassium from the solid phase in that micro-environment.

Furthermore, irrespective of whether it is of any significance in plant nutrition, the release of ions from primary minerals has an important pedogenic function in that it is the point of initiation of many nutrient cycles, hence any factor which accelerates this process is significant in ecosystem function. It is in this context that the effect of the rhizosphere microflora on the nutrition of plants needs to be discussed. This is conveniently done under several headings.

Uptake of dissociated ions from the soil solution

In short term uptake experiments, several workers (Barber 1969; Bowen & Rovira 1966) have shown that the presence of microorganisms on plant roots greatly modifies the absorption and utilization of phosphate by the roots. Not only is the rate of uptake affected, but the pattern of translocation and the direction of metabolism are altered also. The effects are most pronounced when the external supply of phosphate is low, and are not always directed towards increased uptake and incorporation by the plant. It is evident that the rhizosphere microflora influences nutrient absorption even from the solution phase.

Mineralization of organic matter

It has been suggested (sect. 6.3) that the rhizosphere microflora should have a greater ability than the general soil population to effect rapid biochemical changes, and there is some evidence to support this contention. It has been found, for example (Estermann & McLaren 1961), that the enzymatic activity of barley roots is increased by the presence of a root surface microflora, and as a result of this the plant's ability to utilize urea and some organic phosphates is increased. Working with simulated rhizosphere soil, Thompson and Black (1970) measured greater amounts of phosphate in the presence of plant roots than in their absence, the difference being presumed to be the result of greater net mineralization of organic matter in the rhizosphere. Both bacteria and fungi, some of which are stimulated in the rhizosphere (Greaves & Webley 1965), produce the phosphatase necessary to hydrolyse inositol polyphosphates (see sect. 5.2.1).

Ammonifying bacteria characteristically respond to the proximity of living roots, and R/S ratios in excess of 50/1 are not uncommon for this group. This should lead to greater mineralization of organic nitrogen in the rhizosphere, a prediction which can be confirmed by incubating samples of rhizosphere and non-rhizosphere soil in the laboratory, and comparing the amount of NH_4-N released after a standard period of time. Such results do not however conform with field evidence, which indicates that less mineralized nitrogen is present in soil under crops than in fallow, even when allowance is made for the amount of nitrogen taken up by the plants. This apparent anomaly was explained by the studies of W. V. Bartholomew and F. E. Clark (1950), using the heavy isotope of nitrogen (^{15}N) as a tracer, which showed that although the net amount of nitrogen mineralized in cropped soil is only about half that in uncropped soils, the total quantity mineralized is greater under the crop. The difference is due to the fact that most of the nitrogen mineralized under a plant cover is rapidly assimilated again by the microflora.

This example illustrates the general principle discussed in Chapter 5, namely the availability of plant nutrients is the resultant of the opposing processes of mineralization and immobilization. Theoretically, the overall effect in the rhizosphere could be detrimental as often as it is beneficial, and examples of this are known. Thus nitrifying bacteria may be suppressed by root exudates (Rovira & Davey 1974), leading to lower nitrification rates in the rhizosphere. On balance however, the effect is more likely to be beneficial, for the nutrients mainly involved in these microbial transformations are the anions nitrate, sulphate and phosphate. Since these are normally absorbed by plants from the soil solution, it could be argued that the cyclic turnover through organic and inorganic forms, which is mediated by microorganisms, minimizes losses through leaching and provides a continual supply of anions at the root surface.

A rhizosphere effect which may be of particular significance in the nutrition of some forest trees is illustrated by a comparison of the surface microflora of mycorrhizal and non-mycorrhizal roots of yellow birch, a species which forms ectomycorrhizas (see Ch. 7). Table 6.4 shows that there are greater numbers of diverse metabolic groups of bacteria on mycorrhizal roots and this implies a greater availability of nutrients in their vicinity. The increase in methylene blue reducing and glucose fermenting bacteria indicate that certain kinds of oxidizable and fermentable substrates would be broken down more rapidly in the presence of mycorrhizas than in their absence. Especially in the rich humus layers where mycorrhizas abound, the potential ammonifying capacity of their rhizosphere microflora could be of particular importance in the nitrogen nutrition of the trees.

Other interactions in the rhizosphere also have the potential to increase the mineralization of organic matter. D. C. Coleman and colleagues (Coleman *et al.* 1977) reported the results of a microcosm study in which they inoculated partially sterilized soil with bacteria (*Pseudomonas*) alone or in combination with amoebae (*Acanthamoeba*) and nematodes (*Mesodiplogaster*), all isolated from the rhizosphere. They found that up to 35 per cent more phosphate and 50 per cent more ammonium was released by mineralization in the presence of predators.

In summary, while there is evidence that mineralization pro-

Table 6.4 Bacterial populations on mycorrhizal and non-mycorrhizal roots of yellow birch seedlings

	Numbers per gram of dry roots	
Kinds of bacteria	*Mycorrhizal roots*	*Non-mycorrhizal roots*
Methylene blue reducers	5.8×10^8	5.0×10^7
Glucose fermenters		
acid producing	4.6×10^7	5.0×10^5
gas producing	5.8×10^7	1.0×10^4
Ammonifiers	1.2×10^8	1.3×10^7
Fluorescent pigment producers	2.7×10^5	1.0×10^4

Source: After Katznelson *et al.* (1962).

cesses are stimulated in the rhizosphere, much of it is circumstantial, and in any event an increase in the tempo of the mineralization–immobilization cycle will not necessarily result in net mineralization. On theoretical grounds (Ch. 5), the continuous supply of substrates, many of which might be readily utilized as energy sources by a wide range of microorganisms, might be expected to favour net immobilization.

Solubilization of minerals

It has long been known that plant roots can increase the rate of rock weathering by the production of CO_2 and organic acids (see Ch. 5). The combined respiratory activities of roots and microorganisms result in greater CO_2 production from rhizosphere than non-rhizosphere soil; it has been variously estimated that microbial respiration accounts for one-third to two-thirds of this carbon dioxide. Enhanced production of CO_2 in the rhizosphere, where it would dissolve in the soil water to form carbonic acid, should lead to increased solubility of primary minerals, and a consequent increase in the availability of nutrient ions.

A number of low molecular weight organic acids have been identified in the root exudates of plants grown under sterile conditions, and some of these can dissolve phosphate minerals. Some, for example citric, oxalic and tartaric acids, are known to be chelating agents but, while chelation may affect nutrient availability, this may bear little relationship to the effectiveness of these acids as weathering agents (see Ch. 5). In any event, under natural conditions the organic acids present in the root exudates may not persist for long enough to effect dissolution, since they will tend to be used as substrates by rhizosphere microorganisms. Thus the TCA cycle acid, malic acid, can serve as the sole energy source for many rhizosphere bacteria and fungi (Sollins *et al.* 1981).

Plants differ in their ability to dissolve minerals. Thus chemical and X-ray analyses revealed that wheat took up potassium from biotite much more readily than alfalfa (Malquori *et al.* 1975). A. Moghimi and colleagues (Moghimi *et al.* 1978) found that wheat plants absorbed ^{32}P from synthetic hydroxyapatite more efficiently than did maize or peas. They further showed that the 'rhizosphere products', that is root exudates plus microbial metabolites, of wheat plants could themselves solubilize hydroxyapatite while lowering the pH of the medium. It was subsequently shown (Moghimi *et al. loc. cit.*) that the active component was 2-ketogluconic acid, which comprised about 20 per cent of the rhizosphere products. Since the dominant sugars in the rhizosphere products were glucose and fructose, it seems likely that 2-ketogluconic acid was formed by rhizosphere bacteria, since this acid is generated from glucose via the Entner-Doudoroff pathway which does not operate in seed plants.

According to K. Cromack and his colleagues (Cromack *et al.* 1977), one of the main acids involved in biological weathering is oxalic, which is released in copious quantities by fungi thereby enabling them to accumulate calcium in excess of their metabolic requirement as the sparingly soluble calcium oxalate. In so doing, they accelerate weathering by releasing P, Al, Fe and K as stable metallo-oxalic acid complexes. Since the phenomenon of calcium oxalate accumulation by fungi is widespread, it is inferred that the solubilization of minerals initiated by their releasing oxalic acid is likewise common. Such a mechanism could provide an explanation for the changes in mineral structure which are undoubtedly effected by plant roots and their associated mycoflora. For it to operate to a significant degree in the rhizosphere, however, the fungi involved must be able to secure an adequate supply of suitable substrates (energy sources) from root exudates.

While there seems little doubt that larger quantities of low molecular weight organic acids are found in the rhizosphere than in the general body of the soil, there seems to be no data on the relative contributions of roots and microorganisms. One cannot therefore postulate a plant–microbe interaction of significance in plant nutrition solely on the premise that relatively high concentrations of organic acids occur in the rhizosphere. On the other hand, the demonstration of a preferential stimulation in the rhizosphere of microbes that produce such acids would be strong evidence in support of such an hypothesis. Stimulation of bacteria which produce 2-ketogluconic acid (sect. 5.2.1) has indeed been found in the root region of crop plants, by D. M. Webley and R. B. Duff (1965), population densities reaching 10^6 per gram fresh weight of roots and R/S ratios being of the order of 300/1; furthermore, the relative proportion of 2-ketogluconic acid producers increased 50-fold from 0.025 per cent of the total bacterial population in non-rhizosphere soil to about 1.25 per cent in the vicinity of roots.

Supporting evidence for the role of rhizosphere microorganisms as agents of biological weathering derives from the finding that roots of coniferous seedlings and their associated microflora can mobilize potassium from the crystal lattices of micas and felspars (Voigt 1965; Boyle *et al.* 1967). Furthermore, fungi such as *Aspergillus niger* can themselves alter biotite (Boyle & Voigt 1973), causing the release of potassium and other ions. These findings accord with reports of a seasonal variation in microbial population density in the rhizosphere of some tree species, a variation paralleled by changes in the concentration of available potassium in the rhizosphere soil (Kulaj 1962). Electron microbeam scanning (Tan & Nopamornbodi 1979) also revealed higher concentrations of K, along with P, Ca and S, in the rhizosphere of *Pinus taeda* than in the soil 0.2–0.3 mm from the root surface.

The evidence for the alteration of mineral structure presented by G. K. Voigt and his co-workers was based on X-ray diffraction analysis. This technique was also employed by Cecconi *et al.* (1975), in conjunction with chemical analysis, to show that wheat plants could take up potassium from biotite flakes in nutrient solution, and in the process convert the biotite to vermiculite. Another analytical technique, namely the examination of infrared spectra, was used by Silverman and Munoz (1970) to demonstrate that *Penicillium simplicissimum* could alter a variety of igneous rocks in a glucose–mineral salts medium, releasing substantial amounts of Si, Al, Fe and Mg into solution. There seems little doubt that changes in mineral structure can be effected by living organisms. While such biological mobilization of nutrient ions from highly insoluble sources may have little relevance in crop agronomy, its ecological significance could be considerable especially in relation to pedogenesis and mineral cycling in undisturbed ecosystems.

The presence of a rhizosphere microflora appears also to enhance the ability of plants to absorb phosphorus from insoluble fertilizers such as tricalcium phosphate. Whether this can be ascribed to a preferential stimulation of phosphate-dissolving bacteria in the rhizosphere is uncertain, however, for the evidence on this point is contradictory. The solubilization of phosphate minerals is theoretically feasible, and may be achieved by several different mechanisms. One possibility is that bacteria produce more organic acids in the rhizosphere than are present in root exudates alone. This could lead to a greater displacement of PO_4^{3-} ions from adsorption sites and/or the chelation of cations, as described above when outlining the chemical effects of roots on the soil. Another possible mechanism is the utilization of relatively insoluble phosphorus sources – phytates and hydroxyapatite – which is a known property of many microorganisms (Halstead & McKercher 1975; McLaren & Skujins 1971). For this to be effective, of course, microbial phosphate must be somehow transferred to the plant since, if it is released into the medium as PO_4^{3-}, it will simply be readsorbed by the soil. A third possibility is that dead bacterial cells or fragments or dead fungal hyphae might be transported to the root surface by mass flow, and the phosphate released there by phosphatase enzymes. Such enzymes are produced by both bacteria and fungi including mycorrhizal fungi (see previous sub-section; also. Ch. 7).

To summarize, much of the evidence is speculative but biological weathering in the rhizosphere may well account for the uptake of phosphate and other ions when these occur at very low concentrations in the soil solution, or in situations where accepted mechanisms do not provide an adequate explanation. As pointed out by Cosgrove (1977), the potential depletion of the world's mineral phosphate reserves should act as a stimulus to this field of research. While serious shortages of phosphate fertilizers are not expected to occur until well into the next century, their increasing cost may make it obligatory to pursue such investigations sooner. It must be kept in mind that crops recover only a small proportion of the phosphorus added in fertilizers, the greater part accumulating in soil as fixed inorganic P. While some of this accumulated phosphorus is still available to crops and pastures, much of it is not, hence any means of increasing its availability would be of great economic benefit to mankind. One such means might be to enhance microbial solubilization of fixed P in the rhizosphere.

Energy cost to the plant. An energy network diagram, depicting a hypothetical role for the rhizosphere microflora in the solubilization of primary minerals, is presented as Fig. 6.5. Microorganisms in the rhizosphere, utilizing sugars and low molecular weight organic acids in root exudates as substrates, are shown as supplementing their sources of readily available ions (exchangeable and soluble) by dissolving primary minerals, thereby increasing the concentration of dissociated ions in the soil solution whence both plants and microbes draw their nutrient supplies.

If this is accepted as a plausible model, it is pertinent to ask what fraction of plant photosynthate is utilized by the rhizosphere microflora. According to A. D. Rovira (1979), most of the earlier estimates of the amounts of organic materials released by roots were based on studies of plants growing in axenic nutrient solutions, and indicated that less than 1 per cent of the photosynthate produced was 'exuded' into the rhizosphere. However, microorganisms themselves are known to affect exudation (sect. 6.2), and subsequent investigations of the amount of carbon released by plants growing in soil have revealed that a much larger proportion of photosynthate is lost from roots, up to 20 per cent in some instances (Barber & Martin 1976). While much of this extra carbon is in the form of CO_2, sloughings and lysates contribute substantially especially over the time-span of a whole growing

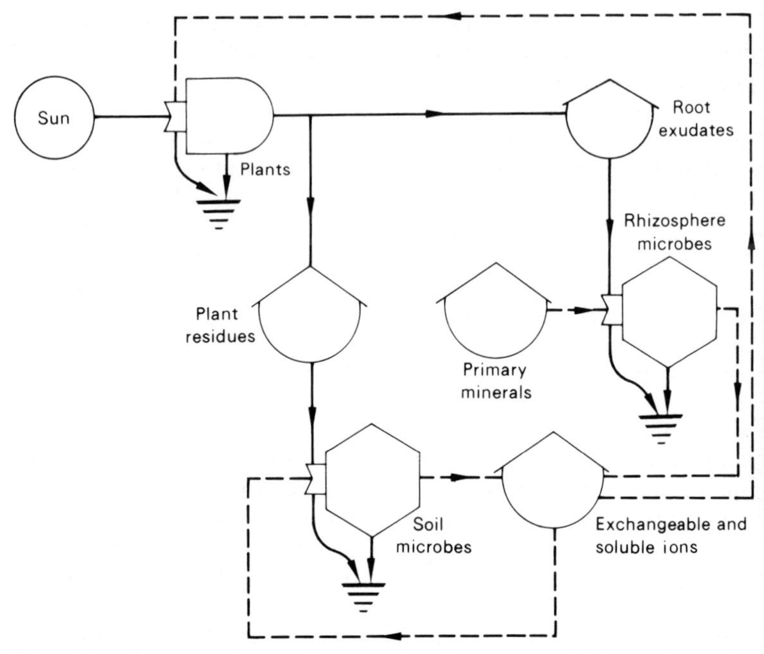

Fig. 6.5 Energy network diagram showing the postulated role of the rhizosphere microflora in the initiation of nutrient cycles. The loss of photosynthate as root exudate enhances microbial growth and leads eventually to a preferential stimulation of acid-producing microbes in the rhizosphere. This in turn facilitates the solubilization of primary minerals which act, through a positive feedback mechanism, to increase the productivity of the system. In immature ecosystems, this could well be a major cause of successional change.

season. J. K. Martin (1978) estimated that nearly two-fifths of the carbon translocated to the roots of wheat was released into the soil, and he attributed this mainly to the lysis of root cell walls rather than to exudation. One would expect CO_2, secretions and exudations to dominate in the early stages of root growth, lysates and sloughings later. Experimental evidence supports this contention (Martin 1978; Warembourg & Billes 1979) and indicates that the chemical composition of organic substrates released from roots changes throughout ontogeny. Short term labelling experiments with $^{14}CO_2$ (Warembourg & Billes *loc. cit.*) further reveals that two phases of CO_2 evolution occur, an initial one due primarily to root respiration followed shortly by a second phase emanating from the microbial respiration of substrates originating from the roots.

The amount and nature of photosynthate 'diverted' towards rhizosphere microorganisms, and which might be utilized to accelerate the dissolution of primary minerals according to the model depicted in Fig. 6.5, must be known if the significance of biological weathering is to be removed from the realm of speculation. Nye and Tinker (1977) have considered this question in relation to the exudation of polycarboxylic acids which are among the most likely compounds produced in the rhizosphere that are capable of causing phosphate desorption. If the equivalent of about 1 per cent of plant dry weight is present in root and/or microbial exudates as citric acid, for example, it can be calculated that a plant of 100 g dry weight might produce enough of this acid to displace 200 mg, which is about the amount of phosphorus used by the plant in attaining that size. Much of this desorbed phosphate would be reabsorbed by the soil or diffuse away from the root, and Nye and Tinker have therefore concluded that, although useful quantities of phosphorus might be taken up by plants via this mechanism, this could happen 'only by a series of assumptions which in total appear unlikely'. If however, the equivalent of 20 per cent of plant dry weight is exuded or converted to polycarboxylic acids in the rhizosphere, then the process is feasible. If in addition any significant fraction of the 'rhizosphere products' appears as chelating acids then the model seems even more realistic.

Rhizospheric nitrogen fixation

Many reports of increased nitrogen fixation have been made following attempts to establish large populations of *Azotobacter* in the rhizosphere (sect. 8.3.1), and inoculation of seed with *Azotobacter* is regularly practised in Russian agriculture. Even in the absence of inoculation, some investigators have found *Azotobacter* to be preferentially stimulated in the rhizosphere, while others have found no evidence of a rhizosphere effect. Climatic or soil differences may account for these conflicting results, at least in part.

There is some evidence that rhizospheric nitrogen fixation may be significant in the nitrogen economy of some grassland and forest ecosystems (Parker 1957; Hassouna & Wareing 1964; Richards 1964). Substantial incorporation of ^{15}N in root tissues of coniferous tree seedlings exposed to air enriched in $^{15}N_2$ (Richards 1973; Bevege *et al.* 1978) may be taken to indicate nitrogen fixation by microorganisms living on or near the root surface. Loose associations are known to exist between various species of N_2-fixing bacteria and the roots of certain tropical grasses (Döbereiner

1974), and there is evidence that nitrogen fixation takes place in the endorhizosphere, that is, in the root cortex (Döbereiner & Boddey 1981).

The potential contribution of rhizospheric nitrogen fixation to the biogeochemical cycle of nitrogen is discussed in Chapter 8. In the present context, it should be kept in mind that, even when agronomically significant amounts of nitrogen are fixed in this manner, many of the crop responses resulting from inoculation with free-living nitrogen fixers are undoubtedly due to factors other than increased supply of nitrogen to the plant. Microorganisms are known to produce both free cytokinins and cytokinin-active nucleosides and excrete them into the culture media (Greene 1980). Cytokinins are growth regulating substances which stimulate cell division in the presence of auxin, and the symptoms associated with plant–microbe interactions frequently resemble the effects of exogenously applied cytokinins. This leads to the suggestion that these hormones may be responsible for some or all of the stimulus to plant growth which has been shown to occur following the inoculation of grasses with nitrogen-fixing bacteria (Gaskins & Hubbell 1979).

6.6 Notes

1. The evidence for active secretion mediated by metabolic energy is, except for root cap mucilages arising from the golgi apparatus, largely circumstantial. Although exudation may occur preferentially at certain sites on the roots, it does not require special structures such as glands. Again, although sucrose is the main sugar translocated to roots it is not commonly found in exudates, which leads to the inference (Warembourg & Morrall 1978) that exudation involves more than just leakage from the phloem.
2. When secondary walls develop in epidermal cells, farther from the root tip, no more secretions or mucilages are released, only exudates.
3. Mucigel is a product of the interaction of roots, soil and microbes and has morphological and biochemical properties quite distinct from those of mucilages produced by roots or microorganisms. It serves an important function in maintaining contact between roots and soils when the roots shrink during daytime water stress. This is thought to provide a mechanism for continuous uptake of water and nutrients by plants (see also sect. 6.5).
4. An exception occurs with ectomycorrhizas (Ch. 7) where short lateral roots may be completely ensheathed by fungal mycelium.

7

Mycorrhizal symbioses

Root-infecting fungi are considered to be parasitic if they cause disease in the host plant, or mycorrhizal if they cause no damage to the host except under unusual circumstances. Separation of root-infecting fungi into such groups is convenient for the purpose of discussion, but the boundaries are not sharp. Mycorrhizal fungi themselves form a heterogeneous group. Some are closely related in their general behaviour to pathogenic root-infecting fungi: indeed, several well-known root disease fungi form mycorrhizas with orchids. At the opposite end of the spectrum some mycorrhizal associations appear to be little more than specialized cases of the rhizosphere effect.

The name **mycorrhiza** is derived from the Greek and means, literally, 'fungus root'. It was first applied in 1885 by A. B. Frank to the composite organs of the Cupuliferae. Similar structures were soon described from many other angiosperms, and also from many conifers. They are characterized by a complete sheath of fungal tissue which encloses the ultimate rootlets of the root system, together with an intercellular infection of the epidermis and cortex (though penetration of the cortical cells sometimes takes place to a limited extent). Such mycorrhizas are called ectotrophic, or ectomycorrhizas. They form a well defined group, distinct in morphogenesis and histogenesis from most other kinds of association between fungi and roots. The term mycorrhiza has however gradually been extended to embrace certain kinds of fungus–root associations in which the hyphae regularly penetrate the cortical cells of the host. These mycorrhizas are called endotrophic, or endomycorrhizas, and in such forms the external mantle of mycelium is usually lacking. Some endomycorrhizas do have a well developed sheath however, for example the so-called arbutoid mycorrhizas, and these constitute a link between the two major kinds of infection.

The mycorrhizal association is a widespread phenomenon, occurring in nearly all families of flowering plants as well as in some bryophytes and pteridophytes. The use of the adjectives, ectotrophic and endotrophic, has fallen into disfavour, the terms ectomycorrhiza and endomycorrhiza, as proposed by Peyronel *et al.* (1969), being now preferred.

7.1 Ectomycorrhiza

This is a common mycorrhiza of forest trees in the families Pinaceae, Betulaceae, Fagaceae, Dipterocarpaceae and Myrtaceae. Infection is a normal and regular event in nature, the mycorrhizal fungi being mainly higher Basidiomycetes.

Structure of ectomycorrhizas

The structure of ectomycorrhizas is an expression of the morphogenetic interaction between plant and fungus. Most mycorrhizal root systems display two classes of lateral roots, 'short roots' of small diameter and determinate length, and 'long roots' which are larger and capable of elongating indefinitely. Short roots (the so-called feeder roots) are ephemeral, while long roots have greater longevity and form the framework of the rootsystem.

Colonisation of short roots by mycorrhizal fungi results in the formation of mycorrhizas of recognizable morphology (Fig. 7.1). By comparison with uninfected rootlets, they are visibly swollen. In most families – both angiosperm and gymnosperm – ectomycorrhizas branch racemosely. The genus *Pinus*, however, is exceptional in having dichotomously branched mycorrhizas. Pines are peculiar in another respect also, the differentiation into long and short roots being sharper than in most other trees.

A typical ectomycorrhiza has many short branches, so that groups of lateral rootlets are formed which may be completely enclosed by mycelium. In section (Fig. 7.2), the rootlet is seen to be bounded by a layer of fungal pseudoparenchyma which may comprise one-quarter of the total volume of the mycorrhiza and two-fifths of its total dry weight. This is the **sheath** or **mantle**, the surface of which is fairly smooth, although some hyphal connections to the surrounding soil may be visible. The inner portion of the sheath is connected to hyphae which ramify between the cells of the epidermis and of the outer cortex, forming a mycelial network called the **Hartig net**. The fungus never penetrates the endodermis in normal mycorrhizas. Hyphae occasionally penetrate the cells of the cortex, but the degree of penetration is not extensive when compared to that found in endomycorrhizas.

(a)

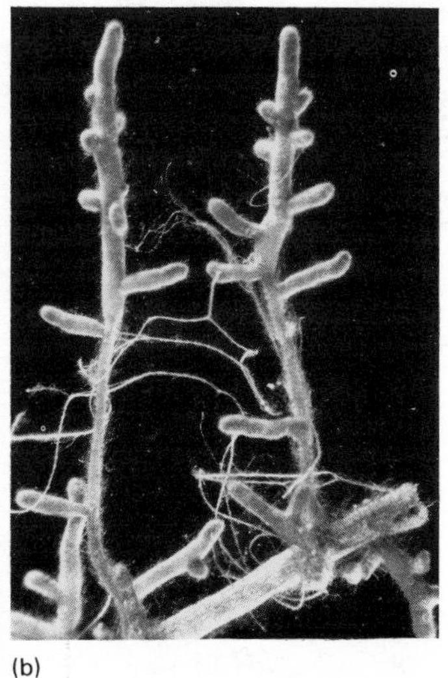

(b)

Fig. 7.1 Gross morphology of ectomycorrhizal root systems. (a) *Pinus*, showing the dichotomous branching of infected roots typical of this genus (b) *Eucalyptus*, showing the racemosely branched mycorrhizas found in most genera other than *Pinus*. Note the absence of root hairs on the mycorrhizal roots, and the presence of numerous mycelial strands and rhizomorphs. (a) Photograph of the litter layer in a pine plantation, kindly supplied by Dr. G. D. Bowen. (b) From Chilvers and Pryor (1965).

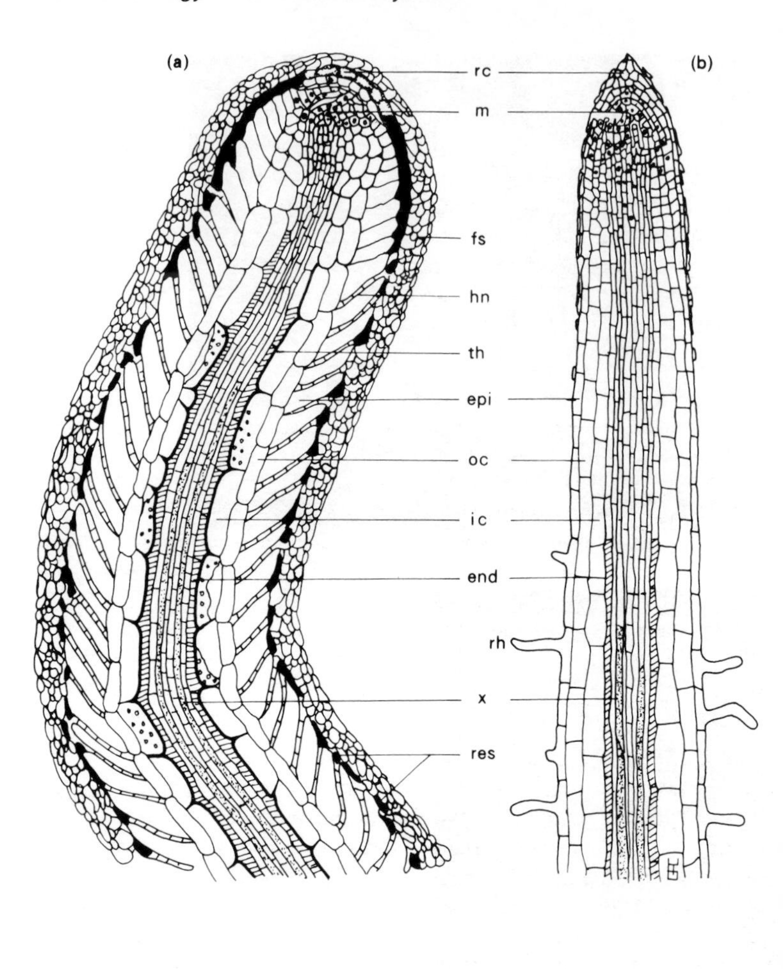
(a)
(b)
rc
m
fs
hn
th
epi
oc
ic
end
rh
x
res

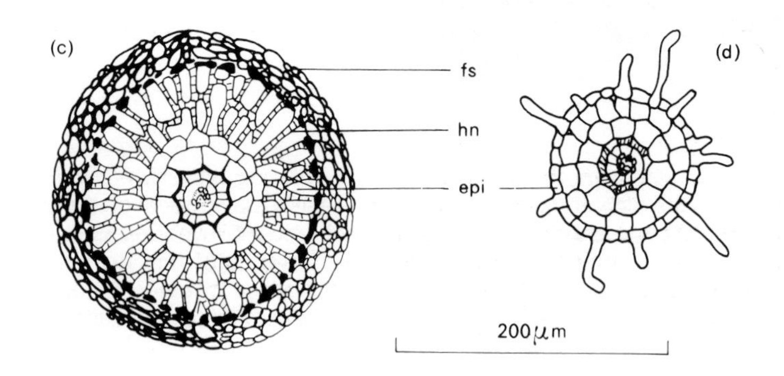
(c)
(d)
fs
hn
epi
200 μm

Long roots may also be subject to colonization but with less apparent morphological modification than occurs in short roots. Where infection of long roots does occur, it is usually limited to the presence of the Hartig net: the sheath is normally (though not always) lacking. Infection of long roots in *Pinus* is of great significance however, since it provides a means whereby the short roots become infected as they emerge from the pericycle. Once the root system is infected, via a young short root, from a source of inoculum in the soil such as a germinating basidiospore, the fungus appears spread intercellularly throughout the cortex of the main root axes (Robertson 1954).

Ectendomycorrhiza and pseudomycorrhiza

These terms are used to designate fungal associations with roots which are characterized by intracellular infection of the host cortex, a coarse Hartig net and a sparse mantle (Moser & Haselwandter 1983). Ectendomycorrhizas are accepted as a distinct form of mycorrhizal symbiosis produced by apparently related strains of fungi, and found mainly on seedlings of *Pinus* and *Larix* in nurseries. Pseudomycorrhizas are regarded as mild pathological infections caused by certain weakly pathogenic fungi that occur normally in the rhizosphere, and which realize their disease-producing potential in the absence of ectomycorrhizal fungi where the seedlings are suffering from some form of physiological stress.

The fungal symbionts

A detailed treatment of the taxonomy of ectomycorrhizal fungi is given by Miller (1982). Most are basidiomycetes but there is evidence that ectomycorrhizas may also be formed by some ascomycetes and zygomycetes. At least twenty-five basidiomycete families are represented, the genera with the broadest host range and most cosmopolitan distribution occurring in the orders Agaricales (*Cortinarius, Amanita, Tricholoma, Boletus, Suillus*), Russales (*Russula, Lactarius*), Hymenogastrales (*Rhizopogon*), Sclerodermatales (*Scleroderma, Pisolithus*) and Aphyllophorales (*Thelephora*).

Fig. 7.2 Anatomy of ectomycorrhizas of *Eucalyptus* compared to uninfected roots. (a), (b) median longitudinal sections of mycorrhiza and non-mycorrhizal root, respectively. (c), (d) transverse sections of mycorrhiza and uninfected root, respectively. rc, root cap; m, meristematic region; fs, fungal sheath or mantle; hn, Hartig net; th, thickened walls of inner cortex; epi, epidermis; oc, outer cortex; ic, inner cortex; end, endodermis; rh, root hair; x, lignified protoxylem; res, collapsed residues of cap cells. (From Chilvers & Pryor 1965.)

Seven families of Ascomycetes are believed to contain ectomycorrhizal fungi, including the imperfect species *Cenococcum geophilum* (syn. *C. graniforme*), which is probably a member of the Eurotiales. The Tuberales is considered by Miller (*loc. cit.*) to be an order of hypogeous, mycorrhizal fungi.

Two species in the Endogonaceae (Mucorales), a family of Zygomycetes which otherwise forms only endomycorrhizas, apparently produce ectomycorrhizas; both belong to the genus *Endogone*.

There have been numerous additions to the list of proven ectomycorrhizal fungi in recent years, and more species will no doubt be added as investigations proceed. Most of the pure culture studies of these fungi have concerned species which form sporocarps regularly and hence can be readily identified. There are however grounds for believing that there are many fungi capable of forming ectomycorrhizas which fruit rarely or not at all, and which, prior to their isolation from infected roots, had never been cultured. As pointed out by Lamb and Richards (1971), these fungi may be more important in mycorrhizal associations in nature than the few dozen species that have been intensively studied in the past.

Specificity of infection

In contrast to the legume–rhizobium symbiosis (Ch. 8), the mycorrhizal association is not one of close specificity. *Cenococcum geophilum* (*graniforme*) and *Pisolithus tinctorius* are among the least specific, being found on a wide range of species both gymnosperm and angiosperm. *Rhizopogon* spp. appear to be restricted to forming mycorrhizas with members of the Pinaceae, while *Suillus grevillei* (syn. *Boletus elegans*) is perhaps the most specific, associating only with *Larix* and a few other coniferous species. Between these extremes are many fungal species of intermediate host range. Most host species seem capable of forming ectomycorrhizas with a number of different fungi, and a single host tree may associate with several species of fungi at the same time. The question of specificity assumes some importance in relation to the introduction of exotic trees in reforestation projects, and it sometimes becomes necessary to introduce the appropriate mycorrhiza-forming fungi, in order to ensure the success of the venture. Various *Eucalyptus* species, and *Pinus radiata*, are widely used as exotics in plantation forestry throughout the world. Malajczuk *et al.* (1982) have recently shown that several eucalypts and *P. radiata* can all form ectomycorrhizas with a group of fungi having a broad host range, but in addition there are some fungi that are host-specific to either *P. radiata* or *Eucalyptus*. Furthermore, there is a tendency for the host-specific fungi to replace those with a broad host range as the trees mature. Such observations point to the need for caution in selecting mycorrhizal associates of tree seedlings for use in forest plantations. While 'close specificity is not a common characteristic of either ectomycorrhizal hosts or fungi' (Harley & Smith 1983), recent studies have raised the possibility that specificity between partners in the ectomycorrhizal symbiosis may be more significant, in terms of evolutionary adaptation, than hitherto believed. Thus R. Molina and J. M. Trappe (1982a) have discerned relationships between particular coniferous genera and certain sections of the genus *Rhizopogon*, whch may indicate coevolution of both host and fungus. If other such examples are found, generalizations about the lack of specificity in ectomycorrhizal infections may be premature.

Formation and abundance of ectomycorrhizas

Recognizable ectomycorrhizas usually appear on seedlings within weeks of germination. The sequence of events is now fairly well understood, though certain details remain to be clarified. It is well illustrated by the study of mycorrhiza formation *in vitro* between the fungus *Piloderma croceum* and Norway spruce, *Picea abies* (Nylund & Unestam 1982). It begins with the selective stimulation of the fungus by root metabolites, followed by the aggregation of hyphae around the root to form a hyphal envelope. This envelope is quite distinct from the sheath or mantle, which forms later, in being constructed of sparsely branched hyphae not closely appressed to the root surface. The third phase involves the penetration of hyphae between the root epidermis and cortex, apparently by mechanical means. Next follows Hartig net formation which is triggered by a morphogenetic change in the fungus, leading to the formation of 'labyrinthic' tissue comprising densely packed, irregularly shaped cells. The Hartig net spreads throughout the cortex, surrounding all its cells but leaving plasmodesmatic connections intact; the endodermis is never invaded nor is the undifferentiated tissue behind the root apex. Finally, the labyrinthic tissue extends to the root surface, covering it with a mantle several layers thick. Eventually, the entire surface of the root is enveloped by the mantle, including the uninfected apex.

This description of the formation of an individual ectomycorrhiza is nicely complemented by Chilvers and Gust's (1982) account of the development of the mycorrhizal root system of *Eucalyptus st-johnii*. Primary infection, from a source of inoculum outside the root, results in the formation of a simple mycorrhiza. During forward growth of the infected root, proliferation of mycorrhizal apices occurs as subsidiary branches emerge from the initial mycor-

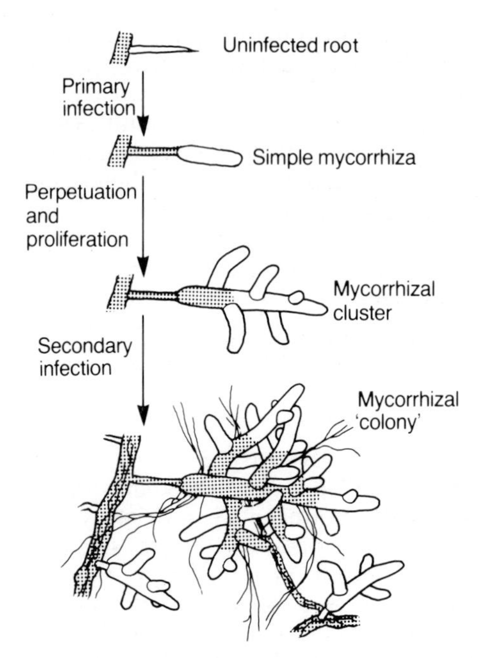

Fig. 7.3 Diagrammatic representation of the organization of ectomycorrhizal apices (simple mycorrhizas) into mycorrhizal clusters and colonies. Shaded and unshaded areas denote older and younger tissues respectively. (From Chilvers & Gust 1982.)

rhizal axis. Subsequently, secondary infections occur from other sources of inoculum, such as hyphal strands, produced elsewhere in the root system. This pattern of development leads to the recognition of three levels of organization within mycorrhizal populations, namely the mycorrhizal apex or tip, mycorrhizal clusters, and mycorrhizal 'colonies' (Fig. 7.3).

Infection and morphogenesis

Although the sequential development of ectomycorrhizas is quite well understood, the actual mechanisms whereby infection is initiated, and their characteristic morphology arises, are complex and our knowledge of them far from complete. The stimulation of fungal growth in the rhizosphere, leading the formation of the hyphal envelope, is presumed to be due to the exudation of growth-promoting substances by the roots, in particular the so-called M-factor (Melin & Das 1954). This substance is also produced by the roots of many other plants, some of which cannot form ectomycorrhizas. Its effect on hyphal growth can be duplicated by nicotinamide dinucleotide (Melin 1962, 1963), although other evidence cited by Slankis (1973) points to its being a cytokinin. Whatever its nature, the M-factor would appear to be a common root metabolite, yet this observation does not accord with its selective action towards mycorrhizal fungi: common soil saprophytes and parasites are unaffected, only prospective mycorrhiza formers being able to react to its presence (Nylund & Unestam 1982).

High levels of auxin are found in ectomycorrhizas and these are of both plant and fungal origin (Slankis 1973). The interaction of auxins and other growth regulators produced by one or both partners is thought by Slankis to be responsible for the changes in root morphology, such as the bifurcation of short roots in *Pinus*, that occur during mycorrhiza formation. However, the ability of short roots of *P. pinaster* to dichotomize was shown by Faye *et al.* (1981) to have a genetic basis, so that the characteristic morphology of ectomycorrhizas may not necessarily be the result of fungal infection. There is also evidence that the internal structure of ectomycorrhizas is due partly to intrinsic factors related to the ageing of root apices (Harley & Smith 1983). Inhibitory substances such as polyphenols (Marks & Foster 1973) and antibiotics (Meyer 1974), produced by host roots, also influence the infection process. These inhibitors may be part of the host's defence barrier against soil-borne infections of all kinds but which the mycorrhizal fungi are adapted to overcome.

Penetration of the fungal hyphae into the middle lamella of the cortex is predominantly if not entirely the result of mechanical pressure generated by osmotic pressure due to the breakdown of glycogen, a fungal storage carbohydrate (Marks & Foster 1973). There is no evidence of enzymatic degradation of cortical cell walls, which indicates that any lytic enzyme activity of the mycosymbiont is suppressed by the host. The infection process ends with the formation of labyrinthic tissue in the cortex, brought about presumably by host cell metabolites. This phenomenon takes place gradually, and the reaction on the part of the host may be induced by the presence of the fungus (Nylund & Unestam 1982). Although both partners to the symbiosis are presumed to have developed a mutual compatibility through the processes of evolutionary adaptation, there seems to be no specific recognition mechanism involved in the initiation of infection, such as is evident in many pathogenic infections or the legume–rhizobium symbiosis (Ch. 8).

Soil and plant factors

Ectomycorrhizas are formed typically by trees of temperate forests growing on brown earth or podzolized soils. They are generally better developed in acid than in neutral or alkaline environments, and are usually more abundant in soil with mor humus than in mull. F. H. Meyer however, reported in 1973 that mycorrhizal development in *Fagus*, as measured by the percentage of mycorrhizal tips, was greater in mull (88%) than in mor (55%); but it should be noted that the absolute number of mycorrhizal tips per unit volume of soil was two orders of magnitude larger in the latter.

An inverse relationship generally exists between nutrient availability and the degree of mycorrhizal infection in *Pinus*: the greater the availability of nutrients, the fewer the mycorrhizas. The effect of increased availability of nitrate is clearly shown in Table 7.1. This table also shows the effect of soil reaction on mycorrhiza development. It has often been assumed that the poor development of mycorrhizas on trees growing in neutral or alkaline soils is due solely to the fact that the mycorrhizal fungi are acidophilic. However, the data tabulated show that mycorrhizas develop more profusely at pH 7.5 where the soil nitrate level is low, than at pH 5.8 where a much greater amount of nitrate is present. Subsequent investigations by Theodorou and Bowen (1969) indicated that high nitrate levels inhibit the infection process, irrespective of soil pH, and that the nitrate effect is enhanced at high pH (8.0) because fungal growth in the rhizosphere is inhibited by the alkaline conditions. In 1981, W. L. Bigg (cited by Harley & Smith 1983) also reported that the nitrate ion could inhibit mycorrhiza formation.

There have been two schools of thought concerning the effect of nutrient supply on mycorrhiza development. A. B. Hatch examined

Table 7.1 Effect of soil nitrate level and soil reaction on mycorrhiza development in *Pinus*

Treatment	*Soil nitrate* (ppm N)	*Soil pH*	*Mycorrhiza* (%)†
Nil	12.7	6.3	25.83 (19.5)
Lime	21.3	7.5	16.86 (8.7)
NH_4NO_3	71.4	5.8	7.21 (2.0)
Lime + NH_4NO_3	115.3	6.9	8.19 (4.3)
LSD for $p < 0.05$	12.3	0.2	5.26
LSD for $p < 0.01$	16.7	0.3	7.25

Source: After Richards and Wilson (1963).

† Transformed data (arcsine transformation) – true values shown in brackets.

this relationship in 1937 and concluded that the predisposing factor in mycorrhiza formation was plant nutrient status, especially that of N, P and K: where the internal concentrations of these nutrients were high and properly balanced, few mycorrhizas would form. Negative correlations between the nitrogen status of seedling pine roots and the degree of mycorrhizal infection have been found by later investigations (e.g. Richards & Wilson 1963, Richards 1965). On the other hand, E. J. Björkman claimed in 1942 that the production of mycorrhizas was conditioned primarily by the presence of free soluble sugars in the root tissues; this occurs whenever carbohydrate synthesis exceeds carbohydrate utilization, and is reflected in a positive correlation between the intensity of infection and the concentration of reducing sugars. Richards (1965) attempted to reconcile the two hypotheses by showing that, in both his and Björkman's data, mycorrhizal infection was better correlated with the ratio of soluble reducing carbohydrates to total N than to reducing substances alone.

The methods of carbohydrate analysis used by Richards, and by Björkman in his 1942 experiments, were relatively crude, and probably did not differentiate between reducing sugars and other reducing substances. In 1970 however, Björkman repeated his earlier experiments using better analytical techniques, with essentially the same results. In 1977, D. H. Marx, in collaboration with A. B. Hatch and J. F. Mendicino, made a detailed re-examination of this matter, and found that variations in the intensity of mycorrhizal infection in *Pinus taeda* seedlings inoculated with *Pisolithus tinctorius* could be attributed largely to variations in the sucrose content of short roots just prior to inoculation. Since these workers paid careful attention to the methods of extraction, purification and analysis of carbohydrates, their findings lend general support to the hypothesis that ectomycorrhizas form under those environmental conditions conducive to the accumulation of free sugars in root tissues.

Ecophysiology of ectomycorrhizal fungi

Knowledge of the cultural behaviour of these fungi began with the researches of the Swedish mycologist Elias Melin in the early 1920s. The results of these and related studies have permitted certain generalizations to be made concerning the ecology and physiology of mycorrhizal fungi. The majority of investigations have concerned their carbon nutrition, and have used standard microbiological laboratory techniques. It should be appreciated that these cultural methods are highly artificial, and far removed from the real world of the soil ecosystem. With this reservation in mind, it seems clear that most ectomycorrhizal fungi grow best on a medium containing simple sugars, and lack the capacity to utilize complex polysaccharides.

Hacskaylo (1973) reviewed studies of the carbohydrate physiology of ectomycorrhizal fungi, and pointed out that conclusions about the breakdown and utilization of various carbon compounds were sometimes invalid, because researchers had not made allowance for the adaptive nature of many enzymes. This characteristic was highlighted by the work of Norkrans (1950), who used small quantities of glucose as a 'start sugar' to demonstrate adaptive growth of ectomycorrhizal fungi on complex carbon compounds. R. J. Lamb (1974) employed this technique to study the ability of twenty-one mycorrhizal fungi of conifers (including some unidentified species isolated from ectomycorrhizas) to utilize a variety of carbon sources. His results, like others before, confirmed the preference of such fungi for simple sugars, the hexoses glucose, mannose and fructose, and the disaccharides cellobiose, trehalose and sucrose all supporting vigorous growth when providing the sole source of carbon. Adding glucose in low concentration (0.1 g $litre^{-1}$) promoted adaptive growth by many of the fungi on the polysaccharides dextrin, glycogen, inulin, starch and cellulose. Several fungi showed adaptive growth after a lag period of incubation (4–6 weeks) even in the absence of start glucose. Lamb also found that pectin could be used as a sole carbon source by all twenty-one fungi, including *Suillus grevellei* and *S. luteus*. Contrasting results were later published by Lindeberg and Lindeberg (1977) who found that neither of these species, nor three other mycorrhizal fungi tested, were able to grow on pectin even in the presence of start glucose.

Strain differences within species may account for some of the divergent results that have been reported, and it is safe to assume that, with few exceptions, ectomycorrhizal fungi show very limited cellulolytic or ligninolytic activity, that the monosaccharides glucose, mannose and fructose are universally good sources of carbon, and that the utilization of disaccharides and the simpler polysaccharides varies within and between species.

It should be borne in mind that most of the pure culture studies of mycorrhizal fungi have concerned species which form sporocarps regularly and hence can be readily identified. There are now grounds for believing that there are many fungi capable of forming ectomycorrhizas that fruit rarely or not at all, but which can be isolated from mycorrhizas by appropriate techniques (Zak & Bryan 1963; Zak & Marx 1964). Some of these unidentified fungi, which are widespread in plantations (Zak & Bryan *loc cit*. Lamb & Richards 1971), along with *Pisolithus tinctorius, Thelephora terrestris, Rhizopogon luteolus* and *R. roseolus*, which are among the dominant mycorrhiza formers in forest nurseries (Marx & Bryan 1970; Lamb 1974), appear to be capable of utilizing quite a wide range of carbon sources.

In addition to the usual requirement for a simple energy substrate, some mycorrhizal fungi need accessory growth factors as well. E. Melin and his associates have shown that pure cultures of most mycorrhizal fungi are partially or completely dependent on thiamine (vitamin B_1) or one of its constituent moieties, pyrimidine or thiazole. Certain species are also deficient in other vitamins. In addition, unidentified growth-promoting substances have been found in root exudates, not only of the coniferous host, but in the exudates of several herbaceous angiosperms (see above under 'Infection and morphogenesis'). Amino acids may also exert a stimulating effect, but the nitrogen requirements of mycorrhizal fungi do not differentiate them from other basidiomycetes. In common with many other fungi and bacteria, they tend to make better growth on ammonium and simple organic nitrogen compounds than on nitrate; some indeed appear to lack the enzyme nitrate reductase. Nor is it likely that they can use the complex organic nitrogen compounds of soil organic matter: Lundeberg (1970) tested the ability of both mycorrhizal and non-mycorrhizal fungi to absorb nitrogen from ^{15}N-labelled humus agar, and found that none of the former could though several of the latter were able to do so.

Neither their carbohydrate physiology, nor their nitrogen nutrition, nor even their requirement for growth factors, set ectomycorrhizal fungi apart as a distinctive ecological group of soil organisms. Although generalizations about their nutritional physiology should not be made solely on the basis of laboratory experiments, the evidence suggests that in nature most of them are probably incapable of leading an independent saprophytic existence in the soil. This conclusion derives from the knowledge that most ectomycorrhizal fungi (unlike many litter-decomposing basidiomycetes) are unable to decompose cellulose or lignin, and require simple sugars as sources of carbon; furthermore, they have poor competitive saprophytic ability, being readily overgrown in culture by fast growing sugar fungi. It thus seems unlikely that ectomycorrhizal fungi can obtain the sugars essential for growth from any source other than the living roots of their hosts. Exceptions to this generalization exist, however, and attention has already been drawn to the ability of some common mycorrhizal fungi in nurseries and plantations to utilize complex polysaccharides. It cannot be inferred from this observation however, that such fungi have more than a limited saprophytic ability, or that they are merely facultative symbionts. In the absence of host roots, they may well survive in the soil in a quiescent state.

Harley and Smith (1983) have emphasized that the nutritional physiology of the fungus in the mycorrhizal condition, where hyphal extension is not taking place rapidly, may be very different from the behaviour displayed by the same species under laboratory conditions conducive to rapid vegetative growth. It seems that the ecophysiology of these ectomycorrhizal fungi is similar to that of many other fungal inhabitants of the inner rhizosphere, and this fact, along with their taxonomic diversity, 'poses in acute form the question of what common property enables them to produce composite organs of similar morphology, histology and function.'

7.2 Endomycorrhiza

A rational discussion of endomycorrhizas is much more difficult than one of ectomycorrhizas, because they do not form such a natural unit and are much more diverse in their morphology. The traditional recognition of two groups of mycorrhizas distinguished between those in which the fungus penetrates the host cells and those where it does not. This system of classification is no longer regarded as satisfactory by some authorities because of the very diversity of endomycorrhizas, not only in their structure but also in respect of the taxonomic status of both fungal symbiont and host plant. J. L. Harley and S. E. Smith (1983) have warned against too readily discarding the older viewpoint, stressing the very great similarity in the structural relationships between host cell and fungus in all endomycorrhizas, and pointing out that this resemblance extends to the ultrastructure of intercellular infection in nitrogen-fixing symbioses (Ch. 8).

The long-established division of mycorrhizas into ecto- and endomycorrhizas is retained in this book: the fact that some of the latter have fungal mantles or sheaths is considered of lesser import than the fact that they are all characterized by an intracellular infection of the cortical cells. On the basis of this definition, ectendomycorrhizas (sect. 7.1) ought to be considered as a form of endomycorrhiza. These symbiotic associations however occur most frequently in seedlings of coniferous trees which are typically ectomycorrhizal in the adult stage. Furthermore, they display some of the structural characteristics of each type of mycorrhiza. Following the recommendation of Harley and Smith (1983), the term ectendomycorrhiza is used in a purely descriptive sense and implies no functional significance. This leaves five distinct kinds of fungus–root symbioses which may be classified as endomycorrhizas:

(a) vesicular-arbuscular mycorrhizas;
(b) arbutoid mycorrhizas;
(c) monotropoid mycorrhizas;
(d) ericoid mycorrhizas;
(e) orchid mycorrhizas.

The attributes of these and, for comparison, ectomycorrhizas are shown in Table 7.2. The various forms of endomycorrhiza will be decribed in turn, with more attention being given to vesicular-arbuscular mycorrhizas because of their potential agronomic and hence economic importance.

7.2.1 Vesicular-arbuscular mycorrhizas

This is by far the most abundant kind of mycorrhiza; indeed T. H. Nicholson (1967) described it as 'a universal plant symbiosis'. Infected roots do not usually show the marked morphological peculiarities common to most ectomycorrhizas. Many look superficially like uninfected roots, quite undistorted and with root hairs. Some however lack root hairs and have a recognizable morphology. They are called **vesicular-arbuscular** (v-a) mycorrhizas, after two kinds of organs – vesicles and arbuscules – which occur in infected tissues (Fig. 7.4). They are found in practically every taxonomic group of plants, and the list of species not infected is probably shorter than the list of those which are. They are common in many staple crop plants, having been recorded from date palms, coffee, tea, cocoa, rubber and citrus. Those gymnosperms which do not have ectomycorrhizas are usually found to have v-a mycorrhizas, for example the coniferous families of the southern hemisphere, Araucariaceae and Podacarpaceae; many members of the Cupressaceae also form v-a mycorrhizas. In addition, they occur in several angiosperm families of great economic importance, such as the Fabaceae, that is Papilionaceae (including those legumes nodulated by the N_2-fixing bacterium, *Rhizobium*), the Rosaceae and the Gramineae. Ferns, lycopods and bryophytes also form v-a mycorrhizas. Some species in a number of genera, *Acacia*, *Casuarina* and *Populus* for example, regularly produce both v-a mycorrhizas and ectomycorrhizas.

Structure and development of v-a mycorrhizas

Infection of roots in v-a mycorrhizas is not continuous but rather

Table 7.2 The characteristics of the important kinds of mycorrhiza. The structural characters given relate to the mature state, not the developing or senescent state

	Kinds of mycorrhiza						
Character	*Vesicular–arbuscular*	*Ecto-mycorrhiza*	*Ectendo-mycorrhiza*	*Arbutoid*	*Mono-tropoid*	*Ericoid*	*Orchid*
Fungi septate	–	+	+	+	+	+	+
Fungi aseptate	+	(+)†	–	–	–	–	–
Hyphae enter cells	+	–	+	+	+	+	+
Fungal sheath present	–	–	+ or –	+	+	–	–
Hartig net formed	–	+	+	+	+	–	–
Hyphal coils in cells	+	–	+	+	–	+	+
Haustoria dichotomous	+	–	–	–	–	–	–
Haustoria not dichotomous	–	–	–	–	+	–	+ or –
Vesicles in cells or tissues	+ (or –)	–	–	–	–	–	–
Achlorophylly	– (or +)	–	–	– (or +)	+	–	+
Fungal taxon	Phyco	Basidio Asco Phyco	Basidio Asco?	Basidio	Basidio	Asco (Basidio)	Basidio
Host taxon	Bryo Pterido Gymno Angio	Gymno Angio	Gymno Angio	Ericales	Mono-trop-aceae	Ericales	Orchid-aceae

Source: From Harley and Smith (1983).

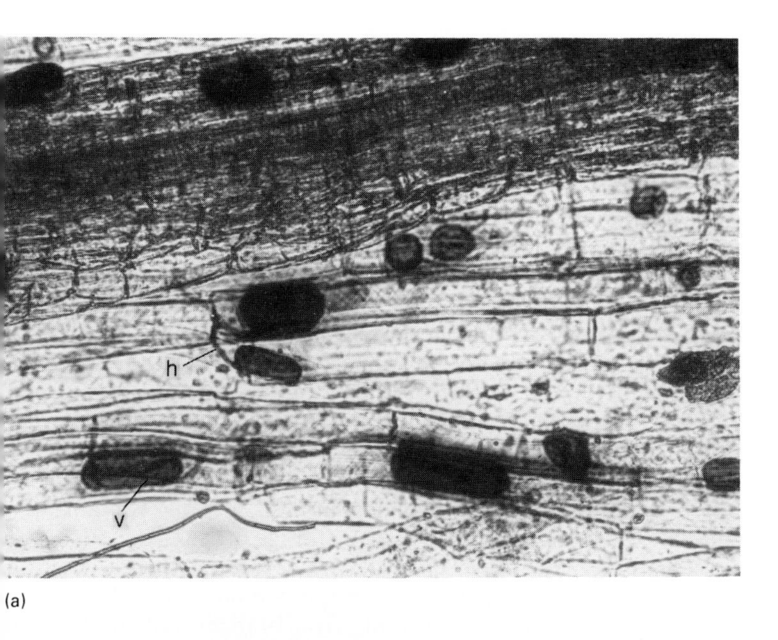

(a)

(b)

Fig. 7.4 Vesicular-arbuscular mycorrhiza. (a) whole mount of mycorrhizal root of *Atriplex*, showing darkly stained fungal hyphae (h) with terminal and intercalary swellings or vesicles (v) (x150). (b) longitudinal section of onion root infected with *Glomus mosseae* showing trunk (at) branches (ab) of arbuscules (x180). (a) Photograph kindly provided by Dr. D. I. Bevege. (b) From Cox & Tinker (1976).

discrete areas of infection are evident, each having a hyphal connection (or connections) with the extramatrical mycelium in the soil. The infecting hyphae may branch before entry and form several appressoria on the root surface. Within the root, the hyphae ramify within and between the cells of unspecialized tissues, such as the root cortex. They do not enter the stele or tissues containing chlorophyll. Within infected cells, coils of hyphae may be formed, or the cells may contain complex branched hyphal systems known as **arbuscules**. Arbuscules are the result of profuse dichotomous branching of hyphae which finally are no longer distinguishable as such but appear as a granulated mass of protoplasm intimately mixed with the protoplasm of the host cell. They are found mainly in the inner cortex and are thought to be involved in nutrient transfer between the symbionts. Intercalary or apical swellings are often found on the main hyphae, and are called **vesicles**. They are sometimes very large and thick walled, and distort the cells or intercellular spaces in which they develop. Vesicles contain large amounts of oil and probably serve as storage organs. There is never any sheath formed around the root surface, although occasionally a loose weft of hyphae may be present, extending up to about 1 cm into the surrounding soil; this results from branching of the infecting hyphae before entry, as described above. Microscopically, the external mycelium is frequently dimorphic with slender, thin walled, septate branches arising from characteristic angular projections on coarser, irregularly thickened, non-septate hyphae. The extramatrical mycelium may also carry spherical or ellipsoid spores usually about 100–300 μm in diameter, though sometimes attaining a diameter of 800 μm; these spores may be contained in rudimentary sporocarps. External vesicles are sometimes present, too. A diagrammatic picture of a v-a mycorrhiza is given in Fig. 7.5.

Specificity of infection

Lack of specificity in v-a mycorrhizas is, if anything, even more characteristic of this symbiosis than it is of ectomycorrhizal associations. 'In general . . . a vesicular-arbuscular fungus isolated from one species of host plant can be expected . . . to infect any other species which has been shown to be capable of forming vesicular-arbuscular mycorrhizas' (Harley & Smith 1983). Notwithstanding this generality, a few of the fungal symbionts of v–a mycorrhizas are thought to have a more restricted host range.

Identity of the endophyte

At the turn of the century, P. A. Dangeard gave the name *Rhizophagus* to a fungus causing v-a mycorrhizas on poplar (*Populus*). In 1923, B. Peyronel reported finding hyphal connections between

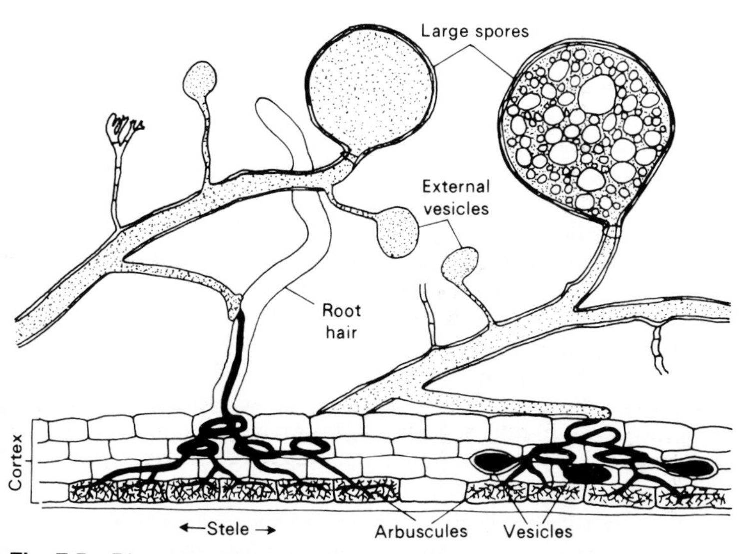

Fig. 7.5 Diagrammatic representation of vesicular-arbuscular mycorrhiza. (From Nicolson 1967.)

mycorrhizal roots of various alpine plants and fruiting bodies of a fungus then known as *Endogone* in nearby soil, and in 1953 and 1956 B. Mosse confirmed Peyronel's observation by finding fructifications of *Endogone* attached to extramatrical mycelium on the roots of strawberry and other fruit plants. J. W. Gerdemann in 1955 described a wet-sieving technique for isolating *Endogone* spores from soil, and showed that v-a mycorrhizas were produced on a wide range of plants following inoculation with these spores. Since that time, it had become apparent that spores of the family Endogonaceae are widespread in soils.

Endogone is a zygomycete of the order Mucorales, some of which produce zygospores in rudimentary sporocarps consisting of a matrix of sterile hyphae, and some of which have free zygospores. The zygosporic species were one of two main groups of *Endogone* recognized by Nicholson and Gerdemann (1968), the other being characterized by the formation of chlamydospores. There are also species in the Endogonaceae which produce both zygospores (or azygospores, i.e. parthenogenic zygospores) and chlamydospores, often within the same sporocarp. Some mycologists once regarded chlamydospores to be azygospores but the consensus of opinion is that the two spore types are quite distinct.

The development of satisfactory methods of recovering spores v-a mycorrhizal fungi from soils, coupled with increased interest in mycorrhizal research, has shown that the Endogonaceae is one of the most common families of soil fungi. As more and more spores were described, it became apparent that the genus to which they were all assigned, namely *Endogone*, was in fact a heterogeneous assemblage of genera and species. J. W. Gerdemann and J. M. Trappe revised the genus in 1974 and several genera are now recognized as containing mycorrhiza-formers. The name *Endogone* is reserved for those species which form zygospores in sporocarps, and there is evidence to suggest (Harley & Smith 1983) that most if not all species of *Endogone* (*sensu stricto*) do not form v-a mycorrhizas at all but are either free-living or ectomycorrhizal.

Four genera recognized by Gerdemann and Trappe form v-a mycorrhizas: *Gigaspora* and *Acaulospora*, which produce azygospores free in the soil, and *Glomus* and *Sclerocystis* which bear chlamydospores; in *Glomus* the chlamydospores occur free or in sporocarps, while in *Sclerocystis* they are borne in a single, uniform layer in a sporocarp. Typical spores and sporocarps of these four main genera of v-a mycorrhizal fungi are illustrated in Fig. 7.6.

More than one fungus may be involved in v-a mycorrhiza formation with a given plant species. This is well illustrated by the work of D. I. Bevege, who in 1971 described ten distinct spore types (representing seven putative species) associated with the mycorrhizal roots of *Araucaria cunninghamii* over a wide geographic range. At that time, he assigned them all to the genus *Endogone*, recognizing some as chlamydospores and some as zygospores or azygospores. From his descriptions, it is apparent that these ten spore types could be assigned to one or other of the genera *Gigaspora*, *Acaulospora* and ***Glomus***, as later defined by Gerdemann and Trappe (*loc. cit.*). Representatives of each of these genera were shown by Bevege, by means of spore inoculation trials, to form v-a mycorrhizas with *A. cunninghamii*.

Life history of the fungus. Repeated attempts to isolate and maintain v-a mycorrhizal fungi in axenic culture have not yet been successful, and consequently they are considered to be ecologically obligate symbionts having little or no saprophytic ability. Vegetative hyphae may survive for some time in senescent or dead roots (Tommerup & Abbott 1981) but any hyphal growth made by v-a mycorrhizal fungi in soil is thought to be dependent on nutrient and energy reserves either stored in spores or accumulated in hyphae from infected roots.

The inability to culture the fungus axenically makes the study of its life cycle difficult. D. I. Bevege (1971), however, devised a technique which permitted him to study the life history of a v-a mycorrhizal fungus in association with the roots of its host, under microbiologically controlled conditions. His observations, which have not been published, cast considerable doubt on the validity of establishing genera on the basis of spore morphology. He grew *A. cunninghamii* seedlings in glass boxes containing a peat–sand medium packed behind a synthetic felt liner and kept moist with nutrient solution. Plants were grown with their roots confined to the space between the liner and the glass wall of the box, and each was inoculated with twelve spores (azygospores?) of a species of *Gigaspora* which is a common endophyte of v-a mycorrhizas of *A. cunninghamii*. This arrangement enabled the development of both roots and fungus to be monitored with a stereo-microscope, and recorded photographically at appropriate intervals of time.

Five weeks after inoculation, many spores were germinating, the germ tubes emerging through the spore wall near the bulbous suspensor. Within 2 weeks of germination, sporangia began to form on short lateral branches to the germ tubes. These sporangia disintegrated after a further 4 weeks, releasing 6–20 sporangio-spores (20–25 μm diameter) into the medium, where they failed to develop further and eventually lysed. Hyphae developing from the germ tubes meanwhile ramified over the root surface, producing infection appressoria 10 weeks after inoculation. Three weeks later, spores 250–350 μm in diameter, and identical with those used as inoculum, formed on hyphae emerging from infected roots several centimetres from the site of inoculation (Fig. 7.7). Some of these spores later germinated, about 8 months after the original inoculum had done so.

Shortly after the formation of the azygospores, clusters of up to twenty vesicles were produced on hyphae emerging from infected roots. Vesicles were spherical-clavate (25 × 18 μm) with either smooth or verrucose walls, and had oily contents; they emptied with age. About the time that the secondary azgospores began to germinate, that is, about eight months after the primary azygospore inoculum germinated, chlamydospores formed in sessile clusters of seven or eight on short hyphal side branches. These spores were much smaller than azygospores, being only 120–200 μm in diameter, and had a single-layered wall 3.5 μm thick in contrast to the three-layered wall of azygospores (these had a 7.5 μm-thick outer wall and two membranous inner walls). Spores identical to these chlamydospores are regularly associated with azygospores of *Gigaspora* in the rhizosphere of *Araucaria* in the field. On the basis of their morphology, they would have to be assigned to the genus *Glomus*.

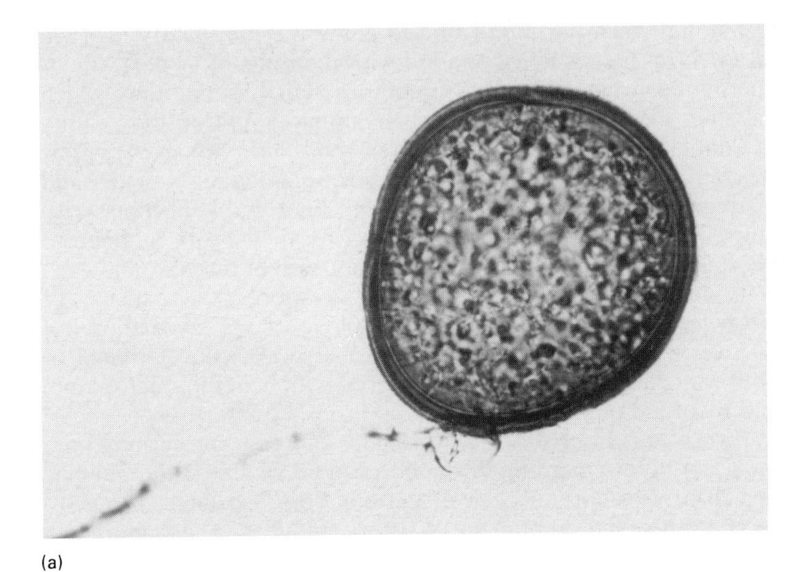

(a)

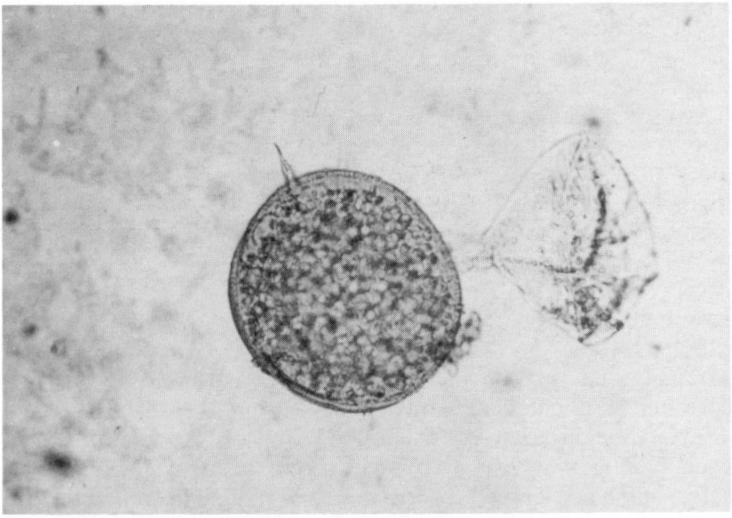

(b)

Fig. 7.6 Representative spores of the four main genera of Endogonaceae known to form v-a mycorrhizas. (a) and (b) azygospores of *Gigaspora* and *Acaulospora*, respectively; (c) chlamydospore of *Glomus*; (d) cross-section of sporocarp of *Sclerocystis* revealing a single layer of spores. (From Hall & Abbott 1981.)

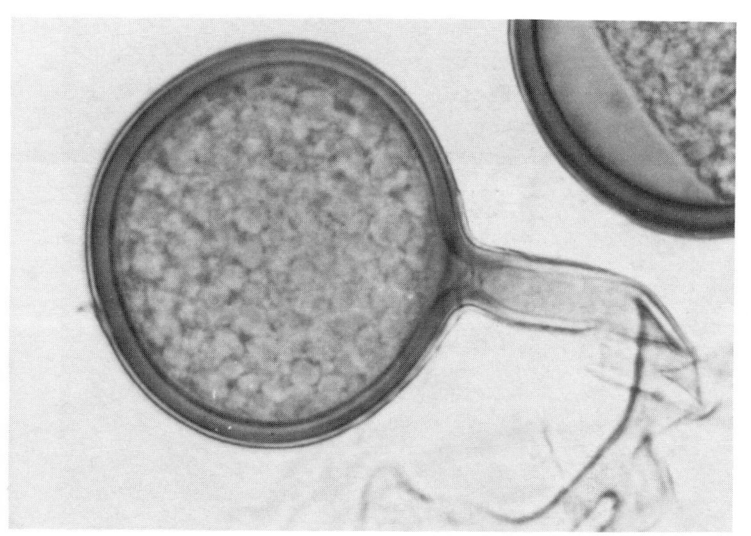

(c)

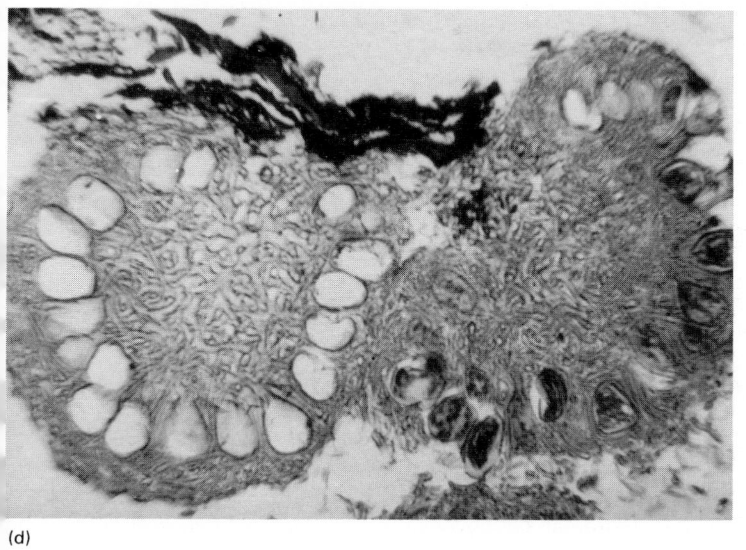

(d)

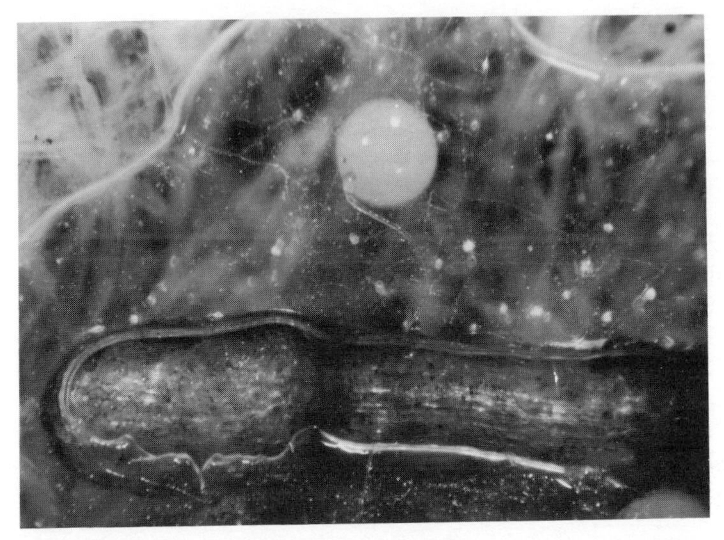

(a)

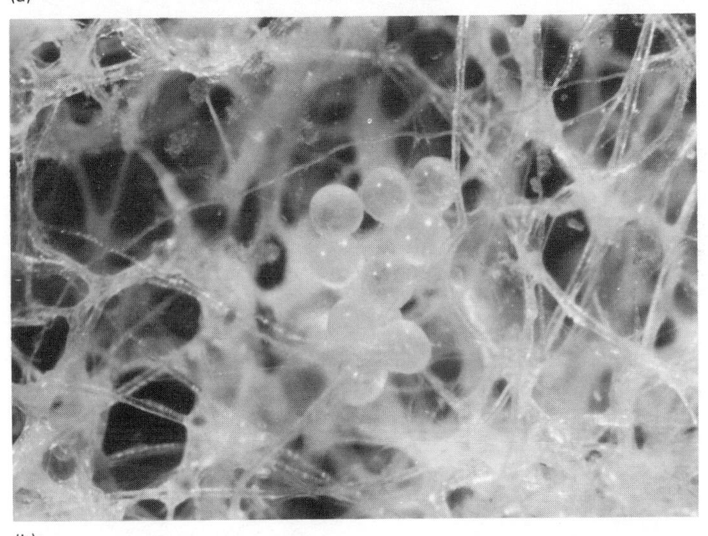

(b)

Fig. 7.7 Extramatrical development of *Gigaspora araucareae* forming v-a mycorrhizas with the conifer, *Araucaria cunninghamii*. (a) Bulbous-based azygospore about 350 µm diameter along with the rootlet from which the subtending hypha has grown; (b) group of sessile chlamydospores approximately 130 µm in diameter. The network of coarse fibres is the material lining the observation boxes. For description of life-history, see text. (From Bevege 1971.)

These observations of D. I. Bevege effectively cover the whole life cycle of the fungus, and demonstrate that two spore types which are so distinct as to be classified as members of different genera, are in fact produced on the same mycelium and must therefore belong to one species. This fungus was shown by Bevege to produce v-a mycorrhizas with *Araucaria* with the formation of arbuscules, vesicles, chlamydospores and azygospores within the fine roots. He proposed that it be designated a new species, *Endogone araucareae*, but the name *Gigaspora araucareae* now seems more appropriate.

Factors affecting the formation of v-a mycorrhizas

The factors affecting endophyte development have not been so widely studied as in ectomycorrhizas. Partly this is due to the difficulty of designing a satisfactory method of assessing the degree or intensity of infection. The preferred method involves a determination of the percentage of root length infected by the fungus, based on a sample of infected root segments of predetermined length (Table 7.3); the precision of method is not necessarily high. Other aspects of fungal development, such as the amount of extramatical hyphae and the size of the spore population in the soil, also pose difficult sampling problems. Nevertheless it is necessary to acquire quantitative data of this kind in order to have a basis for interpreting the ecological significance of the symbiosis.

Both environmental and biotic factors are known to affect the formation of v–a mycorrhizas and the degree of infection of the

Table 7.3 Intensity of mycorrhizal infection in *Araucaria cunninghamii* roots

Source of root	*Root type*	*Percentage infected*[†]	*Vesicles per cm*	*Spores per cm*
Fertilizer trial age 4 yrs	Short	98.0	9.6	0
	Long	48.8	8.7	0
Underplanting age 7 yrs	Short	73.6	10.8	0
	Long	39.9	10.0	0
Pot trial age 2 yrs	Short	88.9	15.8	3.5
	Long	32.6	55.5	2.3

Source: After Bevege (1971).

[†] Percentage infection in long roots is based on the total number of 'short root equivalents', found by dividing the total length of long roots in the sample by the length of the average short root (approx. 1 mm).

host cortex. Large differences between plant species growing under the same conditions have often been discerned. A plot of percentage infection against time is typically sigmoidal (Harley & Smith 1983), the duration of the lag phase being determined by one or more of the following factors: root density, growth rate of fine roots, number of spores per unit volume of soil, spore germination percent, and hyphal growth rate. The interaction of these biotic factors can have a significant effect on the response of crop plants to inoculation since, as several workers (e.g. Daft & Nicolson 1969; Rich & Bird 1974) have emphasized, early infection is all-important in this context. Environmental factors affecting the formation of v-a mycorrhizas include nutrient supply and balance, light, temperature and soil pH. There is generally an inverse relationship between soil phosphate supply and the intensity of infection but the effect of phosphorus may be influenced by the supply of other nutrients, especially nitrogen. Factors which might increase the soluble carbohydrate concentration in roots, such as increased light intensity, also increase mycorrhizal infection, but the underlying relationships are not as well understood as they are in ectomycorrhizas.

The organization of v-a mycorrhizas is complex, and the several phases of intramatrical and extramatrical endophyte activity – hyphae, vesicles, arbuscles and spores – may react to the carbohydrate and nutritional status of the host in different ways. Furthermore, a knowledge of plant physiology suggests a high degree of positive feedback involving growth, nutritional status and carbohydrate availability. The interrelationships among the many participating processes, both host and endophyte, make it difficult if not impossible to determine causal relationships between environmental factors and the formation and development of v-a mycorrhizas.

Infection and morphogenesis

Infection is initiated by hyphae originating from a germinating spore or from a nearby mycorrhizal root. A generalized account of the process has already been given when describing the life-cycle of *Gigaspora araucareae*. The rudimentary mycelium arising from a spore appears to be stimulated by the proximity of a susceptible root into branching to form delta-like complexes of fine (2–7 μm diameter) non-septate hyphae. When the lateral hyphae, or sometimes the main hypha, contact the root, they produce appressoria from which infection pegs protrude through the epidermis. Infection appears to occur most readily in newly differentiated tissue just behind the root tip, and occasionally it takes place via a root hair. After initial penetration, the hypha passes into the outer cortex, where it branches and spreads longitudinally in both directions for a distance of some 5–10 mm from the point of entry, so that a discrete region of infection is formed. As indicated earlier, infection is not systemic but consists of a series of such disjunct infection units. Each new rootlet must therefore be infected afresh from the soil, either from another source of primary inoculum or from hyphae emerging from the parent root. Secondary infections from the extramatrical hyphae, which may be quite extensive, are common.

Within the cortical cells, the hyphae branch dichotomously to form arbuscules. In some species, the appearance of arbuscules is preceded by the formation of hyphal coils in the outer cortex. The fine structure of an arbuscule is illustrated in Fig. 7.8. The mechanism by which the fungal hypha penetrates the cell wall is not known for certain. Barbara Mosse (1962) presented evidence that pectinases facilitated entry into the root but their presence is not essential to infection in *A. cunninghamii* (Bevege 1971). According to Harley and Smith (1983), it is likely that initial infection can occur only in the younger parts of the root before the cell walls are fully formed. These authors suggest that the ability of hyphae to penetrate host cells and tissues is not due to enzymatic activity on the part of the fungus but rather lies in its ability to inhibit the host enzymes needed for cell wall synthesis.

Studies with the election microscope have greatly enhanced an understanding of host–endophyte interactions in the formation of arbuscules. The penetrating hypha does not breach the plasmolemma but invaginates it: even the ultimate branch tips of the arbuscule, which are less than 1 μm in diameter, remain surrounded by the host plasmalemma. Between this and the hyphal wall lies the 'interfacial matrix', which is continuous with the periplasm of the host and comprises scattered polysaccharide fibrils and membranous vesicles; the latter seemingly derive from the host plasmalemma. The arbuscule eventually occupies the greater part of the cortical cell (Fig. 7.8) and provides a very large contact area for exchange of nutrients and metabolites between fungus and root.

Both the host cell and the arbuscule are physiologically very active, as evidenced by the appearance of their nuclei and organelles under the electron microscope. Arbuscules are not long-lived however, and after a few days or at most two weeks, they degenerate into nebulous clumps of tissue which J. M. Janse observed as early as 1897 and called sporangioles. It was formerly believed that the presence of sporangioles indicated that the

(a)

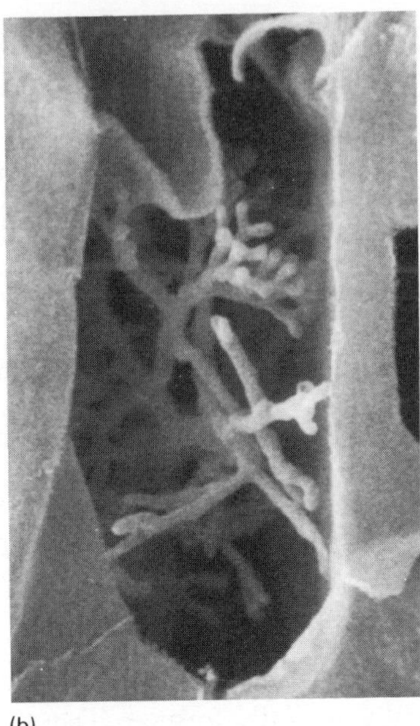
(b)

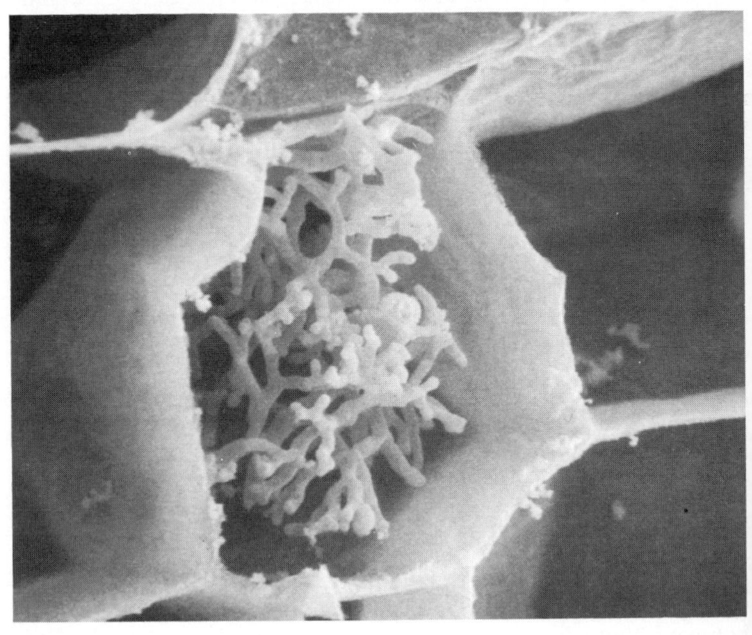
(c)

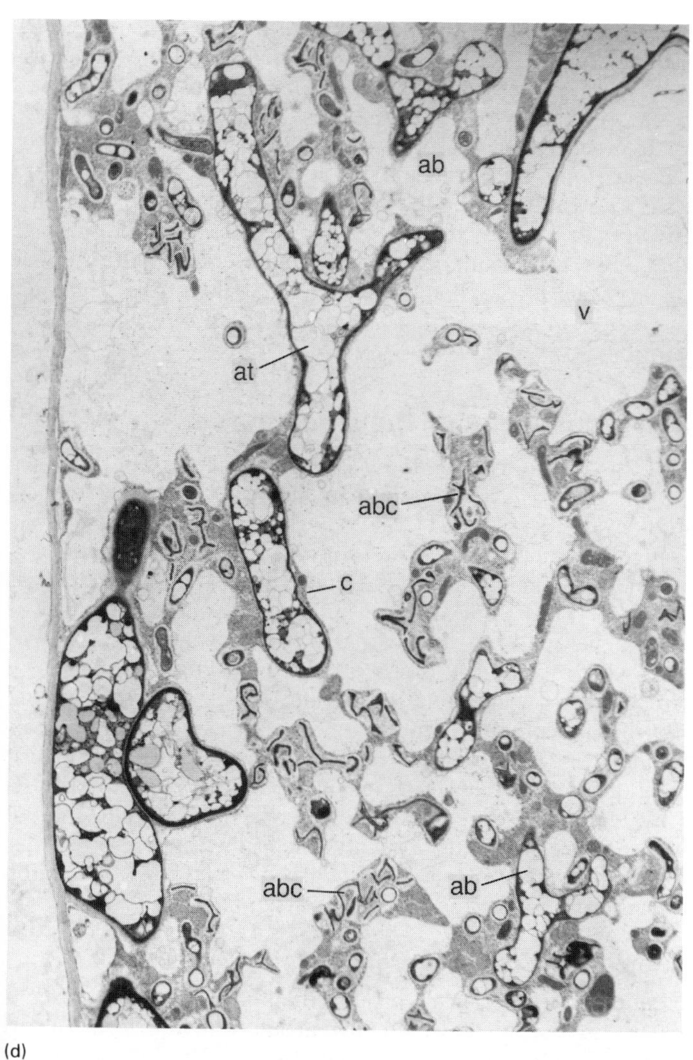

(d)

Fig. 7.8 Fine structure of arbuscules of v-a mycorrhizas (a) (b) (c) scanning electron micrographs of cells of *Liriodendron tulipifera*, showing stages in the development of arbuscules of *Glomus mosseae* (x2000), (d) transmission electron micrograph of mycorrhizal root of onion infected with *Glomus mosseae*, showing sections through arbuscules (x3900): at, trunk; ab, branches; abc, collapsed branches indicating the onset of senescence; the arbuscule is located in the host vacuole (v) and is surrounded by host cytoplasm (c). (a,b,c) from Kinden and Brown (1975) (d) photograph kindly supplied by Dr. G. Cox of the Electron Microscope Unit, University of Sydney.

arbuscule was 'digested' by the host as a means of acquiring the nutrients taken up from the soil by the fungus. It is now held that bidirectional transfer of materials takes place across the interfacial matrix of young and mature arbuscules; Harley and Smith (1983) suggest that this might be supplemented by similar activity at the interface surrounding intracellular coils or intercellular hyphae.

Less is known about the mechanism of formation of that other definitive organ of v-a mycorrhizas, the vesicle. It is evident however that these intercellular or intracellular swellings of the hyphae contain large amounts of lipid. They are found mainly in the middle or outer cortex in the older parts of infection units, and serve as a means of energy storage.

7.2.2 Mycorrhizas of the Ericales

Three of the five kinds of endomycorrhiza recognized in the introduction to section 7.2. are found in the order Ericales, namely arbutoid, monotropoid and ericoid mycorrhizas. They exhibit a wide range of structure and function, yet resemble in some ways ectomycorrhizas on the one hand and v-a or orchid mycorrhizas on the other. Thus both arbutoid and monotropoid mycorrhizas have well developed fungal sheaths and Hartig nets, and many of the fungi which form arbutoid mycorrhizas on one host can form ectomycorrhizas on another. Again, the infected cells of ericoid and arbutoid mycorrhizas contain fungal coils as do orchid mycorrhizas (and to a lesser degree, v-a mycorrhizas), and the fungi of monotropoid mycorrhizas transport carbohydrates into their hosts as do the endophytes of orchid mycorrhizas. The occurrence of mycorrhizas so variable in form and function within the one order of flowering plants poses some intriguing questions about the origin and evolution of the mycorrhizal symbiosis.

Ericoid mycorrhizas

These endomycorrhizas are typical of the family Ericaceae, except for the tribe Arbutoideae which have arbutoid mycorrhizas (q.v.). Many well known genera in the tribes Rhododendroideae (*Rhododendron*), Ericoideae (*Erica, Calluna*) and Vaccinioideae (*Vaccinium, Pernettya*) possess ericoid mycorrhizas. So too do many genera in the related families Epachridaceae (*Monotoca, Leucopogon, Epachris*) and Empetraceae. Most of these plants have relatively fine roots (very fine in some species), and they are typically woody shrubs or small trees found mainly in heaths on acid or peaty soils. The fungus is usually restricted to the outermost layer of cortical cells (which in some species may constitute the entire cortex); hyphae never penetrate the endodermis and reach the stele.

Infection begins with fine, septate hyphae forming a loose weft over the root surface but no sheath is formed. The hyphae pass through host cell walls without first developing appressoria; intercellular penetration seems to be aided by fungal pectinases. Within the cortical cells, the fungi form extensive coils which are surrounded by the host plasmalemma and an interfacial matrix of pectic material. Although the greater part of the root system in these plants is mycorrhizal, infection is not systemic in the cortex; rather the hyphae spread along the root surface and branch to infect cortical cells at discrete intervals. The fungus and the infected plant cell appear to remain physiologically active for a period of some weeks, but eventually both deteriorate. Unlike the cessation of activity in v-a mycorrhizas, however, where degeneration of the arbuscule occurs while the host cell remains intact, in ericoid mycorrhizas the cell degenerates followed by the hypha.

Ericoid mycorrhizas have been synthesized on various susceptible host species by inoculation with fungi isolated from infected roots. These fungi, which were first brought into pure culture by K. D. Doak in 1928, are sterile, slow-growing, dark mycelia. Pearson and Read (1973) found a lack of specificity among the fungal endophytes isolated from species of *Vaccinium, Erica, Rhododendron* and *Calluna*. When growing symbiotically under aseptic conditions, one of the isolates produced apothecia which are ascocarps (see sect. 1.7.2) of the type characteristic of a certain group of Ascomycetes; it was subsequently described by Read (1974) as *Pezizella ericae*. It seems likely that the endophytes of ericoid mycorrhizas may all be Ascomycetes of the genus *Pezizella*, and there may indeed be only one species participating. Harley and Smith (1983), however, draw attention to evidence implicating Basidiomycetes as well in the formation of ericoid mycorrhizas.

Arbutoid mycorrhizas

Mycorrhizas of the tribe Arbutoideae (family Ericaceae) were first described from *Arbutus unedo* by M. Rivett in 1924; the host plants are mostly woody shrubs and trees. Root systems of the Arbutoideae, unlike those of most ericaceous plants, are typically heterorhizic, the short roots being converted to mycorrhizas with a well defined sheath and a Hartig net in the outer cortex. From the Hartig net, the fungus penetrates the cortical cells where it forms extensive coils of hyphae.

The fungal symbionts of arbutoid mycorrhizas are Basidiomycetes (Table 7.2). Many of the species which enter into this kind of symbiosis with green plants of the genera *Arbutus* and *Arctostaphylos* form ectomycorrhizas with coniferous trees (Zak 1976; Molina & Trappe 1982b). The ability of the same fungus to cause intracellular infection of their ericaceous hosts and intercellular infection in conifers reflects contrasting host–endophyte responses, the biochemical basis of which remains to be determined. The physiological attributes of arbutoid mycorrhizas are presumed to be similar to those of ectomycorrhizas, and this being so, they would be expected to confer the same ecological advantages on their hosts.

Monotropoid mycorrhizas

The Monotropaceae is a family of achlorophyllous plants which are completely dependent on their mycorrhizal fungi for supplies of carbon and energy. The ultrastructure and development of the mycorrhizas of *Monotropa hypopitys* were described by Duddridge and Read (1982). Its roots form a 'root-ball' throughout which a fungal mycelium ramifies, enclosing the ectomycorrhizal roots of neighbouring green plants. The root-ball is the survival organ of *Monotropa* during the winter months, and gives rise to flowering shoots (scapes) from adventitious buds each spring. As the root grows, a sheath and Hartig net are formed, and peg-like haustoria push into the epidermal and cortical cells from the hyphae, apparently at the expense of glycogen stored in the sheath. Initially the host cell walls invaginate to enclose these fungal 'pegs' but later the pegs penetrate the walls and emerge into the cells. As the scapes approach senescence, the pegs burst at their tips, releasing their contents into sacs enclosed by the host plasmalemma. These sacs contain groups of organelles (endoplasmic reticulum, mitochondria) of host origin and later, as the pegs collapse, fungal protoplast. At this stage, hyphae from the sheath invade and colonize the senescent cortical cells. Thus the structure and function of the monotropoid mycorrhiza changes in concert with the seasonal development of the host plant.

As long ago as 1881, F. Kamienski observed that the roots of *Monotropa* were completely enveloped in fungal tissue, and surmised that the plant depended for its nourishment on hyphal connections with nearby trees. This was confirmed by E. J. Björkman (1960) who injected ^{14}C- and ^{32}P-labelled orthophosphate into ectomycorrhizal spruce and pine trees in the field, and several days later detected ^{14}C and ^{32}P in the tissues of *Monotropa* growing beneath them.

7.2.3 Orchid mycorrhizas

A detailed description of orchid mycorrhizas is beyond the scope of this book. The account which follows is taken from Harley and Smith (1983) which should be consulted for further information.

The family Orchidaceae contains many thousands of species widely distributed around the world. In most flowering plants, the seedling stage during which the germinate is dependent on seed reserves is relatively short, but in orchids the seeds are very small (only a few micrograms in weight) and have very limited carbohydrate reserves, and consequently the period during which the seedling is unable to photosynthesize is prolonged. Indeed some orchids lack chlorophyll throughout their whole lifetime. Orchid embryos are virtually undifferentiated and, though they imbibe water and swell, do not develop unless they are supplied with soluble carbohydrate or are infected with suitable mycorrhizal fungi. Some species require an exogenous supply of vitamins and other growth factors as well as sugars, if they are to grow asymbiotically. The monosaccharides D-glucose and D-fructose, and the disaccharide sucrose, which are readily metabolized by all angiosperms, are suitable energy sources. The fungal disaccharide trehalose is also probably satisfactory for most orchid seedlings, but another common fungal metabolite, mannitol (a polyhydric alcohol), does not support the further development of some orchid embryos. While there is no doubt that mycorrhizal infection is not obligatory for the successful culture of orchid seedlings, under natural conditions the necessary soluble sugars must be provided by the fungal symbiont. The carbon thus derived permits the seedling to continue its development, perhaps for years, before any green leaves are formed. Even after they have begun to photosynthesize, it is not known for certain whether orchids remain dependent, wholly or in part, on carbohydrate from their fungal symbionts. Mature orchids usually have mycorrhizal roots irrespective of whether or not they contain chlorophyll, and there is indirect evidence, for some terrestrial forms at least, that they must continue to be supplied with carbon by their symbiotic fungi, for they are able to maintain a subterranean existence for many years and continue to flower periodically, even if growing in heavy shade. It is the net movement of carbon from substrate through fungus to host that sets orchid mycorrhizas apart from all other mycorrhizal symbioses, with the exception of monotropoid forms.

The fungal symbionts

The fungi of orchid mycorrhizas are Basidiomycetes. They are relatively easy to isolate from roots and many belong to the form genus *Rhizoctonia*. J. H. Warcup (1981) induced many *Rhizoctonia* isolates of Australian orchids to form perfect stages, and assigned them to various genera including *Thanatephorus* (*Corticium*) and *Ceratobasidium* which contain strains pathogenic to crop plants. Other well known plant parasites, such as *Armillariella* (*Armillaria*) *mellea*, and many wood-rotting fungi including *Coriolus versicolor* and *Fomes* spp., are also found to be mycorrhizal with orchids. Most of these fungi produce exoenzymes which hydrolyse one or more of the complex polysaccharides starch, pectin, cellulose or lignin.

Specificity

As with other kinds of mycorrhizal symbioses, there is a range of specificities in orchid mycorrhizas. *Caladenia* spp. are normally associated with the fungus *Sebacina vermifera* whereas *Dactylorchis purpurella* can be infected by many different fungi. Similarly, most fungal symbionts can infect a variety of host orchids, and even *S. vermifera* is not restricted to association with *Caladenia*. In considering specificity in orchid mycorrhiza, Harley and Smith (*loc. cit.*) have emphasized the point that this attribute has usually been judged on the effect of a fungus on early seedling growth rather than on its ability to form mycorrhizas with adult plant roots.

Infection and anatomy

There are two phases of infection in a mycorrhizal orchid, namely primary infection of the germinating seedling and secondary infection of new roots of the adult plant. In germinating, the orchid embryo absorbs water, swells, bursts the testa and produces a few epidermal hairs. Fungal hyphae penetrate the walls of these hairs or those of epidermal cells near the suspensor of the embryo. Infection spreads throughout the basal region of the protoderm (the meristematic tissue which developes into the underground stem or corm), and extensive hyphal coils or 'pelotons' form within the cells. The pelotons are surrounded by the host cytoplasm and plasmalemma. They remain active for a few days only, then degenerate; finally, the identity of the hyphae is lost and their walls become diffuse. At this stage, the host cell may become reinfected and the process repeated several times.

There is considerable variation in the degree to which adult, chlorophyll-bearing orchids are mycorrhizal. Some species are reported to be only sporadically infected but there is a clearly recognizable infection cycle in others. Reinfection of new roots of adult orchids takes place from the soil or, less frequently, from perennating tubers or rhizomes. This occurs regularly after dormancy and especially in spring, with infection spreading during the summer months and becoming most evident in autumn.

There is no external sheath of fungal tissue, nor any Hartig net, in orchid mycorrhizas. From a functional standpoint, Harley and Smith characterize the symbiosis as a three-phase system: the external hyphae in the soil, absorbing carbon compounds and other nutrients; an association of active hyphae and host cells; and a digestive phase in which the hyphae collapse and die, followed by death of the host cells.

7.3 Nutrient and metabolite interchange between symbionts

If mycorrhizal associations have any regulatory role in nature, it is likely to be through their effect on one of the major functional processes in ecosystems, namely energy flow or nutrient circulation. For example, if the presence of mycorrhizas increases the rate of transfer of chemical elements from lithosphere to biosphere, one might be justified in assuming that this particular plant–microbe interaction was in fact significant in ecosystem regulation.

Effect of ectomycorrhizas on yield and nutrient uptake

Since most ectotrophic mycorrhizal fungi are unable to utilize complex polysaccharides, they are unlikely to promote rapid breakdown of litter and humus and so affect the nutrition of their hosts in this way. There are however exceptions to this generality, for example several species of *Tricholoma* have the capacity to utilize cellulose as a carbon and energy source (Norkrans 1950). It is not inconceivable that, among the many unidentified fungi that have been isolated directly from mycorrhizas during the past two decades, there are others with litter decomposing capabilities; certainly many of them can utilize a wide range of carbon sources in pure culture (sect. 7.1). Furthermore, many mycorrhizal fungi produce phytases (Theodorou 1971; Ho & Zak 1979), and the sheaths of ectomycorrhizas in nature show phosphatase activity (Bartlettt & Lewis 1973; Alexander & Hardy 1981), which might give them access to the insoluble inositol phosphates of litter (sect. 5.2.1). If this is so, then it is not unreasonable to suppose further that trees harbouring such symbionts might be to some degree independent of normal mineralization processes. This

Table 7.4 Effect of ectomycorrhiza on yield and nutrient uptake of *Pinus elliottii*

Fungus	*Dry weight of seedlings* (g)†	*Nutrient content of seedlings* (mg)†		
		N	P	K
Nil (uninoculated)	0.26	2.44	0.22	2.27
Rhizopogon roseolus	1.39	13.03	1.37	12.54
Suillus granulatus	1.58	15.26	1.64	14.80
E8.22‡	2.81	38.92	3.86	35.60

Source: After Lamb and Richards (1971).
† Differences between treatments highly significant in all cases.
‡ Unidentified fungus isolated from roots of mature *P. elliottii*.

hypothesis remains to be substantiated, whereas there is much accumulated knowledge which indicates that mycorrhizal fungi can influence the nutritional status of their hosts in other ways.

There is considerable experimental evidence to support the view that ectomycorrhizas function very efficiently as nutrient-absorbing organs. This is illustrated by Table 7.4, which shows clearly the superiority of mycorrhizal over non-mycorrhizal plants, when both are grown in a peat–sand substrate at low levels of available nutrients. This table also indicates that certain unidentified fungi may be more effective symbionts, in the seedling stage at least, than some of the better known mycorrhizal fungi. There is also evidence that strains of the same fungal species may differ in their effectiveness (Marx 1979). If the stimulus to seedling growth afforded by ectomycorrhizas persists until the trees reach merchantable size, then considerable economic benefit might be derived by manipulating the symbiosis to achieve earlier or higher yields in afforestation projects.

One of the first to demonstrate the beneficial effects of ectomycorrhizas was A. B. Hatch, who in 1937 pointed out that infection of roots by mycorrhizal fungi increased the effective root surface for the absorption of nutrients. This was brought about in three ways: the life of infected roots was prolonged, their degree of branching was enhanced, and their diameters were increased. A second effect of mycorrhizal infection of great significance in nutrient absorption, is that a layer of fungal pseudoparenchyma is interposed between the root surface and the soil. When considering nutrient interchange between host and fungus, it must therefore be kept in mind, as J. L. Harley (1969) has stressed, that a typical ectomycorrhiza consists of four interconnected tissue systems:

(i) an external mycelium in the soil,
(ii) the fungal sheath or mantle,
(iii) the hyphae ramifying between the epidermal and cortical cells of the host and forming the Hartig net, and
(iv) the host cortex and stele.

E. Melin and H. Nilsson, in a series of papers published during the 1950s, described an elegant technique to demonstrate that hyphae emanating from the mycelial sheath can transport nutrients to the roots of the host. They grew pine seedlings in aseptic sand cultures and then introduced pure cultures of mycorrhizal fungi in shallow dishes placed on the surface of the sand (Fig. 7.9). In the course of time the hyphae grew over the edge of the dish, through the sand, and eventually formed mycorrhizas on the rootlets. By inserting isotopically labelled nutrient solutions in the dishes, they demonstrated the movement of PO_4^{3-}, Ca^{2+}, NH_4^+ and

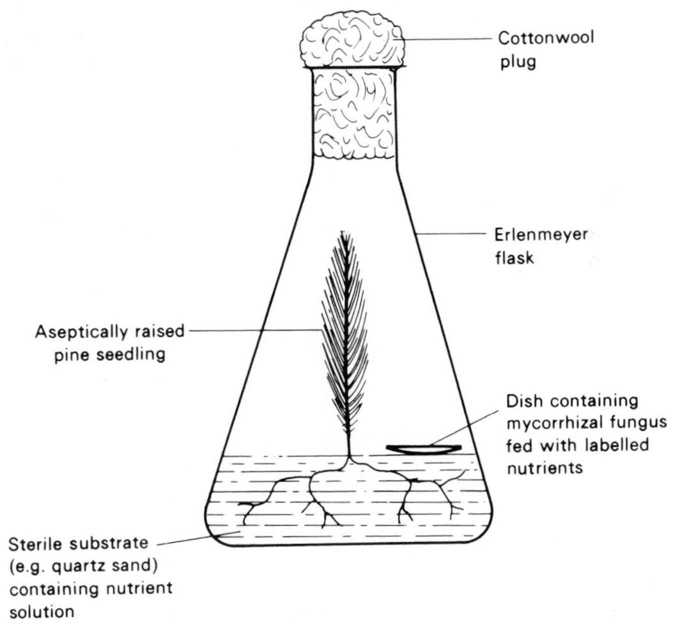

Fig. 7.9 Technique used by E. Melin and H. Nilsson to demonstrate transport of nutrients to pine seedlings through extramatrical hyphae of ectomycorrhizas. For explanation, see text.

amino-N through the hyphae to the roots and into the tops of the seedlings. Just what contribution translocation of nutrients through fungal hyphae from zones of uptake remote from the root surface makes in nature has never been properly assessed, although Stone (1950) was able to show that Monterey pine seedlings with an extensive system of extramatrical hyphae absorbed soil phosphate more rapidly than those with less well developed hyphal connections. In coarse textured mineral soils, hyphae often ramify several centimetres away from mycorrhizas. In the litter layer however, while there may be extensive development of extramatrical hyphae at the L–F horizon interface, this is by no means universal. Skinner and Bowen (1974) observed that soil conditions had a marked effect on the amount of extramatrical mycelium produced by any given strain of fungus, and furthermore they found large differences between strains. In some circumstances, there is little growth of hyphae from *Pinus radiata* mycorrhizas in litter, up to 0.1 mm only; even this may be ecologically significant, of course.

Mechanisms of ion absorption

J. L. Harley and his students at Oxford have elucidated many of the mechanisms of nutrient absorption, by using excised mycorrhizas of beech (*Fagus sylvatica*) in short-term experiments (see, e.g. Harley & McCready 1952a, b; Harley *et al.* 1956; Harley & Wilson 1959). Rates of absorption of infected roots are about five times as great for PO_4^{3-}, and about twice as great for K^+, as in uninfected roots. As in the study of nutrient uptake by excised roots of non-mycorrhizal plants, it has been found that absorption is an aerobic process, and the process of ion accumulation is linked to respiration, occurring only at the expense of metabolic energy. Mycorrhizal roots are also similar to uninfected roots in that their absorption rate is influenced by temperature, and will not proceed in the presence of metabolic inhibitors. Furthermore, both mycorrhizal roots and non-mycorrhizal roots show the property of selective absorption, such as preferential uptake of K^+ from a mixture of metallic cations.

In summary therefore, Harley's studies have shown the process of nutrient uptake by mycorrhizas is similar to that of other absorbing organs, though seemingly more efficient. The presence of the fungal sheath, however, complicates this simple view of mycorrhizas as highly efficient ion-absorbing organs. A large proportion of the ions absorbed are retained initially in the sheath or mantle, as much as 90 per cent in the case of PO_4^{3-}. Release of accumulated PO_4^{3-} to the whole system is a gradual process, dependent on temperature and oxygen supply. Apparently the transfer of PO_4^{3-} from sheath to core is, like uptake from an external solution,

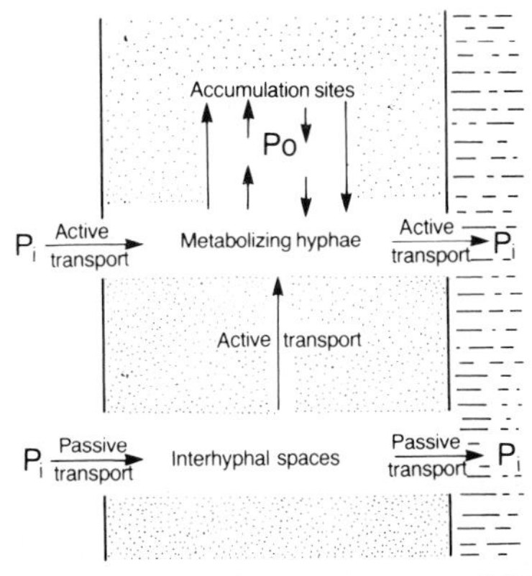

Fig. 7.10 Diagrammatic view of the passage of phosphate through the sheath of ectomycorrhiza. (Adapted from Harley 1969.)

dependent upon the expenditure of metabolic energy in respiration. It is most rapid when PO_4^{3-} is absent from the external medium.

Harley's postulated scheme for the uptake of phosphate from the soil, and its passage through the sheath, is shown in Fig. 7.10. Phosphate is absorbed actively by the sheath and the extramatrical hyphae, and accumulated in the sheath largely as polyphosphate (Harley & McCready 1981). Once the accumulation sites are fully occupied, phosphate may diffuse directly to the host tissue provided the concentration gradient between the external solution and the plant cells is sufficiently steep. At the phosphate concentrations which normally pertain in the soil solution however, this is unlikely to occur, so that ecologically the diffusional pathway is insignificant. Instead, it is highly probable that, under natural conditions, the host draws its phosphate from the fungal sheath by mechanisms involving remobilization of polyphosphate and active transport to the cortical tissues.

It must be emphasized that the pattern of nutrient absorption in mycorrhizas is not qualitatively different from that which exists in uninfected roots, only quantitatively so. The latter are quite capable of accumulating and storing nutrients in the cortex before passing them on to the shoot. As J. L. Harley and D. H. Lewis (1969) have succinctly stated, 'mycorrhizas differ essentially from the root by possessing an extra fungal cortex into which very rapid accumulation is possible'.

Other workers, for example Morrison (1962a), using intact pine seedlings, have confirmed the findings of the Oxford group on the uptake of PO_4^{3-} by excised beech mycorrhizas. These whole-plant experiments clearly show the greater ability of mycorrhizal plants to take up PO_4^{3-} from the substrate, and reinforce the view that the sheath acts as a primary reservoir of accumulation from which PO_4^{3-} passes steadily into the host. A somewhat different pattern of absorption has been found for sulphate however, both with intact pine seedlings and excised beech mycorrhizas, indicating that this anion is not metabolized by the fungus, but has a free pathway through the sheath into the host tissue (Morrison 1962b, 1963).

In the study of ion absorption by mycorrhizas, greatest emphasis has been placed on phosphate. This is partly because of the ready availability of the radioisotope ^{32}P and its suitability for short term uptake studies, and partly because phosphorus plays a leading role in metabolic processes. However, in 1937 A. B. Hatch drew attention to the fact that ectomycorrhizas could benefit their hosts by promoting the uptake of any limiting nutrient, and J. L. Harley has frequently stressed the fact that, since the fungal sheath physically isolates root tissues from the soil, any element which passes through the sheath may be affected by it (see, e.g. Harley & Smith 1983, p. 202).

Ectomycorrhizal fungi can use ammonium and some simple forms of organic nitrogen, including amino compounds, in pure culture but their capacity to utilize nitrate is limited and varied (sect. 7.1). Nitrate is not likely to be abundant in the typical habitats of ectomycorrhizal trees, which are acid soils often high in phenolic derivatives (see sect. 5.2.2). The complex organic nitrogen compounds found in humus are unsatisfactory sources of nitrogen for mycorrhizal fungi in culture (Lundeberg 1970). The most important source of nitrogen for ectomycorrhizal plants in nature is ammonium, and the absorption of this ion by excised beech mycorrhizas has been studied by Carrodus (1966, 1967). Uptake of NH_4^+ is stimulated by exogenous sugars because it is rapidly assimilated into amino compounds, especially glutamine. In this it differs quantitatively, though not qualitatively, from phosphate uptake for which process there are usually adequate carbohydrate reserves for the effect of exogenous sugars not to be readily apparent in short term experiments. In nature, the fungal storage carbohydrates in the sheath – trehalose, glycogen and mannitol (see p. 301) – are mobilized to provide glucose and fructose as substrates for respiration and carbon sources for biosynthesis, during the uptake and assimilation of ammonium. Beech mycorrhizas do not absorb nitrate rapidly and have little or no capacity to assimilate it Carrodus 1967; Smith 1972) but whether this applies to ectomycorrhizas in general is not known for certain.

G. D. Bowen and C. Theodorou in 1967 emphasized one aspect of mycorrhiza formation which, although pointed out by Hatch thirty years before, is often overlooked. This is the longevity of mycorrhizas in comparison with uninfected roots. They showed, by means of a root scanning technique, that some uninfected portions of the root system of *P. radiata*, namely the zones of cell elongation, were just as efficient at absorbing ^{32}P as were mycorrhizas (Fig. 7.11). Mycorrhizas however persist in one position in the soil for considerable periods of time whereas the highly efficient uninfected portion of the root 'moves' through the soil as the root elongates. The limiting factor in phosphate uptake from soil is not usually the plant's ability to absorb phosphate, but rather the rate at which phosphate ions are brought to the root surface. Whether this occurs mainly by diffusion or mass flow, the distance ions will move increases with time. Therefore an organ which is active in nutrient uptake for long periods has the potential to absorb ions from a much greater volume of soil than an ephemeral structure.

Mycorrhizal longevity is significant not only for phosphate aborption but also for the uptake of ammonium and potassium ions, which are absorbed on soil colloids and have low mobility relative to demand. The value of ectomycorrhizas might therefore be enhanced in mineral soils by hyphal strands growing towards exchange sites and primary mineral sources. This effect might not be so pronounced in litter hirizons where, as already noted, hyphal outgrowths from the sheath may be short and few. Harley and Smith (1983) point out, however, that in litter layers the nutrient sources tend to be brought to absorbing surfaces by processes such as leaf fall, comminution and mixing of litter by invertebrates, and percolating water. 'In such a situation the intense exploitation by mycorrhizas will lead to efficient trapping of nutrients without extensive extramatrical mycelium'.

Effect of v-a mycorrhizas on plant growth and nutrient status

Because the endophyte cannot yet be grown in pure culture, it is difficult to design experiments which would provide a critical test of the effect of mycorrhizas on plant growth. Nevertheless, evidence has accumulated that they enhance the growth of the host under certain circumstances. Barbara Mosse in 1957 tested the effect of inoculating apple seedlings, grown in autoclaved soil, with sporocarps of the mycorrhizal fungus; inoculated seedlings were larger than uninoculated and had generally higher nutrient contents. In 1959 G. T. S. Baylis raised seedlings of *Griselinia littoralis* (Cornaceae) in non-sterile soil and transplanted them to

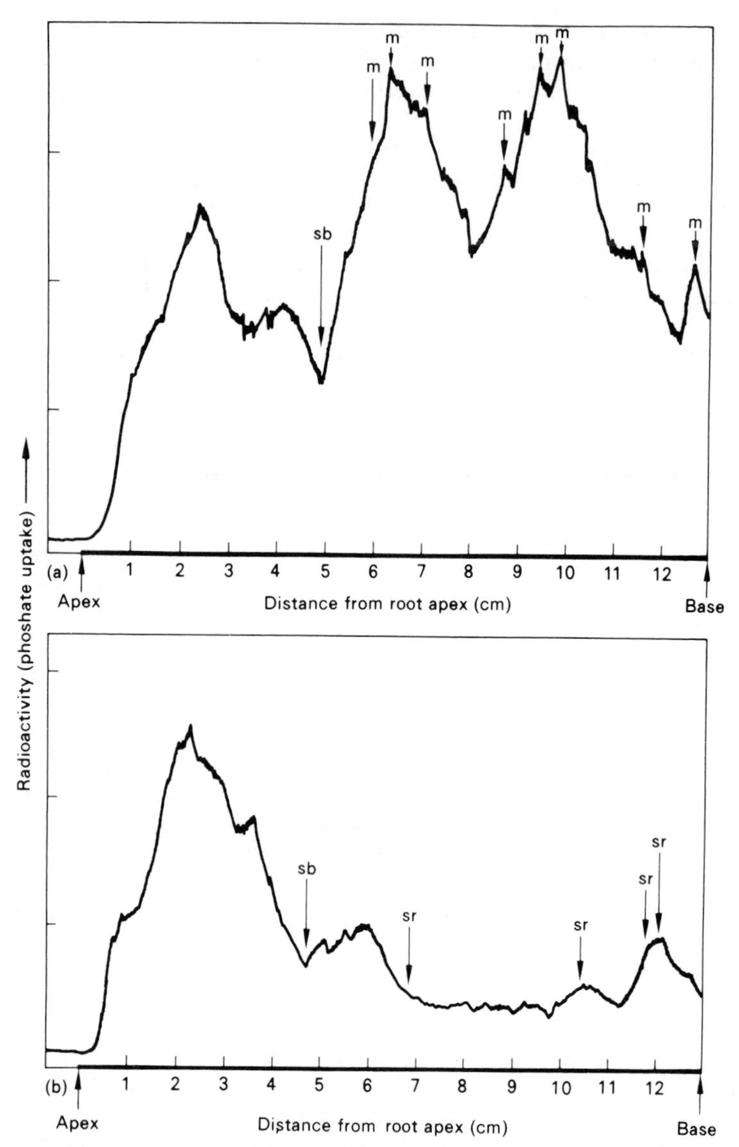

Fig. 7.11 Root scan of mycorrhizal (a) and uninfected roots (b) of *Pinus radiata*, showing how ectomycorrhizas permit the root system to retain a high capacity for absorbing ^{32}P from a 5×10^{-6} M solution of KH_2PO_4 for a relatively long period of time. m, position of mycorrhiza; sr, position of uninfected short root; sb, commencement of suberization. (From Bowen & Theodorou 1967.)

autoclaved soil deficient in phosphorus; a proportion of the transplants became mycorrhizal, grew better and took up more phosphorus than uninfected plants. Some care is needed when interpreting the results of experiments carried out in autoclaved soils, since it has been shown that heating may produce toxins which can be destroyed by a number of different microorganisms. Especially in experiments such as those just described, where the method used to infect seedlings would have introduced other soil microbes as well as the mycorrhizal fungus, it can be argued that the beneficial results obtained are due primarily to the destruction of toxins produced during sterilization. J. W. Gerdemann in 1964 used soil which had been steam-sterilized ('partially sterilized') rather than autoclaved, and also added leachings from sporocarps to his control plants in an attempt to introduce at least some of the microorganisms which were present as contaminants of his inoculum. These experiments showed that inoculation with fungal sporocarps increased the yield and phosphorus uptake of maize seedlings growing in phosphate deficient soils.

Benefit to the host seems more likely to result from mycorrhizal infection when conditions for growth are least favourable. For example, M. J. Daft and T. H. Nicholson (1966, 1969) produced v-a mycorrhizas on tobacco, tomato and maize seedlings in sand culture by inoculating them with spores of *Endogone* (*Glomus*), and found that growth was stimulated most under conditions of low phosphate availability; yield differences were greatest when phosphorus was supplied as bone meal and least when it was in the form of dicalcium phosphate. Even where inoculation failed to stimulate growth, it still resulted in increased phosphorus uptake. Further evidence of improved nutrition brought about by these mycorrhizas comes from T. M. Morrison and D. A. English, who in 1967 reported that excised mycorrhizal roots of kauri pine (*Agathis australis*) absorbed much more ^{32}P from a solution of H_3PO_4 than uninfected roots, and presented evidence to show that phosphorus uptake was metabolically dependent.

The enhanced growth and better phosphorus status of hoop pine seedlings resulting from infection by *Gigaspora araucareae* is illustrated in Table 7.5; non-mycorrhizal seedlings showed symptoms of gross phosphorus deficiency. Notwithstanding such positive responses to inoculation, there have been a few reports of growth being depressed by mycorrhizal infection (e.g. Baylis 1967); where this occurs it is probably due, in the opinion of J. L. Harley (1969), to carbohydrate utilization by the fungus.

Response mechanisms

The spore isolation methods introduced by Gerdemann have

Table 7.5 Effect of vesicular-arbuscular mycorrhiza on yield and nutrient status of 2-year old *Araucaria cunninghamii* seedlings growing in a steam-sterilized krasnozem soil

Parameter	*Mycorrhizal status*		*Significance of difference*[†]
	Uninfected	*Infected*	
Dry weight (g)	7.8	72.6	***
P concentration, roots (%)	0.050	0.122	***
N concentration, roots (%)	0.90	0.89	ns

Source: After D. I. Bevege (1971).
[†] Three asterisks indicate that the differences were significant at the 0.1% level of probability; ns = not significant.

permitted the establishment of v-a mycorrhizas under microbiologically controlled conditions, and this has facilitated the study of mechanisms by which phosphorus absorption is enhanced. Bowen *et al.* (1975) used radioactive tracer techniques to study short term uptake of phosphorus by endomycorrhizas of clover, onion and hoop pine. They found that detached mycorrhizal roots of clover and onion took up more than twice the phosphate that excised uninfected roots did from 5×10^{-6} M $KH_2\,^{32}PO_4$ solution. When the root systems of intact hoop pine (*A. cunninghamii*) seedlings were tested, unsuberized mycorrhizal roots absorbed 1.6–1.8 times more ^{32}P than similar uninfected roots. By the use of respiration inhibitors, absorption was shown to be metabolically mediated. Autoradiographs of whole roots revealed that the mycorrhizal portions were the most radioactive, while autoradiography at the cellular level showed that most of the activity was in fungal tissues, with very little accumulation of ^{32}P in neighbouring cortical cells. The higher rate of uptake by infected parts of the root system is therefore probably due to uptake by the fungus and not to the stimulation of normal uptake mechanisms of the host cells following upon infection. The high affinity of v-a mycorrhizas for phosphate, relative to uninfected roots, has been confirmed with tomato plants by Cress *et al.* (1979).

Expressing uptake in terms of 'inflow' per unit length of root provides a realistic means of comparing the efficiencies of infected and uninfected roots for absorbing immobile ions such as phosphate (Harley & Smith 1983). On this basis, Sanders and Tinker

(1973) showed that the inflow of phosphate into mycorrhizal roots of onion was several times greater than into non-mycorrhizal roots. The explanation for the increased uptake efficiency of mycorrhizas is that it is due in large part to the growth of extramatrical hyphae into the surrounding soil. Bowen *et al.* (1975) used microautoradiography to demonstrate the accumulation of ^{32}P in extramatrical hyphae. They also found that removal of the external hyphae from mycorrhizal roots reduced phosphate uptake, which indicates that translocation of phosphate through the hyphae is likely to be more rapid than diffusion through the soil.

Another feature of the v-a mycorrhizal symbiosis is the apparently greater longevity of infected roots. In *A. cunninghamii*, active uptake of phosphate by mycorrhizal short roots continues even after they have become suberized (Bowen *et al.* 1975). Moreover, the mycorrhizas of this species show renewed growth from their apices, the fresh (unsuberized) tissues becoming infected by extramatrical hyphae emanating from the older tissues; consequently they may persist as efficient absorbing organs for two or three years. This should greatly increase the volume of the absorption zone around them in the same way as has been postulated for long-lived ectomycorrhizas.

Once absorbed by v-a mycorrhizas, phosphorus is stored as inorganic polyphosphate granules in the fungal vacuoles, as in ectomycorrhizas. Movement of phosphate from fungus to root and thence to shoot is rapid (Bowen *et al.* 1975); transfer probably takes place mainly via the arbuscules which, as indicated in sect 7.2.1, provide a very large area of interface with the infected cortical cell.

As with ectomycorrhizas, research into the nutritional physiology of v-a mycorrhizas has centred on phosphorus. Experiments have revealed increased uptake of ^{45}Zn by hoop pine mycorrhizas (Bowen *et al.* 1974) and this element too can be transported through the extramatrical hyphae of white clover (Cooper & Tinker 1978). Absorption of ^{35}S is also enhanced by mycorrhizal infection but this may be an indirect effect consequent upon improved phosphorus status of mycorrhizal plants (Rhodes & Gerdemann 1978).

The evidence for increased phosphorus uptake by plants carrying v-a mycorrhizas, especially under conditions of low phosphate availability, is indisputable. Whether they can exploit relatively insoluble sources of soil phosphate is in doubt, however, even though the mycorrhizal fungi may be capable of synthesizing phosphatase enzymes. The ecological significance of these endomycorrhizas would therefore seem to stem primarily from improved phosphorus nutrition of the host plant, brought about primarily by the development of an extensive extramatrical mycelium. This attribute is likely to be especially important for woody shrubs and trees, particularly if infection promotes root longevity, since as K. P. Barley has noted, such plants generally have lower rooting densities than herbaceous species.

Effect of ericoid mycorrhizas on host nutrition

The biology of ericoid mycorrhizas is now fairly well understood, as a result of a series of papers by D. J. Read and his colleagues, dating from 1973. Much of this work is summarized in Stribley and Read (1975), Read (1978) and Read (1982).

Most ericaceous species are heathland plants of acid, nutrient-deficient soils where nitrogen mineralization is restricted and ammonium is the major, if not the sole, source of inorganic N. In this environment, mycorrhizal infection leads to an increase in the concentration and content of N and P in plants such as *Calluna vulgaris* and *Vaccinium macrocarpon*. This is brought about partly by the development of an extramatrical mycelium, and partly because the absorption system of fungal hyphae has a much higher affinity for NH_4^+ than does that of roots. These effects of ericoid mycorrhizas on host nutrition are most evident under conditions of low to moderate supply of soil ammonium: at very low or very high levels of NH_4^+, there is often no beneficial effect; indeed when nitrogen is very deficient, mycorrhizal plants may be smaller than uninfected ones. In the latter circumstance, Stribley and Read have suggested that the carbon drain to the fungus in heavily infected plants may outweigh any nutritional advantage gained by the host.

There is evidence that mycorrhizal plants of *Vaccinium*, when grown aseptically, are able to utilize amino acids as their sole soil of nitrogen, and translocation of ^{15}N-glutamine from the endophyte *Pezizella ericae* to its host occurs in this genus. Strains of *P. ericae* can use various salts of inositol phosphate in culture, from which it is inferred that mycorrhizal plants have access to organic forms of soil phosphate. Orthophosphate is also absorbed and translocated to the host by extramatrical hyphae, and phosphorus is stored in the fungal tissues as polyphosphate granules.

Transfer of carbon from host to fungus in ectomycorrhizas

It might be inferred from E. J. Björkman's hypothesis that relates mycorrhiza formation and development to the soluble carbohydrate content of roots (sect. 7.1), that the host plant provides the fungus with a source of carbon for energy, biosynthesis and growth. Such a correlation does not, however, indicate a cause and effect relationship. The first direct evidence that photosynthate moved from plant to fungus was provided by E. Melin and H. Nilsson in 1957. They raised Scots pine (*P. sylvestris*) seedlings in axenic culture with *Rhizopogon roseolus* or *Boletus* (*Suillus*) *variegatus*, allowed them to photosynthesize $^{14}CO_2$, and found that the products of photosynthesis were rapidly transported to the root tips and to the mycelium of the mycorrhizal fungus. A small proportion of this (approx. 10%) may have been the result of dark fixation, since ^{14}C was found in the control plants which had been decapitated to prevent them from photosynthesizing. It is now known that photosynthate is preferentially diverted to the symbiotic organ. This was clearly shown by Bevege *et al.* (1975) who compared the translocation of photosynthetically fixed carbon to mycorrhizal and non-mycorrhizal roots of the same *P. radiata* seedlings, 24 hours after exposing them to a 2-hour pulse of $^{14}CO_2$: mycorrhizas contained eight times as much ^{14}C per unit weight as uninfected roots.

Further definitive evidence of the movement of sugars from host to fungus in ectotrophic mycorrhizas comes from the studies of D. H. Lewis and J. L. Harley, reported in 1965. The main soluble carbohydrates found in beech mycorrhizas are the sugars glucose, fructose, sucrose and trehalose, and the sugar alcohol mannitol, the latter two compounds being widespread as storage carbohydrates in fungi generally. They could not be detected in uninfected beech roots, which indicates that their existence in mycorrhizas is due to the presence of fungal tissue. Mycorrhizas placed in a solution of ^{14}C-labelled glucose accumulated mainly trehalose and glycogen, in a solution of labelled fructose they accumulated mannitol, and when fed with labelled sucrose, they accumulated all three storage carbohydrates (Table 7.6). Irrespective of which of the three sugars were fed to mycorrhizas, ^{14}C

Table 7.6 Percentage distribution of ^{14}C, absorbed as sucrose from an external solution, in excised beech roots

Type of root	*Insoluble carbohydrates*†	*Soluble carbohydrates*				
		Trehalose	*Sucrose*	*Glucose*	*Mannitol*	*Fructose*
Mycorrhizal	54.0	17.2	2.6	0.9	24.4	0.9
Uninfected	22.5	0	47.5	19.0	0	11.0

Source: Data of D. H. Lewis and J. L. Harley. From Harley (1965).
† Mainly glycogen.

activity occurred almost entirely in fungal storage carbohydrates, about two-thirds of them in the sheath and the remainder in the Hartig net. Uninfected roots, on the other hand, accumulated only glucose, fructose and sucrose; these sugars could also be found in the core tissue of mycorrhizas. To test the hypothesis that simple sugars move from host to fungal tissue and are there converted to fungal storage carbohydrates, Lewis and Harley prepared agar blocks containing ^{14}C-labelled sucrose, and attached mycorrhizal roots to them after first removing a ring of sheath from the base of the mycorrhiza, so that the only fungal tissue in direct contact with the agar was the Hartig net. After an appropriate period, the distribution of ^{14}C-labelled carbohydrates among the components of the mycorrhizas was determined. A typical result is shown diagrammatically in Fig. 7.12. In the basal part of the core the ^{14}C label was found mainly in sucrose, with slight activity in trehalose due to the presence of the Hartig net. In the middle zone, where the sheath was present as well, sucrose and mannitol were about equally labelled, with smaller amounts present as trehalose and the two hexoses. The apical section was dissected into sheath and core before analysis. In the core, most of the activity was in sucrose, as in the basal part, again with some activity in trehalose

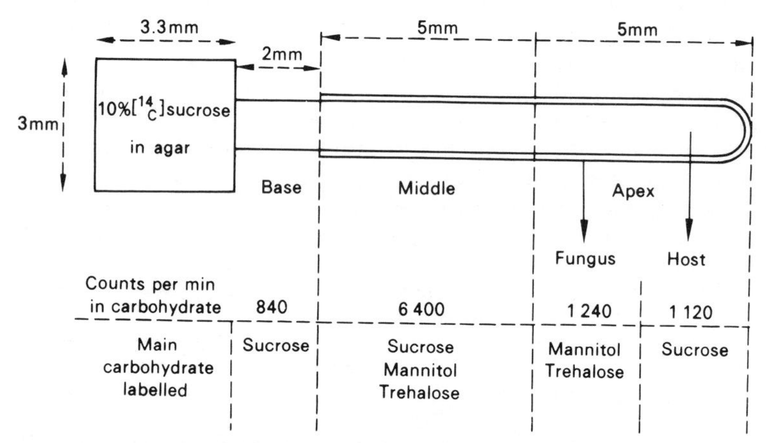

Fig. 7.12 Transfer of carbohydrate from host to fungus in ectomycorrhizas of *Fagus sylvatica*, as shown by the experiments of D. H. Lewis and J. L. Harley. The upper part of the diagram indicates the method; the centre, the amount of labelled carbon in the carbohydrates of various regions after 20 hours; the bottom, the most highly labelled carbohydrates present after 20 hours. For further explanation, see text. (From Harley 1965)

and hexoses. In contrast, the apical sheath showed most activity in the fungal carbohydrates mannitol and trehalose, while sucrose activity was greatly reduced.

The fact that the ^{14}C label remained predominantly as sucrose in both basal and apical cores indicates that sucrose can move through the host tissue of beech mycorrhizas. From there it moves to the fungal tissues of the Hartig net and sheath where it is hydrolysed to its constituent moieties (glucose and fructose) by extracellular fungal enzymes, before being absorbed and converted to mannitol and trehalose. Furthermore, there is virtually no reciprocal flow of carbohydrates from fungus to host since, compared to mycorrhizal roots, uninfected beech roots show little or no ability to absorb and utilize trehalose and mannitol.

Lewis and Harley's hypothesis is that the fungus of ectomycorrhizas absorbs sugars from the host and converts them into reserve carbohydrates peculiar to itself, and in this way maintains a concentration gradient favouring continued movement of sugars from host to fungus. Furthermore, since re-utilization of these fungal storage carbohydrates by the host is insignificant, the fungus gains an additional advantage in that it does not have to share its energy reserves with the host. This state of affairs, that is, where the recipient converts the transferred carbohydrate into a form unavailable to the donor, exists also in the lichen symbiosis, and indeed appears to be a common feature of mutualistic symbioses between autotrophs and heterotrophs (Smith *et al.* 1969). The hypothesis is strengthened by the finding of Bevege *et al.* (1975) that the roots of *Pinus radiata* seedlings exposed to $^{14}CO_2$ in the light had nearly half the labelled carbon as trehalose and about one-fifth as mannitol.

Harley and Smith (1983) have pointed out that the thesis of unidirectional movement of carbohydrate from host to fungus has proved so attractive that many workers have overlooked the possibility that some carbon might flow in the opposite direction. Although, as indicated above, non-mycorrhizal beech roots have little ability to absorb and utilize trehalose and mannitol, they are able to take up these fungal carbohydrates to some extent and to metabolize trehalose slightly. Furthermore, there is no reason to suppose that any glucose or fructose which might be present in the sheath or Hartig net, could not be absorbed by cortical cells. In fact, Reid and Woods (1969) showed that ^{14}C from labelled glucose, applied to mycelial strands of *Thelophora terrestris* emanating from mycorrhizas of *P. taeda*, was translocated to the host. On the other hand, Harley and Smith (*loc. cit.*) reported that when D. H. Lewis fed ^{14}C-glucose to beech mycorrhizas, only trehalose and mannitol in the sheath became labelled and a negligble amount of ^{14}C appeared in the host tissues.

It would seem that some movement of carbon from endophyte to host does occur but not a great amount. Glutamine, for example, may be transferred thus, and utilized by the host for synthesizing amino acids in the tricarboxylic acid cycle. Such a transfer represents the return of carbon previously transported to the fungus from the plant. As pointed out by Lewis (1975), the recycling of carbon as amino-compounds should not be equated with the unidirectional flow of carbohydrate from the photoautoroph to the chemoheterotroph.

Carbon transfer in endomycorrhizas

Some general information on carbon movement between symbionts in orchid mycorrhizas has been given in sect. 7.2.3; no further details are provided here. This discussion is restricted to vesicular-arbuscular and ericoid mycorrhizas.

Compared to ectomycorrhizas, much less is known about the transfer of carbonaceous materials from plant to fungus in v-a mycorrhizas. Since the association seems to be obligate for the fungus, it is reasonable to assume that its energy source is soluble carbohydrate in the host root. Support for this view comes from the observation that the growth of extramatrical hyphae is greatly stimulated once infection takes place. I. Ho and J. M. Trappe were the first to show, in 1973, that ^{14}C-labelled photosynthate could be found in the external hyphae of endomycorrhizal plants. This was subsequently confirmed by Bevege *et al.* (1975) and Cox *et al.* (1975). Chemical analyses have failed to reveal the presence of specific fungal carbohydrates such as mannitol and trehalose, which are so characteristic of other mycorrhizas; only a small amount of the storage carbohydrate, glycogen, is present. The lipid content of mycorrhizas may however be considerably higher than that of uninfected roots (Bevege *et al.* 1975; Cooper & Lösel 1978), and it has been suggested that lipids may have the same storage function in v-a mycorrhizas as do polyols, glycogen and trehalose in other mycorrhizas.

Bevege *et al.* used ^{14}C-labelling techniques to show that the synthesis of amino and other organic acids, proteins and cell wall materials represents an additional 'sink' for photosynthate in v-a mycorrhizas, and Harley and Smith (1983) have calculated that this fungal biosynthesis and growth would increase the respiratory demand of roots for carbohydrate by more than 10 per cent. Although many aspects of the carbohydrate metabolism of v-a mycorrhizas remain to be elucidated, there is no reason to doubt that net transfer of carbon from host to endophyte occurs in this symbiosis.

Knowledge of carbohydrate movement in ericoid mycorrhizas is also fairly recent. When $^{14}CO_2$ is photosynthesized by *Vaccinium*, the label is found as glucose, sucrose and fructose in the roots of uninfected plants, and as these sugars together with mannitol and trehalose in the roots of mycorrhizal plants (Stribley & Read 1974). This labelling pattern is similar to that found in ectomycorrhizal plants and indicates carbohydrate transfer from autotroph to heterotroph. As in that symbiosis, a small amount of carbon is recycled to the host plant in the form of amino acids or amides.

Processes involved in nutrient and metabolite transfer

Bidirectional transfer of molecules and ions can only occur when active cells of both partners are in juxtaposition. This occurs when mycorrhizas have reached the mature state. At this stage in the mutualistic association, transfer between the cells of the two symbionts takes place across an interfacial zone, bounded on one side by the host plasmalemma and on the other by the fungal plasmalemma. Between the plasmalemmas is a matrix comprising the host cell wall (considerably modified in endomycorrhizas, less so in ectomycorrhizas) and material derived from the host, and the fungal cell wall which is relatively unmodified. In the words of Harley and Smith (1983), 'an "apoplastic space" separates the fungal plasmalemma from the host plasmalemma and through this compounds and ions must pass as they are transferred from one organism to another'. Although the collapse of fungal hyphae, which is a regular feature of endomycorrhizas, may contribute to nutrient and metabolite transfer, this is believed to be quantitatively much less important than movement between living cells.

Harley and Smith (*loc. cit.*) recognize three possible mechanisms whereby carbohydrate might be transferred from host to endophyte in all kinds of mycorrhiza:

(1) possible leakage into the apoplast, followed by active transport into the fungus;
(2) fungus-induced disruption of host cell wall formation, such that precursor molecules accumulate in the apoplast and are selectively absorbed by the fungus;
(3) hydrolysis of host cell wall constituents (e.g. pectin) by the fungus and accumulation of the products in the apoplast, where they are selectively taken up by the fungus.

One or other of these mechanisms may dominate at different stages in the development of any particular mycorrhiza.

These authors also present a schema for explaining phosphate transfer from fungus to host, involving the metabolism of polyphosphate in which form phosphorus is stored in fungal tissues. Polyphosphate breakdown releases orthophosphate which sets up or maintains a concentration gradient sufficient to cause passive leakage of inorganic phosphate into the apoplast, whence it is actively absorbed by the host. Two separate mechanisms are postulated to explain how the concentration of inorganic phosphate in the fungal pool might be increased:

(1) simple hydrolysis of polyphosphate to orthophosphate;
(2) enzymatic linkage of polyphosphate degradation with the active uptake of hexoses leaked into the apoplast from the host, followed by the release of inorganic phosphate during subsequent metabolism of the hexose phosphate so formed.

If a real understanding of mycorrhizal symbiosis is to be attained, it is important that the biochemical mechanisms underlying host response to infection be elucidated. Only then is it likely that the manipulation of these mutualistic associations for increased biological productivity can be successfully achieved.

7.4 Mycorrhizas in nutrient cycling and energy flow

Energy network diagrams for non-mycorrhizal and mycorrhizal ecosystems are presented in Fig. 7.13. The details of the mineralization–immobilization cycle have been omitted for the sake of clarity. Plants and soil microorganisms are shown as drawing on the same inorganic nutrient pool, and the two nutrient cycles are assumed to be competitive. However, in the mycorrhizal system the symbiotic fungus is interposed between the nutrient pool and the plant. In both systems, productivity is determined by the operation of a work gate through which energy flow from the sun is controlled by the supply of nutrients from the inorganic pool in the soil. The competitive flow towards the heterotrophic soil microbes (i.e. the immobilization rate) is governed at another work gate by the energy flow from plant residues. This is a function of the rate at which the litter reservoir is replenished which in turn depends on the rate of energy capture by the producers.

The presence of an additional module in the mycorrhizal system (Fig. 7.12b) has important implications for the processes of energy flow and nutrient cycling. The fungus acts as a physiological sink for plant sugars, utilizing them for biosynthesis and growth, and/

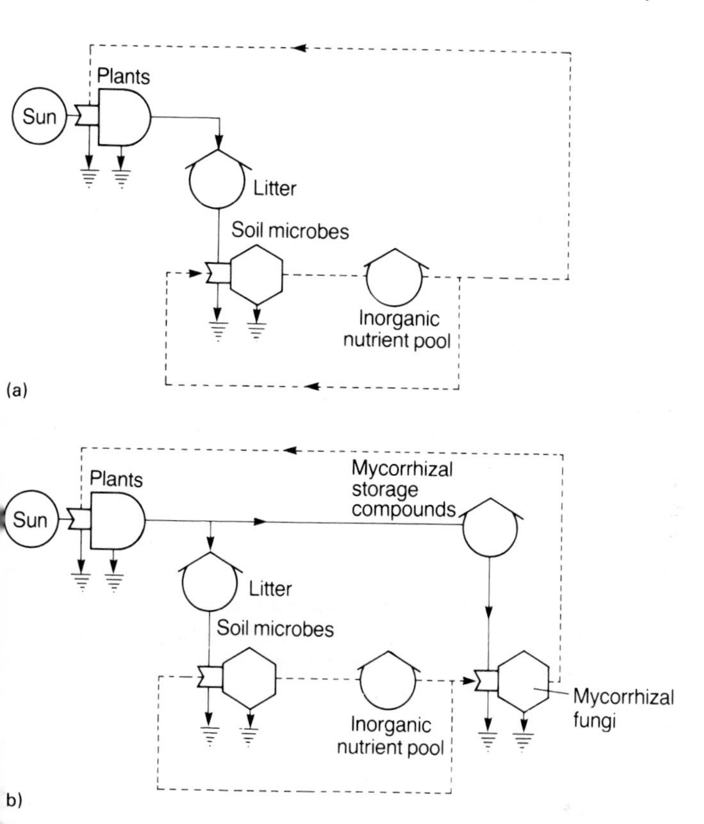

Fig. 7.13 Energy network diagram of hypothetical ecosystems. (a) Non-mycorrhizal system. (b) Ectomycorrhizal system.

or converting them into storage compounds which the host is unable to metabolize. The continual movement of assimilates into the fungal symbiont activates the work gate which regulates nutrient flow into the mycorrhiza and thence to the host. Being independent of plant residues as an energy source, the mycorrhizal fungus competes successfully with free-living soil microorganisms for mineral nutrients. Especially in low fertility soils, where the intensity of infection is usually greatest, the role of mycorrhizas is seen as speeding the passage of essential elements through the nutrient cycle, and minimizing losses from the system through leaching. Productivity is enhanced through the ability of mycorrhizas to increase the concentration of limiting nutrients at the work gate which controls the flow of solar energy into the plants.

As indicated in section 7.3, increasing efficiency of ion uptake is only one of several mechanisms whereby the fungal endophyte confers an ecological advantage on its host. Other factors already mentioned include the greater longevity of mycorrhizas which permits them to exploit a larger volume of the soil solution, and the production of surface phosphatases which might give them access to phytates not readily available to uninfected roots. In addition to these factors, the production and excretion of organic acids, especially oxalic acid (Cromack *et al.* 1979; Malajczuk & Cromack 1982), might enable them to dissolve fixed forms of phosphorus in acid soils, namely iron and aluminium phosphates, by chelating Fe and Al (but see sects. 5.2.1 and 6.5).

Energy cost to the host

Whatever the means by which the plant is presented with an increased supply of any limiting nutrient, the mechanism which initiates it is the diversion of photosynthate into fungal storage and growth sinks. The amount so diverted is not known for certain but it is possible to make a rough estimate. In the oak–pine forest studied by G. M. Woodwell and his co-workers at Brookhaven (see Ch. 4), root respiration was estimated not to exceed 424 g m^{-2} yr^{-1} dry matter, or 28 per cent of total autotroph respiration. Ectomycorrhizas are responsible for 40–50 per cent of the root respiration of beech (Harley & Smith 1983). If this value applies also to the ectomycorrhizal oaks and pines, then 11–14 per cent of the total producer respiration (R_a) is accounted for by the mycorrhizal fungi. Since in this forest, R_a represents 56 per cent of gross production, it follows that for every 100 g carbon fixed by photosynthesis, 16 g are consumed by root respiration, and of this amount 6–8 g is respired by the mycorrhizas.

In addition to the respiratory consumption of photosynthate, a further quantity is used for the production of fungal tissues, including sporocarps. The latter can be quite substantial, reliable estimates ranging from 65 to 410 kg ha^{-1} yr^{-1}. In the *Abies amabilis* ecosystem of western Washington State, in the United States, K. A. Vogt *et al.* (1982) found that biosynthesis of mycorrhizal sheaths, sporocarps and sclerotia utilized an amount equivalent to 14–15 per cent of net primary production. Applying this percentage to the Brookhaven oak–pine forest, dry matter production by mycorrhizal fungi would be 15 per cent of 1195 or 179 g m^{-2} yr^{-1} which is 6.6 per cent of gross production (GP). Fogel and Hunt (1979) have presented data which suggest that the proportion of GP utilized by mycorrhizal fungi for biosynthesis is greater than this but theirs is probably an overestimate because not all the fungal tissues included were necessarily mycorrhizal. Nevertheless it is possible that Vogt and her colleagues have underestimated, and assuming this to be so we might double their figure and accept that approximately 13 per cent of GP, or 13 g m^{-2} yr^{-1} for every 100 g of photosynthate produced, is used for biosynthesis by the mycorrhizal fungus. Since net community production (NEP) in the Brookhaven forest was shown to be 22 per cent of GP (sect. 4.7), mycorrhizal production accounts for about three-fifths of NEP. This value seems high, though not perhaps unreasonable, and is accepted for the present purpose. Adding to it the respiratory loss from mycorrhizas (around 7 g), the total annual usage by the symbiotic fungus is 20 g m^{-2}. In summary, then, the 'energy cost' of the symbiosis to the autotrophic component in this oak–pine forest is estimated to be 20 per cent of gross production.

It should be kept in mind that the proportion of total community respiration (R_e) contributed by autotrophs will vary throughout succession, being greater earlier and smaller later. The relative contribution of mycorrhizal fungi to producer (autotroph) respiration will likewise change as succession proceeds. Although there is no experimental evidence to support such a contention, it seems reasonable to suppose that this will be greater in the early stages, when net primary production and the annual demand for nutrients is highest. The Brookhaven forest is in a late stage of succession ($GR/R_e = 1.3$), and in less mature forests the amount of gross production diverted into fungal storgae carbohydrates might be expected to exceed the 20 per cent calculated for this ecosystem.

There is less data on which to base an estimate of the proportion of photosynthate diverted to the endophytes of v-a or ericoid mycorrhizas. As Harley and Smith (1983) point out, it may be inferred from the fact that mycorrhizal infection may sometimes depress the growth rate, that the fungus makes substantial demands on the plant for carbohydrate. The development of hyphae external to the root requires carbon skeletons for biosynthesis, and sugars as respiratory substrates: the increased respiratory demand alone is likely to be more than 10 per cent in v-a mycorrhizas (sect. 7.3). Despite the presence of an extramatrical mycelium, however, 'it is difficult to believe that the demands of their mycorrhizal fungi on the photosynthetic products of their individual plants are proportionately as high as those of ectomycorrhizal fungi' (Harley & Smith 1983).

The expenditure of a substantial fraction of photosynthate to maintain the mycorrhizal symbiosis raises questions about its adaptive significance. In the process of evolution, increased growth rate would seem to be of little adaptive value unless it enhanced survival, which it may do for the ephemeral colonizers of habitats subject to recurrent drought, but hardly for the ectomycorrhizal tree species which dominate the forests of temperate, mesic environments. Adaptation is directed towards survival, not growth rate, and the energy cost of the mycorrhizal habit should be seen in the context of survival rather than in terms of productivity.

Role of mycorrhizas in biogeochemical cycling

As indicated in Chapter 5, the biogeochemical cycle begins with nutrient uptake by roots and ends with the release of inorganic ions from decomposing plant residues. One consequence of the increased uptake and storage of nutrients by mycorrhizas is, therefore, that they contribute substantially to the detrital pool (see Fig. 5.1). Below-ground additions to this pool remain one of the least studied aspects of mineral cycling. The quantities of nutrients entering the decomposer pathway from dead roots can be very large, sometimes far exceeding those added by above-ground litter fall. According to R. Fogel (1980), mycorrhizal mortality accounts for half the annual throughput of biomass and over two-fifths of the nitrogen released annually in a Douglas fir ecosystem. Unlike the decomposition of plant residues on the soil surface, the decay of mycorrhizas takes place within the root system sorption zone, and frequently in the immediate vicinity of other, actively absorbing, mycorrhizas. This reinforces the point made earlier that the ecological role of mycorrhizas is to increase the rate of biogeochemical cycling and to minimize leaching losses to drainage waters.

7.5 Other regulatory functions of mycorrhizas

In addition to their direct effects on nutrient cycling and energy flow, there is evidence that mycorrhizas have other regulatory functions. It is known, for example, that mycorrhizal infection affects water uptake by plants, and the activities of nitrogen-fixing bacteria and root disease fungi.

Water relations

Vesicular-arbuscular mycorrhizas enhance water transport in plants, through a lowering of root resistance (Safir *et al* 1971, 1972) or an increase in stomatal conductance (Levy & Krikun 1980; Allen 1982). Reduced root resistance to water flow is due, according to Hardie and Leyton (1981), to the extension of the root system by extramatrical hyphae.

Ectomycorrhizal seedlings are sometimes held to be more

drought resistant than uninfected plants, and certainly mycorrhizal fungi vary in their ability to withstand low soil water potentials (Mexal & Reid 1973). Using tritiated water, Duddridge *et al* (1980) showed unequivocally that water was transported through rhizomorphs of *Suillus bovinus* into *Pinus sylvestris*, and suggested that rhizomorph-forming fungi might be well adapted as symbionts of plants growing in habitats subject to drought. This could well apply to *Cenococcum geophilum* (syn. *C. graniforme*), which is widely recognized as having drought-resistant strains, and which has long had a reputation for forming mycorrhizas under dry soil conditions.

Soil-borne plant pathogens

There is evidence that mycorrhizas have a protective function against certain root disease organisms (Zak 1964). Many ectomycorrhizal fungi produce antibiotics in pure culture which inhibit the growth of other microbes, including some pathogenic root-infecting fungi. One of these is *Leucopaxillus cerealis* var. *piceina* which forms ectomycorrhizas with short-leaf pine (*P. echinata*), a tree which is susceptible to littleleaf disease caused by the root-rot fungus *Phytophthora cinnamomi*. D. H. Marx and C. B. Davey (Marx 1969a,b; Marx & Davey 1969a,b) have shown that fully developed mycorrhizas formed by this and other fungi on shortleaf and loblolly pines resist attack by the pathogen under controlled conditions in the laboratory, and have isolated and identified the antibiotic agent. They further showed that various morpholotical forms of naturally occurring mycorrhizas of shortleaf pine were resistant to infection; mycorrhizas with incomplete fungal mantles did permit some infection to occur but the pathogen did not invade the Hartig net region. Thus it seems probable that ectomycorrhizas have an additional regulatory function in forest ecosystems as biological control agents against root pathogens.

Marx (1975) has further discussed the protective role of ectomycorrhizas, and pointed out that the endophyte might restrict the activity of root pathogens by ways other than the secretion of antibiotics: by competition for available nutrients in the rhizosphere, by the sheath creating a physical barrier to pathogenic infection, and by encouraging the development of an antagonistic microflora in the 'mycorrhizosphere'.

The effect of v-a mycorrhizas on the root-rot fungus, *P. cinnamomi*, has also been investigated, but with conflicting results (Harley & Smith 1983): in some circumstances, mycorrhizal infection confers protection on the host while in others the incidence of disease is increased. Protection seems more assured against the vascular wilt fungus (*Verticillium albo-atrum*) and the take-all fungus (*Gäumannomyces graminis*), disease resistance in the latter case being attributed to the improved phosphate nutrition of mycorrhizal plants (Graham & Menge 1982). The interactions between mycorrhizal fungi and root pathogens are clearly complex, and the regulatory role of the symbiosis remains to be clarified.

Interaction with nitrogen-fixing bacteria

There are indications (Bevege *et al.* 1978) that nitrogen fixation by free-living bacteria in the rhizosphere is enhanced in the presence of mycorrhizas. More definitive evidence is available for interactions involving symbiotic nitrogen-fixing bacteria and actinomycetes, that is, in the legume – rhizobium and the non-legume – actinomycete symbioses. Improved phosphorus nutrition is thought to be the main reason for enhanced nodulation and increased nitrogenase activity in legumes (Mosse *et al.* 1976; Smith *et al.* 1979); a further contributing factor may be the increased uptake of other ions involved in process of nitrogen fixation, such as Cu^{2+}, Zn^{2+} and Mo^{3-} (Rovira *et al.* 1983).

8

Root nodule symbioses and other nitrogen-fixing systems

The roots of some plants bear nodules which are the visible manifestation of a symbiotic association between these plants and certain bacteria. In contrast to mycorrhizas, whose main function appears to be to use an existing supply of nutrients from the soil more efficiently, this symbiosis is able to 'fix' atmospheric nitrogen, that is to utilize the molecular nitrogen of the atmosphere for biosynthesis and growth. Except for certain free-living prokaryotes, no other organisms can assimilate gaseous nitrogen. Fixation of nitrogen is therefore one of the fundamental reactions of the biosphere (see sect. 8.3).

Two broad groups of angiosperms possess root nodules, one comprising legumes and the other certain non-legumes, the bacterial component being different in the two symbioses.

8.1 The legume–rhizobium association

The legume–rhizobium symbiosis is of great significance in agriculture, for the growth of crop plants is limited more frequently by a deficiency of nitrogen than of any other element. Efficiently nodulated and properly managed leguminous crops can fix between 100 and 175 kg N ha^{-1} yr^{-1}; indeed there are authenticated records of fixation rates, under the stimulus of mowing or grazing, exceeding 500 kg per hectare each year. Highest rates of fixation are usually found in association with herbaceous legumes, but some woody species have considered potential also, for example *Acacia mollissima* when introduced to South Africa is reported to have fixed nearly 200 kg N ha^{-1} yr^{-1} over a thirty year period.

The ability of legumes to maintain or restore soil fertility was known to the ancient Romans, but it is less than 100 years since this has been attributed to the presence of the peculiar nodular structures found on their roots. Around the middle of the nineteenth century it was learned that the growth of most non-leguminous plants depended on the supply of inorganic nitrogen from the soil but this was not necessarily so for legumes. In the 1880s, H. Hellriegel and H. Wilfarth found that when quartz sand, a medium free of nitrogen, was inoculated with a small quantity of soil from an area where legumes grew well and then sown with legume seed, the seeds developed into healthy plants bearing root nodules and gained considerable amounts of nitrogen. Oats treated in the same way did not develop nodules and grew poorly. When a sterile soil inoculum was used the legumes failed to nodulate, made poor growth, and gained no nitrogen (Table 8.1).

On the basis of these results, Hellriegel and Wilfarth (1888) postulated that bacteria in the root nodules assimilated elemental nitrogen from the air and that the pea plants used the nitrogenous compounds formed by the bacteria. This hypothesis gained support when T. Schloesing and E. Laurent found, in 1892, that any gain in the nitrogen content of legumes was balanced by a corresponding loss from the atmosphere. Direct evidence that root nodules were the site of fixation had to await the advent of isotopic tracer techniques, and even then was not obtained until 1952, when W. H. Aprison and R. H. Burris demonstrated that excised root nodules of soybean could fix $^{15}N_2$ for a short period, although they rapidly lost the capacity to do so. The root nodule organism was first isolated by M. W. Beijerinck in 1888. He named it *Bacillus radicicola*, but several species are now recognized, all of which are placed in the genus *Rhizobium*. Specific rank has in the past been conferred on

Table 8.1 Effect of inoculation with soil containing *Rhizobium* on growth of legumes (peas) and non-legumes (oats)

Treatment	*Yield*	
	Oats (g)	*Peas* (g)
No N added		
Uninoculated	0.6	0.8
Inoculated with soil which had grown legumes	0.7	16.4
Inoculated with sterile soil	–	0.9
112 mg N added as NO_3^-		
Uninoculated	12.0	12.9
Inoculated	11.6	15.3

Source: Based on the data of Hellriegel and Wilfarth (1888).

many strains of rhizobia on the basis of their constant association with a particular legume, but recent taxonomic evidence indicates that *Rhizobium*, as now constituted, probably contains only two or three distinctive groups which might properly be regarded as separate species.

Incidence and specificity of nodulation

The Leguminosae, in the wide sense, contains three large groups which are given family status by most taxonomic authorities. Hutchinson (1969) recognizes the families Mimosaceae, Caesalpiniaceae and Papilionaceae and places all three in the order Leguminales. Other taxonomists classify the papilionaceous species as belonging to the family Fabaceae and assign the three families to the order Fabales. Still others see the Leguminosae as a subdivision of the order Rosales. The Mimosaceae and the Caesalpiniaceae are mostly tropical in distribution while the Papilionaceae (Fabaceae) contains both tropical and temperate species. Not all legumes will nodulate, the ability to do so being least pronounced in the Caesalpiniaceae: two-thirds of the genera examined in this family are reported as lacking nodules. In contrast, nodulation is almost universal in the Papilionaceae, 95 per cent of genera being nodulated (Allen & Allen 1981). The Mimosaceae occupy an intermediate position. Nodule shape and size varies considerably from species to species (Fig. 8.1). Annuals usually have large nodules grouped about the taproot or the first order laterals. The nodules of perennial species tend to be smaller and more widely distributed over the root system. New nodules are formed throughout the growing season and old nodules are sloughed off. Nodulation is affected by seasonal conditions and by the growth patterns of the plant so that nodules may be absent at any given time, even from plants which normally nodulate abundantly. As several workers have pointed out, it is probably unwise to generalize about the incidence of nodulation in the Leguminosae, since only a small proportion of the total number of species has ever been examined to determine whether root nodules are present.

D. O. Norris suggested in 1956 that any general description of the legume–rhizobium symbiosis should be based on the fact that the vast majority of legumes are tropical plants, and that their ancestral ecological habitat was probably the strongly leached, acid soils of tropical rainforests. The temperate genera evolved from tropical forebears and some of them, particularly the genera *Trifolium, Medicago, Melilotus* and *Trigonella* of the tribe Vicieae, have become adapted to neutral or calcareous soils of higher nutrient status. Although other workers have challenged Norris'

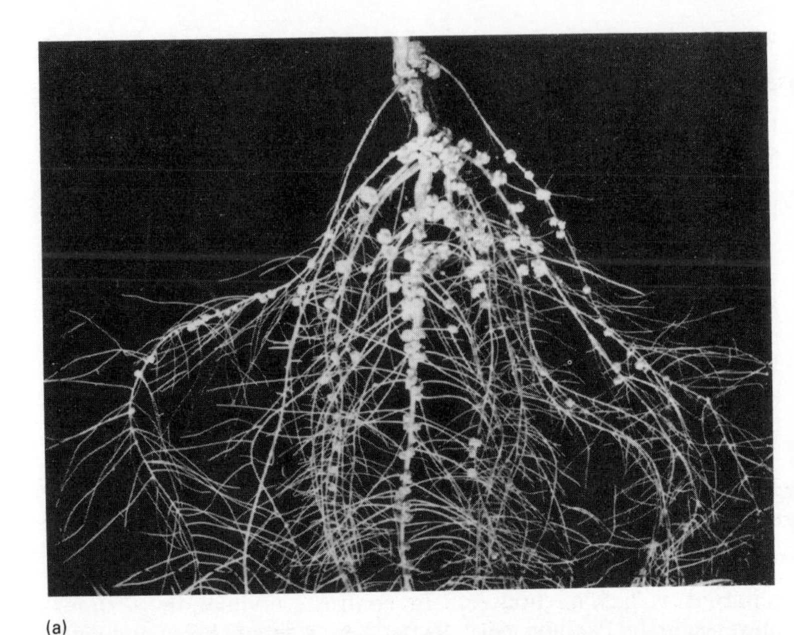

(a)

(b)

Fig. 8.1 Root nodules of legumes, (a) soybean, *Glycine max*. (b) Red clover, *Trifolium pratense*. (From Stewart 1966.)

hypothesis concerning the evolution of the Leguminosae, it is indisputable that the bulk of research done on symbiotic nitrogen fixation in legumes deals with the Trifolieae and the Vicieae, and caution must be exercised in generalizing from the results of this research until such time as more is known about the tropical species. This is particularly true when we come to consider the significance of the legume–rhizobium symbiosis in the geochemical cycle of nitrogen (sect. 8.3.1).

There is a certain degree of specificity in nodulation, and legumes may be arranged in groups, the members of one group generally forming nodules with the same strain of *Rhizobium*, but for the most part not nodulating with rhizobia from other groups. Seven such 'cross-inoculation' groups, and a number of small 'strain specific' groups, have been recognized. The seven major groups are the field pea (*Pisum*), medic (*Medicago*), clover (*Trifolium*), lupin (*Lupinus*), garden bean (*Phaseolus*), soybean (*Soya*) and cowpea (*Vigna*) groups. Within the various cross-inoculation groups, sub-groups may be distinguished which are specific in their requirements for particular strains of rhizobia; for example, three sub-groups are recognized in the clover cross-inoculation group. The scheme is of such limited applicability to the Leguminosae as a whole that it has fallen into disfavour. Nowadays, the last four groups, together with a number of the strain-specific groups, are considered as a 'tropical legume miscellany'. This miscellany of species contains some forming more or less discrete groups, some which will nodulate with a wide range of rhizobial strains, and others which will nodulate only when infected with specific strains of bacteria.

Rhizobium strains not only differ in their ability to form nodules with a given legume, but also in their capacity to fix nitrogen in association with the particular host. The property of nitrogen fixation is a function of both host and bacterium, and a strain of *Rhizobium* may be highly efficient, that is 'effective', on one host yet serve no useful purpose, in other words be 'ineffective', on another. Effective nodules are generally larger than ineffective nodules, and can also be distinguished by their pink coloration, which is due to a red internal pigment, leghaemoglobin. The effectiveness of rhizobia is an important factor to be taken into account when legumes are used in agriculture. When new legumes are being introduced to a region for the first time, care must be taken to ensure that effective strains of rhizobia are present in the soil in sufficient numbers to form nodules on the developing seedlings. In many cases, the native rhizobia are not effective for the introduced plant, and inoculation of the seed with an effective strain of *Rhizobium* can mean the difference between a successful introduction and complete failure of the crop, especially if the soil is low in available nitrogen. The causes of the host – bacterium specificity in nodulation and nodule effectiveness are not known for certain but there is some evidence that both attributes are controlled by genetic factors in the host.

Inherited host factors and nodulation

There have been rapid advances in our knowledge of the genetics of *Rhizobium* in recent years, but this will not be dealt with here; interested readers should consult the conference proceedings edited by A. H. Gibson and W. E. Newton in 1981. Of more immediate concern in the context of this chapter is an understanding of host genetics as it affects nodulation and nitrogen fixation in legumes, knowledge of which has developed more slowly. P. S. Nutman and his colleagues (see Nutman 1981) have identified several major genes in red clover which interfere with the release of bacteria from infection threads, or inhibit the released bacteria from multiplying, or prevent their transformation into bacteroids (see below for a description of the infection process).

D. P. S. Verma and associates have identified nodule-specific proteins of host origin which are induced during nodule development as a response to infection of soybeans by *Rhizobium japonicum*. These polypeptides have been termed 'nodulins' (Verma *et al.* 1981) and are associated with a number of nodule-specific messenger RNA sequences. It is likely that the RNA is responsible for encoding the nodulin sequences, although there is no direct experimental evidence for this as yet. The expression of these nodule-specific host genes is significantly altered in nodules formed by ineffective strains of *Rhizobium*, from which Verma *et al.* (*loc. cit.*) infer that some of them may be involved in the process of symbiotic nitrogen fixation.

There is also evidence of host control over nodule number and nodule size. According to Gibson and Jordan (1983), however, the physiological bases of this and other disruptions to the symbiotic system, such as those described by Nutman and Verma, have not yet been adequately defined.

Effect of edaphic factors on nodulation

Nodulation is affected by a variety of edaphic factors, some of which act via their influence on the nutrition of the host. For example, infection is conditioned by the carbohydrate and nutrient status of the plant, in a manner analogous to their effect on mycorrhiza development in pine seedlings (sect. 7.1). Plants which are grossly nitrogen deficient nodulate sparingly or not at all, while a high level of availability of soil nitrogen depresses nodulation and reduces fixation, the critical level of mineral soil nitrogen varying with the nature of the crop and the environmental conditions. Many other factors are however involved in the nodulation process and these will be discussed later.

Calcium and phosphorus play an important part in the legume–rhizobium relationship. Phosphorus-deficient plants do not nodulate properly and in addition an adequate supply of soil phosphorus helps to maintain the population of nodule bacteria in the soil at a high level. Calcium is important for the nutrition of both legumes and bacteria. Most specialized temperate legumes of the clover, medic or field pea type have a high calcium requirement, and frequently do not nodulate well on acid soils, unless these are limed. In contrast lupins, though confined to temperate regions, have the ability to nodulate on acid soils even in the absence of lime. The effect of lime on the symbiotic system of temperate legumes growing in acid soils is confounded by its effect on the availability of molybdenum. Adequate molybdenum is necessary for the functioning of the fixation process, and its availability is increased by adding lime. Work with subterranean clover on the tablelands of south-eastern Australia has shown, however, that the amount of lime needed to induce nodulation is less than that needed to correct any molybdenum deficiency. Substantial economies can be effected, in this instance, by using molybdenum-superphosphate to correct molybdenum deficiency and drilling in small quantities of lime (about 220 kg per hectare) with the seed (Anderson & Moye 1952). Furthermore, since superphosphate contains sufficient calcium to overcome any calcium deficiency on these soils, the effect of lime in promoting nodulation is not due to an increase in the supply of molybdenum or calcium. The increase in nodulation brought about by liming is in fact caused by an increase in rhizobial numbers resulting from a decrease in soil acidity. When it was realized that it was necessary only to increase soil pH in the rhizosphere to ensure good nodulation, the technique of 'lime-pelleting' of subterranean clover seed was developed: seed pelleted with a small amount of finely ground calcium carbonate produces as vigorous a pasture as that obtained when much larger quantities of lime are applied in the drill rows (Loneragan *et al.* 1955).

It should not be assumed, on the basis of these results with temperate species, that all legumes will nodulate better on acid soils if these are limed. D. O. Norris pointed out in 1965 that the well known legumes of temperate agriculture, such as the clovers, peas, vetches and lucerne, all have highly specialized, fast-growing rhizobia which produce acid when grown on appropriate media in the laboratory, whereas all the tropical legumes have slower growing, alkali-producing rhizobia. Norris postulated that this enables the tropical strains of rhizobia to resist acid soil conditions, whereas unless the rhizosphere of temperate legumes is ameliorated with lime, their acid-producing rhizobia fail to cause nodulation. Lime should not, therefore, be applied indiscriminately to tropical legumes, for not only are their rhizobia well adapted to growing in acid soils but in addition there is evidence that they are more efficient than temperate species in extracting calcium from soils of low calcium status.

The specific role of calcium in nodulation, and the importance of its interaction with hydrogen ion concentration, were clarified in 1958 by J. L. Loneragan and E. J. Dowling. They found that the amount of calcium required for nodulation of subterranean clover was greater than that which produced maximum growth of *Rhizobium*, and also was in excess of plant requirement when nitrogen was supplied as potassium nitrate. They concluded that the effects of H^+ and Ca^{2+} on nodulation could best be explained in terms of their influence on the level of calcium in the plant. C. S. Andrew and D. O. Norris came to a similar conclusion in 1961, after experimenting with several tropical and temperate species. A leguminous tree of the Australian semi-arid zone, mulga (*Acacia anuera*), has maximum nodule development associated with a clearly defined optimum level of calcium in the phyllodes of 0.56 per cent: nodulation declines rapidly as phyllode calcium concentration deviates from the optimum in either direction (Fig. 8.2).

Soil temperature is another environmental variable which has been extensively studied. The temperature of the rooting medium affects both the initiation of infection and the subsequent development of nodules. Tropical legumes have a higher temperature requirement for nitrogen fixation (though not necessarily for nodulation) than temperate species (Gibson & Jordan 1983). Soil temperature exerts its influence on the nitrogen-fixing system through its effect on the symbiotic association. Other edaphic factors probably behave likewise, thus the symbiosis does not necessarily react to environmental variables in the same way as the plant or bacterium growing separately.

Further aspects of the effect of soil properties, and other environmental variables, may be found in the review by A. H. Gibson and D. C. Jordan published in 1983.

Infection and nodule development

Rhizobia are facultative symbionts able to live as normal compo-

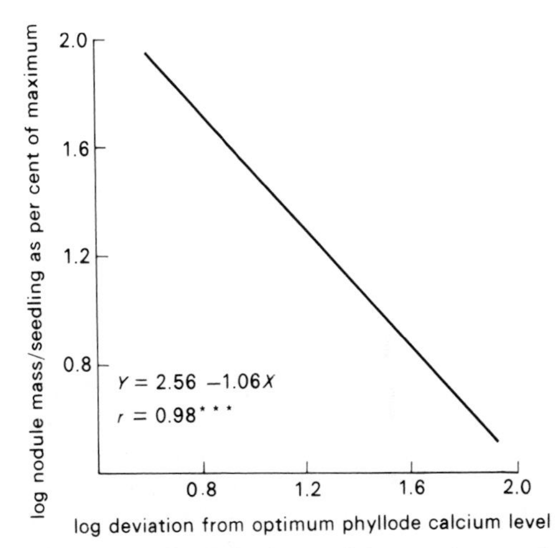

Fig. 8.2 Effect of plant calcium status on the extent of nodulation in *Acacia aneura*. (From O'Hagan, B.J. (1966) MSc thesis, University of New England, Australia)

nents of the soil microflora in the temporary absence of their hosts, but their continued existence as free-living heterotrophs depends on the presence of the host root. Whereas nodule bacteria are little affected by the roots of most non-legumes, they are greatly stimulated in the rhizosphere of their particular host, and R/S ratios almost always exceed 1000/1. The cause of this large and specific rhizosphere effect is unknown. The population of *Rhizobium* in the rhizosphere of legumes may reach 10^6 cells per gram or larger. Such large population densities are not needed to initiate infection in axenic culture, where fewer than 100 bacteria in the whole rhizosphere may be sufficient to start infection. Why such high numbers are necessary in nature remains to be determined, but a variety of microbial interactions may be involved (Parker *et al.* 1977).

The process of nodule development involves several distinct stages, which have been clearly described by P. S. Nutman (1963) and which are illustrated in Fig. 8.3. The first stage includes the events up to and including infection of the root. Nodules seem to be initiated at predetermined sites on the root, usually via root hairs though in some species infection can occur directly through the epidermis. Infection is localized in the zone of very young root hairs, immediately behind the root tip, and only a very small percentage of root hairs becomes infected.

The inner layer of the root hair cell wall consists of cellulose microfibrils while the outer layer is amorphous and comprises pectic substances. Electron micrographs of the pre-infection phase show rhizobia embedded in the pectic layer and this has led to the hypothesis that the bacteria bind to specific sites on the root hair wall. Some workers (e.g. Dazzo & Hubbell 1975) believe that the rhizobia attach to strain-specific glycoproteins (lectins) on the surface of the root hair but the evidence for this is inconclusive. Others (e.g. Robertson *et al.* 1981) suggest that the binding results from rhizobium polysaccharide–plant pectin or polysaccharide–polysaccharide interactions.

Another regular feature of the pre-infection phase is the curling of the root hair. The nature of the substances which cause root hair curling is uncertain but indoleacetic acid (IAA) has been implicated. The amino acid tryptophan is exuded by the host roots and is thought to be converted to IAA by the rhizobia. It should be noted that the exudation of tryptophan, its conversion to IAA, and root hair curling, can all occur whether or not the *Rhizobium* strain involved can nodulate with the host in question. Root hair curling is apparently not an essential prerequisite of infection though it may facilitate it by slowing down the rate of extension growth of the hair near the point of attachment of the bacterium.

The entry of rhizobia into the root hair is accomplished by invagination of the inner wall and its associated plasma membrane to form an infection thread, which is a hypha-like tube composed mainly of cellulose secreted by the host. The enzymes responsible for invagination have not been identified for certain. It was formerly held that a host polygalacturonase was induced by the presence of the extracellular slime layer of the bacterium, and this helped to plasticize the root hair cell wall. Pectinases and cellulases may also be involved, although there is no evidence that rhizobia themselves can produce such enzymes.

The infection thread grows at its tip, which is free of cellulose, towards the base of the root hair cell and thence into the cortex. In most legumes, cell division is stimulated in advance of the growing infection thread. This process was once thought to be initiated by a pre-existing tetraploid cell (Fig. 8.3) but recent evidence indicates that the disomatic state, which is common in nodule tissue, arises after the bacteria have been released from the infection thread, which by this time is ramifying through the newly produced cells. Vascular elements soon differentiate within the nodule cortex and link up with those of the parent root. Not all legumes possess infection threads, and in such hosts the bacteria apparently spread from cell to cell in the root and developing nodule by the division of already infected cells. In those species in which rhizobia do not

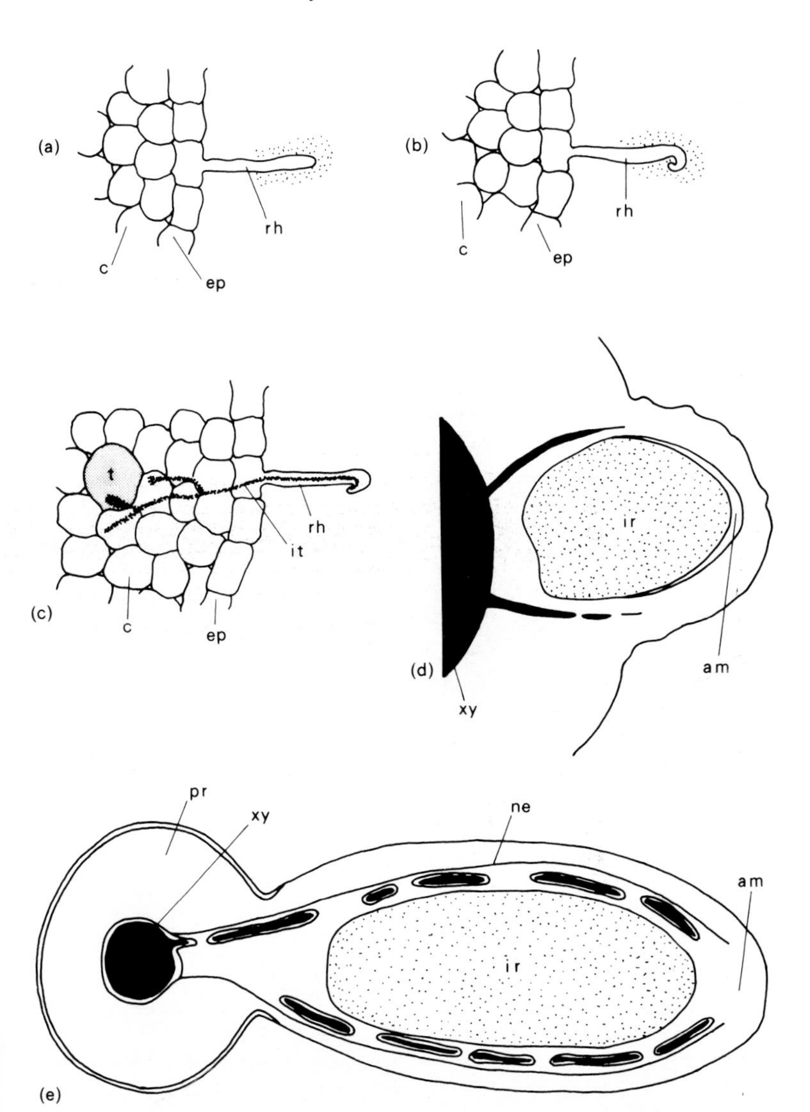

Fig. 8.3 Initiation and structure of pea nodules. (a) Aggregation of rhizobia around root hair. (b) Curling of root hairs. (c) Rhizobia move through infection thread into tetraploid cell. (d) Differentiation of young nodule into central infected region and apical meristem. (e) Longitudinal section through mature nodule: am, apical meristem; c, cortex; ep, epidermis; ir, infected region; it, infection thread; ne, nodule endodermis; pr, primary root; rh, root hair; t, tetraploid cell; xy, xylem. (From Stewart 1966.)

enter via root hairs, the infection process is less well understood.

Bacteria released into the cortical cells are isolated from the host cytoplasm by being enclosed in vesicles pinched off from the plasma membrane lining the infection thread. Within these membrane envelopes, the rhizobia multiply and become transformed to 'bacteroids', which are irregularly shaped and in some legumes even branched to produce X and Y forms. From one to several bacteroids are contained within each membrane envelope, depending on the identity of the host (Fig. 8.4).

The process by which bacteria in the infection thread enter the host cell appears to be similar to the endocytotic process for passage of particles and macromolecules across cell membranes. The bacteria are first adsorbed onto the outer surface of the cell's boundary

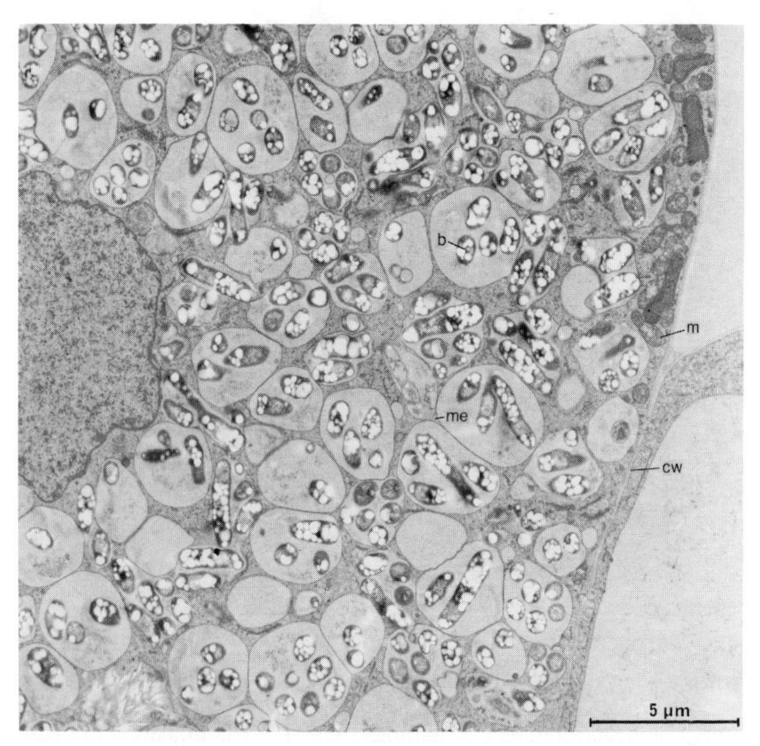

Fig. 8.4 Electron micrograph of mature cell of soybean nodule, showing groups of bacteroids (b) enclosed in membrane envelopes (me). cw, cell wall; m, mitochondrion. Photograph kindly provided by Dr. D. J. Goodchild.

membrane, which then invaginates to enclose them in a vacuole. It has been estimated that each central tissue cell of a mature soybean nodule contains up to 500 000 bacteroids, and that there are some 10 000–40 000 such cells in each nodule.

Bacteroids do not divide within the nodule and have not so far been cultivated on laboratory media. Concomitant with bacteroid formation, the pigment leghaemoglobin is produced, the enzyme nitrogenase is synthesized, and nitrogen fixation begins. After a period of weeks or months (except in those species with perennial nodules) leghaemoglobin is replaced by bile pigments, the bacteriods lyse, and degenerative processes spread throughout the nodule. Rod-shaped bacteria, which have meanwhile lain dormant in the infection thread, invade and multiply in the old nodule which then decays, permitting the bacteria to escape into the soil.

Nodulation of non-legumes by rhizobia

In 1973, M. J. Trinick reported that nodules could be formed by *Rhizobium* on *Trema aspera*, a non-leguminous species of the family Ulmaceae. This plant was later re-identified as *Parasponia andersonii* by Akkermans *et al.* (1978). Isolates from *Parasponia* nodulate several tropical legumes, and rhizobia isolated from a number of such legumes are able to nodulate *Parasponia*. Anatomically, the nodules closely resemble the actinorrhizas of *Alnus* (Trinick 1979), and contain leghaemoglobin-like proteins. The evolutionary significance of this symbiosis remains to be determined.

Molecular biology of nitrogen fixation

The availability of ^{15}N as a tracer enabled ammonia to be established as the first product of N_2 fixation in intact organisms, in this event *Azotobacter* (Wilson & Burris 1947). In 1960, J. E. Carnahan and co-workers perfected the technique of demonstrating N_2 fixation in cell-free preparations, and using this technique, McNary and Burris (1962) were able to show that N_2 fixation was an anaerobic process requiring pyruvate to produce a reductant and ATP for energy transfer. It had been discovered much earlier that the growth of legumes in soils deficient in molybdenum (Anderson 1942) or sulphur (Anderson & Spencer 1950) was inhibited, and from this and other evidence it was inferred that these elements were essential for the nitrogen-fixing process. It is now known that the enzyme complex nitrogenase contains two enzymes, a relatively large Mo–Fe protein which binds and reduces N_2, that is a dinitrogenase, and a smaller Fe–S protein which acts as a dinitrogen reductase; both are highly labile in the presence of oxygen. The complete nitrogen-fixing system requires an election donor (ferredoxin or flavodoxin) to bring about the protonation of N_2, ATP-Mg^{2+} to mediate electron transfer, and a means of regenerating ATP from ADP and P_i (Burris *et al.* 1981).

The site of fixation within the nodule

All effective nodules contain bacteroids, the amount of nitrogen fixed being strongly correlated with the amount of bacteroid-containing tissue. This suggests that the bacteroids themselves fix nitrogen which is then transferred to the host. A positive correlation also exists between the amount of leghaemoglobin and the nitrogen fixing ability of nodules. The pigment changes from red to green when nitrogen fixation ceases in annual plants at the end of vegetative growth or when they are placed in the dark for a few days.

Although the relationship between the presence of bacteroids and nitrogen fixation was known for many years, all efforts to obtain *in vitro* fixation by bacteroids failed; even in excised nodules it continues only briefly. It now appears that this is due to the sensitivity of the nitrogen fixing process to oxygen: the bacteroids require oxygen for respiration but too much oxygen inhibits fixation. Thus F. J. Bergersen and G. L. Turner (1967) found it possible to demonstrate fixation in bacteroids by studying the process in washed bacteroid suspensions obtained from breis (homogenates) extracted by crushing nodules under an atmosphere of argon. The problem of providing sufficient oxygen for bacteroid respiration at oxygen tensions low enough to permit nitrogen fixation to proceed is apparently solved through the agency of leghaemoglobin, which is able to transport oxygen to the bacteroids at many times the rate of simple diffusion. This red pigment is a regular feature of the central tissue of all effective legume root nodules, where it occurs as a partially oxygenated solution bathing the bacteroids in their membrane envelopes (Bergersen & Goodchild 1973). It has the capacity to bind O_2 and release it at the very low O_2 concentrations (0.01–0.1 μM) necessary for maximum nitrogenase activity. Leghaemoglobin thus serves as a macromolecular carrier to facilitate the diffusion of O_2, resulting in increased ATP concentration in the bacteroids with a consequent stimulation of nitrogenase activity. The gas exchange system of the nodule as a whole is enhanced by a radially oriented network of small, uninfected and vacuolated cells connecting the central, N_2-fixing tissue to the soil atmosphere. These interstitial cells are linked by plasmodesmata to the bacteroid-filled cells and presumably have the additional function of metabolite transfer between symbionts.

8.2 Non-legume root nodules: actinorrhiza

Root nodules known as **actinorrhiza** are found in eight angiosperm families other than those of the Leguminosae. According to Hutchinson (1969), these families fall into seven orders. Altogether, some eighteen genera are represented (Table 8.2). These nodulated non-legume genera have a world-wide distribution, from the arctic to the tropics, and (excluding *Rubus*) comprise nearly 300 species. About 40 to 50 per cent of these species are known to possess actinorrhizas though not all have been thoroughly examined. *Rubus* contains but one actinorrhizal species so far as is known, *R. ellipticus*. Except for the herbaceous *Datisca* spp., all these nodulated plants are woody shrubs or trees, and they often occur naturally on infertile soils. Their ecological significance, and their contribution to the geochemical cycle of nitrogen, will be discussed later (sect. 8.3).

Identity of the endophyte

Until recently, most attempts to synthesize nodules in axenic culture, using microorganisms isolated from naturally occurring nodules as inocula, were unsuccessful, and evidence as to the nature of the endophyte was perforce derived from cytological studies. Many conflicting reports came from investigations with the light microscope, but electron microscopy has revealed that the nodule endophyte is an actinomycete, occurring as fine, branched, septate hyphae less than 1 μm in diameter. A prominent feature of most

Table 8.2 Non-leguminous plants with root nodules (actinorrhizas)

Order	*Family*	*Genus*
Coriariales	Coriariaceae	*Coriaria*
Rosales	Rosaceae	*Cercocarpus, Chamaebatia, Cowania, Dryas, Purshia, Rubus*
Myricales	Myricaceae	*Comptonia, Myrica*
Fagales	Betulaceae	*Alnus*
Casuarinales	Casuarinaceae	*Casuarina*
Cucurbitales	Datiscaceae	*Datisca*
Rhamnales	Eleagnaceae	*Eleagnus, Hippophaë, Shepherdia*
	Rhamnaceae	*Ceanothus, Colletia, Discaria*

Source: After Gibson and Jordan (1983).

actinorrhizas, under the light microscope, is the presence in some cells of club shaped or spherical structures known as vesicles; they are prevalent in nodule tissue which is actively fixing nitrogen but are absent from ineffective nodules. Evidence from the acetylene reduction technique (see sect. 8.3) indicates that they are essential for the fixation process. In electron micrographs, the vesicles are seen to be terminal cells of hyphae, 3–4 μm in diameter, with a complex internal structure. They have not been described from all actinorrhizas examined, however, and are notably absent from N_2-fixing *Casuarina* nodules. In some species, irrespective of whether vesicles are produced or not, the hyphae appear to fragment into bacteroid-like segments; these structures were identified as spores by van Dijk and Merkus (1976), and are formed in sporangia which are characteristic of the particular endophyte. In *Alnus glutinosa*, some strains produce abundant sporangia and spores while others produce very few (van Dijk 1978).

In 1970, J. H. Becking proposed that a new family of bacteria be established (Frankiaceae) to include all those actinomycetes which form nodules on the roots of non-leguminous dicotyledons, and he recognized a number of species which he assigned to the genus *Frankia*. D. Callaham *et al.* (1978) isolated and cultivated *in vitro*, the endophyte of *Comptonia* actinorrhizas, and since that time the actinomycete *Frankia* has been isolated from various hosts, and actinorrhizas have been synthesized by inoculating susceptible plants with the isolates under microbiologically controlled conditions. The actinomycete, while relatively slow growing, does not have to be cultivated in a particularly complex nutrient medium.

Incidence and specificity of nodulation

Good nodule development depends on favourable environmental conditions. For most genera, nodulation is best in a substrate with a pH around neutrality. The numbers of nodules per plant decline as the level of combined nitrogen in the rooting medium increases; nodule weight however increases, except at very high levels of available nitrogen, when it too is depressed.

Organisms from one genus will not necessarily infect plants in another genus. Until quite recently, all the evidence pertaining to such specificity was based on inoculation with crushed nodule suspensions. Very little of this earlier work has been confirmed using pure cultures of actinomycetes as inocula. According to information assembled by Quispel and Burggraaf (1981), strains isolated from *Alnus* spp. (Betulaceae) will nodulate species of *Comptonia* and *Myrica* (Myricaceae) but will not form actinorrhizas with species of *Eleagnus, Hippophaë* or *Shepherdia* (Eleagnaceae) nor with species of *Ceanothus* (Rhamnaceae) or *Coriaria* (Coriaceae). Until such time as more data derived from studies with pure culture inocula are available, few generalizations about specificity are possible: at most it could be said that there appears to be nothing like the marked specificity found in the legumes and especially in the Papilionaceae (Fabaceae).

Infection and nodule development

The nodules of non-legumes are perennial. Externally they are fairly distinct from those of legumes. At first they appear as lateral swellings on the roots and later become strongly lobed, producing large nodule clusters up to 5 cm in diameter (Fig. 8.5). Two distinct types were previously recognized (Becking 1977): one with negatively geotrophic roots arising from the nodule lobe apices, found in *Casuarina* and *Myrica*; and one with slow growing, dichotomously branched apices, typified by *Alnus*. According to Gibson and Jordan (1983) however, both types of actinorrhiza have been found on *Alnus* and *Myrica*.

Actinorrhizas, unlike legume nodules, are modified lateral roots. Infection occurs at discrete points on the root and is often associated with the distortion of root hairs. The endophyte penetrates via a crook or fold in the deformed hair, apparently after being encapsulated on its surface by extracellular substances of host origin (Angulo *et al.* 1976; Callaham *et al.* 1979). Entry into the plant tissues appears to be achieved by invagination of the outer cytoplasmic membrane of the root hair cell, as in legumes (Becking *et al.* 1964). As the endophyte advances through the root cortex, a lateral root or 'prenodule' is initiated in the pericycle, and is transformed to an actinorrhiza by invasion and multiplication of the actinomycete in its cortical cells. Anatomically, actinorrhizas are characterized by a well-developed cork layer, an infected cortex and a central stele which shows secondary thickening. The presence of the endophyte in the cortex contrasts markedly with the infected region in legume nodules, which is internal to the vascular system.

The nitrogen-fixing process in actinorrhizas

Conclusive evidence that the nodules are the site of fixation came from the investigations of G. Bond in the 1950s (Bond 1955, 1956, 1967). If the nodules are removed from a plant, it soon develops symptoms of nitrogen deficiency which persist until new nodules are formed. When nodulated plants are allowed to fix nitrogen labelled with ^{15}N, the label is found mainly in the nodules, and if the nodules are removed before the test and they and the rest of the root systems are then gassed separately, only the nodules show this evidence of fixation.

(a)

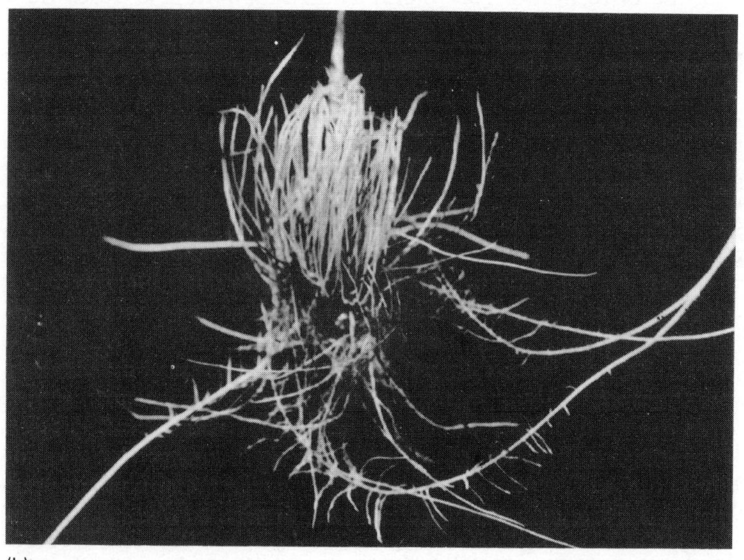

(b)

Fig. 8.5 Root nodules (actinorrhizas) of non-leguminous angiosperms. (a) *Alnus glutinosa*. (b) *Myrica gale*, 10-week-old rootsystem showing negatively geotropic roots arising from the lobes of some nodules. (From Stewart 1966.)

The amount of nitrogen fixed is proportional to the volume of infected tissue and, using appropriate techniques, high levels of nitrogenase activity can be demonstrated in nodule extracts containing a high proportion of vesicles. Presumably, the nitrogenase system is located in the vesicles (Tjepkema *et al.* 1980) but this is not certain, nor is the site of fixation known in *Casuarina* nodules which lack a vesicular stage.

Although the presence of a haemoglobin in actinorrhizas has been reported it is doubtful whether these reports are correct. Indeed, one of the surprising aspects of the physiology of actinorrhizas is the apparent tolerance of the N_2-fixing system to oxygen, as shown by their ability to function effectively at atmospheric O_2 levels. J. D. Tjepkema and colleagues have suggested that the thickened walls of vesicles afford protection against O_2 in a manner similar to that of the heterocysts of cyanobacteria (q.v. sect. 8.3.1.). The corky exterior of *Alnus* actinorrhizas is broken by lenticels which connect the N_2-fixing tissue to the surrounding atmosphere. *Myrica* nodules lack lenticels but their negatively geotrophic roots may be important in maintaining an oxygen supply to the endophyte.

8.3 Biogeochemistry of nitrogen

As a constituent of proteins, nitrogen is of paramount importance to organisms. Despite the fact that there are some 30 000 tons of nitrogen in the air over every acre of the Earth's surface (755 g cm^{-2}), most plants and animals, including a large proportion of the world's human population, suffer from the effects of nitrogen deficiency at some stage of their lives. The great reservoir of nitrogen in the atmosphere is the ultimate source of nitrogen for the biosphere, so that the significance of root nodule symbioses is best gauged by considering their role in fixing gaseous nitrogen relative to that of other agents for removing this element from the air. These agents may be either biological or physical, and their several contributions will be considered in turn.

8.3.1 Biological nitrogen fixation

Biological nitrogen fixing systems involve either symbiotic or free-living soil microbes. The former include root nodule symbioses and several mutualistic associations involving cyanobacteria. Before discussing the relative significance of various diazotrophs (N_2-fixers) in the biogeochemistry of nitrogen, it is first necessary to review the means by which the rate of fixation can be measured.

Measurement of nitrogen fixation

Methods of determining N_2 fixation rates may be based on changes in the total N content of ecosystems or their components, on the incorporation of the stable isotope of nitrogen (^{15}N) by plants or soil, or on nitrogenase activity as measured by the acetylene–ethylene assay (Knowles 1981; Havelka *et al.* 1982).

Kjeldahl analysis. Determination of the total N content of plant samples by the Kjeldahl technique can be used to determine the rate of fixation by leguminous crops, provided appropriate controls are used, e.g. uninoculated plants or a non-nodulating isoline of the same species. In nitrogen deficient substrates such as sand dunes or glacial moraines, where little or no mineral N is available for plant growth, the increase in total soil N over a long period of time may provide a crude estimate of the fixation rate. In very infertile soils during a relatively short time interval however, the soil N content may decline even though the total amount in the soil–plant system increases (Richards & Bevege 1967).

^{15}N analyses. Nitrogen-15 is a naturally occurring, stable isotope of nitrogen, accounting for about 0.4 per cent of atmospheric N_2. By exposing diazotrophic systems to an atmosphere artifically enriched in ^{15}N, fixation can be detected by mass spectrometry as an increase in the $^{15}N/^{14}N$ ratio of plant or microbial tissues (Burris *et al.* 1943). Results are presented as the 'atom per cent ^{15}N excess' in the tissues under test, this being the atom per cent ^{15}N in the sample less the atom per cent ^{15}N in normal air (0.38%); the atom per cent ^{15}N in the sample is the number of ^{15}N atoms expressed as a percentage of the total number of N atoms, i.e. ^{14}N plus ^{15}N. A value of 0.015 atom per cent ^{15}N excess in a single sample serves as a conservative level for establishing fixation (Burris & Wilson 1957). It requires a high degree of technical skill to make precise and accurate estimates of fixation rates by this technique, especially under field conditions, but there are indirect methods based on ^{15}N analysis that are less demanding of technical expertise. One is the so-called ^{15}N-dilution technique in which a low level of $^{15}NH_4$-N or $^{15}NO_3$-N is added to the soil in which a test legume and a suitable non-fixing control plant are growing: fixation by the legume dilutes the label so that the difference in the ^{15}N enrichment of test and control plants can be used to calculate the amount of N_2 fixed.

The 'natural abundance' method depends on the difference which exists in the ^{15}N abundance of soil and air. Presumably as a result of isotopic discrimination over a pedogenic time scale, most soils contain a slightly higher proportion of ^{15}N than is found in atmospheric N_2, the difference amounting to as much as 0.5 to 1.0

per cent (Rennie *et al.* 1976). Furthermore, should any isotopic discrimination occur during the fixation process, it is likely to favour the lighter of the two isotopes. For both these reasons, a plant which is fixing gaseous nitrogen should have a lower ^{15}N content than one which is entirely dependent on soil mineral nitrogen. The difference should theoretically provide a measure of fixation rate but there are several problems associated with its application and it has not yet been adequately evaluated.

Acetylene reduction. This method depends on the ability of nitrogenase to use acetylene (C_2H_2) as an alternative substrate to N_2, reducing it in the process of ethylene (C_2H_4). The rate of ethylene production, as measured by gas chromatography, is taken as a measure of nitrogenase activity. Acetylene reduction by diazotrophs was first reported independently by M. J. Dillworth and by R. Schöllhorn and R. H. Burris in 1966, and a detailed evaluation of the acetylene–ethylene assay was published by R. W. F. Hardy and his colleagues in 1968. Its simplicity and low cost has led to its widespread use since that time but unfortunately its limitations have not always been appreciated.

A major problem is the difficulty of extrapolating from the amount of C_2H_4 produced in the short term (exposures of less than 2 hours' duration are usual) to the fixation of N_2 in the long term (Gibson & Jordan 1983). It is therefore necessary to calibrate the acetylene–ethylene assay against other methods. The theoretical ratio of C_2H_2:N_2 reduced is 3:1 but published observations range from 1:1 to 7:1. One of the reasons for this may be that nitrogenase activity in some legumes declines rapidly in the presence of C_2H_2, due possibly to the cessation of ammonia production (Minchin *et al.* 1983). This results in underestimates of nitrogenase activity based cumulative C_2H_4 production especially in the closed experimental systems which are widely used for this assay.

On the other hand, the acetylene–ethylene assay can overestimate nitrogenase activity in soils because C_2H_2 suppresses the oxidation of endogenously produced C_2H_4. The latter is a minor fermentation product (see sect. 2.2) produced in anaerobic microsites but its normal fate in well aerated soils is to be rapidly oxidized by aerobes. The introduction of C_2H_2 blocks the oxidation of C_2H_4 and it is only then that endogenous C_2H_4 can be detected. Since it is impossible to distinguish between this and the C_2H_4 derived from acetylene reduction, nitrogenase activity is likely to be less than that shown by the assay.

Within the limitations of the methods used to estimate the rate of biological nitrogen fixation, we turn now to a discussion of the relative contributions of the various diazotrophs, both symbiotic and free-living, to the biogeochemical cycle of nitrogen. Non-biological mechanisms for sequestering atmospheric N_2 in the soil–plant system will be discussed in the succeeding section (8.3.2).

Symbiotic fixation

Legumes

The cultivation of leguminous crops is a basic feature of most of the permanent agricultural systems of the world. From the point of view of the agronomist, the legume–rhizobium symbiosis is undoubtedly the most important biological mechanism for adding nitrogen to the soil–plant system. While none would deny its agronomic importance, its significance from the geochemical viewpoint is another matter. G. E. Hutchinson, following a critical review of the available data in 1954, concluded that legumes are responsible for only 3–4 per cent of the nitrogen fixed at the surface of the Earth each year.

As previously indicated (sect. 8.1.), knowledge of the role of legumes in nitrogen fixation is restricted to a few tribes of the family Papilionaceae (Fabaceae). These are cultivated herbaceous legumes capable of fixing nitrogen at rates approximating 100 kg ha^{-1} annually. Under intensive agriculture, even higher rates of fixation have been achieved: about 300 kg ha^{-1} yr^{-1} in the tropics and 600 kg ha^{-1} yr^{-1} in temperate regions (Henzell 1968; Gibson & Jordan 1983). Results for a wide range of legumes studied as part of the International Biological Programme were summarized by P. S. Nutman (1976). A value of the order of 100 kg ha^{-1} yr^{-1} may be accepted, given appropriate management. It would be unwise to assume however that fixation rates attained by these herbaceous species in cultivation apply also to the vast numbers of wild legumes. Taking into account that one-third of the land surface is sparsely vegetated, and that the cultivation of legumes is practised on but a small fraction of the remainder, Hutchinson estimated that the total annual contribution of legumes is slightly less than 0.3 kg ha^{-1} of fixed nitrogen for the Earth's surface as a whole. This estimate might be enlarged if more data were available on nitrogen fixation in the Leguminosae as a whole, for it is not inconceivable that the woody perennial legumes of the tropics and subtropics contribute far more to the nitrogen cycle than the handful of herbaceous species that have been intensively studied. It must be remembered, however, that many tropical legumes apparently do not nodulate, and therefore presumably do not fix nitrogen at all. Furthermore, legumes are more likely to be ecologically dominant in successional than climax plant communities and so may make only a minor contribution to the nitrogen economy of undisturbed ecosystems.

Actinorrhizal plants

Perhaps of greater geochemical significance than legumes are the symbioses involving nodulated non-leguminous plants. The ability of plants such as alder (*Alnus*) to act as soil improvers has been known for almost as long as that of legumes, however quantitative data on their contribution to the nitrogen cycle is meagre. *Gale palustris* (syn. *Myrica gale*), which grows in acid peat bogs, fixes about 9 kg N ha^{-1} yr^{-1} in the field, whereas in artificial culture its nitrogen-fixing ability may surpass that of red clover and other annual legumes (Bond 1951). Under appropriate ecological conditions, non-legumes may achieve quite high fixation rates, for example 62–164 kg N ha^{-1} yr^{-1} for *Alnus crispa* (Crocker & Major 1955), 58 kg N ha^{-1} yr^{-1} for *Casuarina equisetifolia* (Dommergues 1963) and up to 218 kg N ha^{-1} yr^{-1} for *C. littoralis* (Silvester 1977). W. B. Silvester (1977) summarized the literature on N_2 fixation by actinorrhizal plants: an average value for vigorous stands of angiosperms would probably exceed 50 kg N ha^{-1} yr^{-1}. Their contribution to the nitrogen economy of many forest ecosystems is widely recognized, and bearing in mind their broad distribution, it is likely that these actinorrhizal plants will prove to be among the major contributors to the geochemical cycle of nitrogen.

Symbioses involving cyanobacteria

Some diazotrophic cyanobacteria or blue-green algae enter into mutualistic associations with a wide range of eukaryotic organisms from fungi to angiosperms (Gibson & Jordan 1983). Many of these symbioses make a significant contribution to the nitrogen economy of particular ecosystems while some (e.g. the *Azolla–Anabaena* association) may be important agriculturally.

Lichens. A lichen (see Ch. 3) is an association between a fungus, which is usually a discomycete (a sub-group of Ascomycetes), and an alga which may be green or blue-green, i.e. a cyanobacterium. About 8 per cent of all lichens contain cyanobacteria, and it is likely that most of these fix atmospheric nitrogen. Fixation has been demonstrated by the ^{15}N technique in several genera (Bond & Scott 1955). Quantitative data on the contribution of lichens to the nitrogen cycle is lacking, but they no doubt play an important role in vegetating specific habitats such as bare rock surfaces, thereby initiating pedogenesis and primary ecological succession.

In cool temperate, humid coniferous forests dominated by Douglas fir, inputs of atmospheric N_2 up to 10 kg ha^{-1} yr^{-1} have been attributed to lichens (Pike *et al.* 1972). They are abundant on bark of many rainforest trees in the tropics and subtropics and it is conceivable that they play a significant role in the nitrogen economy of rainforest ecosystems. In drier terrestrial arctic and subarctic habitats, lichens – along with epiphytic cyanobacteria – are the major source of nitrogen for biotic communities (Gibson & Jordan 1983). The magnitude of their contribution in the tundra varies greatly with seasonal and environmental conditions but on average is probably about 1–2 kg ha^{-1} annually.

Despite their obvious importance in particular ecosystems, and their widespread distribution, the limitations imposed by exposure to environmental extremes (especially low temperatures and high water deficits) reduce their overall geochemical significance.

The Azolla–Anabaena association. Species of *Azolla*, a small freshwater fern, have a mutualistic relationship with the cyanobacterium *Anabaena azollae*. The fern has small, alternate and overlapping, bilobed leaves, the aerial upper lobe of which contains chlorophyll while the partially submerged lower lobe is nearly colourless. The cyanobacterium is encased in a mucilage-filled cavity in the ventral surface of the upper lobe (Peters *et al*. 1980). *A. azollae*, unlike the free-living species of the genus, is not inhibited from fixing N_2 by high levels of mineral nitrogen so that nitrogenase activity is manifest even in waters containing 100 ppm NH_4-N.

Azolla is widely distributed in temperate and tropical regions but is most common in warm tropical and subtropical waters. It forms dense surface mats in favourable environments, notably in calm, poorly aerated eutrophic waters. It is widely used as a forage and green manure crop and is especially suitable for rice cultivation (Moore 1969). Its N_2-fixing ability varies with seasonal and environmental factors but it is credited with contributing of the order of 100 kg ha^{-1} yr^{-1} in a wide variety of situations (Gibson & Jordan 1983).

Coralloid roots of cycads. The gymnosperm family Cycadaceae contains nine genera and about 90 species, with a distribution restricted to the tropics and subtropics of the southern hemisphere. These plants produce, in addition to their normal roots, clusters of negatively geotropic 'coralloid roots' which arise from the surface of the swollen hypocotyl and taproot. The coralloid roots are infected by the cyanobacterium *Nostoc* (or perhaps *Anabaena*) and occur either at the soil surface or deep within the soil profile. Infected cells form a layer akin to the epidermis of normal roots but overlaid with a secondary cortex.

N_2 fixation in the coralloid roots of cycads has been confirmed by the ^{15}N technique for *Cycas* and *Macrozamia* by F. J. Bergersen *et al*. (1965), and for *Encephalartos* and *Ceratozamia* by G. Bond (1967). Because many of the coralloid roots occur at depth in the soil, where little or no light penetrates, the diazotroph presumably grows and fixes N_2 heterotrophically. The contribution of these symbioses to the geochemical cycle of nitrogen is probably negligible, however they might well be significant in the nitrogen economy of particular ecosystems. *M. communis*, for instance, is a prominent understorey plant of the *Eucalyptus maculata* forests of south-eastern Australia (Bergersen *et al.* 1965) while in the *E. marginata* forests of south-western Australia *M. riedlei* is said to fix 1–8 kg N ha^{-1} yr^{-1} (Grove *et al.* 1980).

Other symbiotic associations

There are other plant–microbe associations which may be of some significance in the nitrogen economy of certain ecosystems, and these are the so-called leaf glands or nodules on leaves. G. Stevenson in 1959 demonstrated fixation of $^{15}N_2$ by leafy shoots of *Coprosma*, a genus of the family Rubiaceae which possesses stipular glands inhabited by bacteria. Three other genera in this family, namely *Psychotria, Pavetta* and *Chomelia*, bear leaf nodules, which are sub-epidermal cavities filled with slime and bacteria. Of these, reliable evidence of fixation has been obtained only for *Psychotria*. W. S. Silver and co-workers found in 1963 that homogenates of the leaves of *Psychotria* fixed $^{15}N_2$; they isolated from the nodules the bacterium *Klebsiella*, and showed that it could fix $^{15}N_2$ in pure culture. Nitrogen fixation is apparently not the only function of the endophyte, because although nodulated plants can grow normally in a nitrogen-free medium, non-nodulated plants are abnormal even when supplied with mineral nitrogen. It has been suggested that the bacteria supply an essential growth factor in addition to fixing nitrogen, and this is supported by the finding that gibberellic acid partially replaces the bacteria in allowing normal growth on combined nitrogen. Leaf nodules also occur in *Ardisia*, a widely distributed tropical genus of the family Myrsinaceae, but the evidence for nitrogen fixation in this instance is contradictory.

Non-symbiotic fixation

Turning now to non-symbiotic fixation, the organisms concerned here are either bacteria or cyanobacteria, that is lower protists. Despite earlier reports that soil-inhabiting yeasts could fix N_2, there is no unequivocal evidence that eukaryotes possess nitrogenase.

Bacteria

Diazotrophic bacteria include representatives of all the major nutritional categories described in Chapter 2, and range from strict aerobes to obligate anaerobes (Table 8.3). Since nitrogenase is oxygen sensitive, aerobic species must have some kind of protective

Table 8.3 Free-living, N_2-fixing bacteria

Nutritional category	*Genus*
Chemoheterotrophs	
Aerobes	*Azotobacter, Azomonas, Azotococcus, Beijerinckia, Derxia, Xanthobacter; Methylobacter, Methylococcus, Methylocystis, Methylosinus*
Anaerobes	*Clostridium; Desulphomaculatum, Desulphovibrium*
Facultative anaerobes	*Arthrobacter; Bacillus; Aquaspirillum, Azospirillum, Campylobacter; Citrobacter, Enterobacter, Erwinia, Escherichia, Klebsiella*
Chemoautotrophs	*Thiobacillus*
Photoheterotrophs	*Rhodospirillum, Rhodopseudomonas, Rhodomicrobium*
Photoautotrophs	*Amoebobacter, Chromatium, Ectothiorhodospira, Thiocapsa, Thiocystis; Chlorobium, Pelodictyon*

Source: Genera based on Postgate (1981) as modified by Gibson and Jordan (1983).

mechanism. In *Azotobacter*, protection is afforded by a very high rate of cellular respiration which maintains intracellular oxygen concentrations at sub-toxic levels. In facultative anaerobes, fixation takes place only at reduced oxygen tensions.

The aerobic chemoheterotrophs include not only the well known genus *Azotobacter* but taxonomically related forms as well. *Azotobacter* has a world-wide distribution, especially in neutral and alkaline soils. *Beijerinckia* is restricted mainly to acid, lateritic soils of tropical and extra-tropical regions. *Derxia* appears to be confined to the tropics.

Azotobacter was first isolated by Beijerinck in 1901. Since then, innumerable papers have been written on its nitrogen-fixing ability, '... their number and volume directly related to the ease and convenience of working with this aerobic organism, and almost unrelated to its real importance in world nitrogen economy' (Henzell & Norris 1962). The contribution of heterotrophic nitrogen fixers to the nitrogen cycle is limited by the availability of a suitable organic substrate. *Azotobacter*, for example, is most abundant in fertile soils with a high organic matter content, although numbers rarely exceed 200 cells per gram. The presence of *Azotobacter* does not however mean that it is making any significant contribution to the nitrogen economy of such soils, since these will normally contain some mineralized nitrogen, such as ammonium, which *Azotobacter* is known to utilize in preference to molecular nitrogen.

The facultative anaerobes, though widespread in soils (*Arthrobacter, Bacillus*) or in decomposing plant residues (*Klebsiella, Enterobacter*), are subject to the same restrictions with respect to energy substrates as *Azotobacter*. This is true also of *Clostridium*, the best known obligately anaerobic diazotroph. *Clostridrium* was the first bacterial genus shown to fix nitrogen in pure culture. Its nitrogen-fixing capacity, like that of *Azotobacter*, is inhibited by small amounts of ammonium or nitrate. Knowledge of its role in nature is quite speculative, for the reason that it is difficult and tedious to work with an organism that is an obligate anaerobe. It is known to be more widespread in distribution than *Azotobacter*, being much more tolerant of acid conditions. Its contribution to the nitrogen cycle could therefore be greater than that of the more readily studied aerobe.

Because of the limitations imposed by their energy requirements, and their preferential use of mineral nitrogen, the agronomic value of free-living, heterotrophic nitrogen-fixing bacteria in cultivated soils is thought to be slight. The same may be true of their geochemical significance; however, as pointed out by H. L. Jensen in 1950, the efficiency of nitrogen-fixing heterotrophs may be greater in the field than in the laboratory. Thus in uncultivated soils, where crops are not removed but where plants are allowed to die and decompose *in situ*, it is conceivable that chemoheterotrophic bacteria, both aerobic and anaerobic, make a substantial contribution to the nitrogen cycle. Although fixation rates of less than 0.3 kg N ha^{-1} yr^{-1} are thought to be representative of the inputs provided by *Azotobacter* and *Clostridium* in unamended agricultural soils (Postgate 1974), there is evidence that fixation rates are an order of magnitude greater in the soil–litter subsystem of forest soils (Silvester 1978; O'Connell *et al.* 1979). Even such small inputs of atmospheric nitrogen may be sufficient, in the absence of disturbance, to balance losses from natural ecosystems by denitrification and leaching.

Several kinds of autotrophic nitrogen-fixing bacteria are known, such as the chemoautotrophic *Thiobacillus ferro-oxidans* and the photoautotrophic green sulphur bacteria (e.g. *Chlorobium*) and purple sulphur bacteria (e.g. *Thiocystis*). It would seem that all photosynthetic bacteria, including the photoheterotrophic non-sulphur purple bacteria (*Rhodospirillum* etc.), are capable of fixing nitrogen. Some are able to fix nitrogen also in the dark, when growing chemoheterotrophically, but their efficiency of fixation is then much lower. It should be kept in mind that the environmental conditions for bacterial photosynthesis are very restrictive. The bacteria require both sunlight and anaerobic conditions, thus are likely to be active only in shallow waters and estuarine muds. Although their local importance in such ecosystems may be considerable, their contribution to the geochemical cycle of nitrogen is probably small.

Associative nitrogen fixation. In 1961, J. Döbereiner recorded the presence of *Beijerinckia* in the rhizosphere of sugar cane, and in 1974 she reported finding high levels of nitrogenase activity in the rhizosphere of *Digitaria decumbens* colonized by *Spirillum lipoferum*, a bacterium which is now classified as either *Azospirillum lipoferum* or *A. brasiliense*. Subsequently, Döbereiner and Day (1976) reported the presence of diazotrophic bacteria in the inner cortical cells of this grass, that is, in the 'endorhizosphere' (sect. 6.1). Many other grasses and cereals have since been implicated in what has become known as **associative nitrogen fixation** (van Berkum & Bohlool 1980). The major site of fixation – cortex or rhizosphere – has not yet been determined with certainty.

It was at first thought that *Azospirillum* was associated only with tropical grasses characterized by the C_4 photosynthetic pathway, but it has since been found in or on the roots of C_3 plants too, not only in the tropics but also in temperate regions. Nor is *Azospirillum* the only diazotroph involved: *Azotobacter paspali* is regularly associated with the roots of some ecotypes of *Paspalum notatum*, while species of *Klebsiella, Enterobacter, Erwinia* and *Pseudomonas* occur in the rhizospheres of various cereals and forage grasses throughout the world (Gibson & Jordan 1983).

Inoculation with *Azospirillum* or other diazotrophs produces variable results. Even when a positive response is evident, there is uncertainty about whether the stimulus to growth arises from N_2 fixation or is due to other causes. As indicated in sect. 6.5, some of the bacteria implicated are known to produce growth factors (IAA, cytokinins and gibberellins) which conceivably stimulate root growth and so enable the plant's root system to exploit the soil more effectively. None the less, inoculation with free-living heterotrophic diazotrophs, especially *Azotobacter*, has been practised for many years in Russian agriculture (Mishustin & Shil'nikova 1971), and evidence from ^{15}N studies indicates that fixed N_2 is rapidly incorporated into grasses (Döbereiner & Boddey 1981) and rice (Watanabe 1981) in amounts which would contribute significantly to the nitrogen economy of these plants. Not all such studies have produced unequivocal results, however (Gibson & Jordan 1983). Whether such fixation as does occur adds much nitrogen to the soil–plant system is even more problematical. There is no doubt that the rates of fixation achieved by free-living diazotrophs in association with plant roots are substantially higher than those attained in non-rhizosphere soil, which is to be expected in view of their dependence on readily available carbohydrates as energy substrates: Döbereiner (1974) claimed that the *Azotobacter paspali–Paspalum notatum* association had the potential to fix as much as 100 kg N ha^{-1} yr^{-1}. There are known to be many ecosystems in which unexplained gains of nitrogen have been recorded, gains of the order of 50 kg ha^{-1} yr^{-1}, for example in soil beneath pure grass swards (Parker 1957) and in pure stands of pines and other conifers (Richards 1964). These ecosystems contain no symbiotic associations which fix molecular nitrogen, yet the measured rates of fixation far exceed the accepted contribution of free-living microorganisms. It follows that either the activities of the non-symbiotic nitrogen fixers are capable of great stimulation in some as yet unexplained fashion (in the rhizosphere, for example) or that nitrogen fixation in such ecosystems is accomplished by hitherto unrecognized processes. Associative nitrogen fixation is now accepted as making a significant contribution to prairie ecosystems (Kaputska & Rice 1978) and, according to Jordan (1981), grasslands such as these depend largely on such rhizospheric fixation to maintain their nitrogen status. It is not improbable that the same is true of some forest ecosystems, since the roots of coniferous tree seedlings can incorporate $^{15}N_2$ in ecologically significant amounts (Bevege *et al.* 1978).

In 1936, J. F. Lipman and A.B. Coneybeare published a detailed analysis of the nutrient budget of the USA and estimated that the mean rate of nitrogen fixation by non-symbiotic bacteria was 6.7 kg ha^{-1} yr^{-1}. Assuming this to embrace associative nitrogen fixation as well, we may accept this figure, as did G. E. Hutchinson (1954), as a reasonable estimate for the Earth's land area as a whole. Assuming that the land occupies 29 per cent of the Earth's surface, we arrive at a figure of 2.8 kg ha^{-1} yr^{-1} as the contribution of free-living and associative heterotrophs to the nitrogen budget of the Earth.

Cyanobacteria

A group of autotrophic microorganisms whose geochemical significance may be greater than any of the free-living bacteria is the cyanobacteria or blue-green algae. The first evidence of nitrogen fixation by these microbes was obtained by B. Frank in 1889. Frank's cultures however were not free of bacteria, and when it was later shown that pure cultures of green algae (Chlorophycophyta) could not fix nitrogen, his results were discounted as being due to bacterial contamination. It was not until 1928 that K. von Drewes was able to prove conclusively that nitrogen fixation occurred in pure cultures of the cyanobacteria *Nostoc* and *Anabaena*. Since then, at least twenty genera and over forty species, from terrestrial, marine and freshwater habitats, from the Antarctic to the tropics, have been shown to fix nitrogen. Fixation in blue-green algae was first confirmed using the ^{15}N technique by R. H. Burris *et al.* in 1943. Nitrogen-fixing ability is widespread in the group but by no means universal. With few exceptions, all filamentous, diazotrophic cyanobacteria have a common peculiarity, they form thick-walled cells called **heterocysts** at intervals along their filaments so that the proportion of species with heterocysts in a mixed algal population is an index of the nitrogen-fixing potential of that population. In these heterocystous forms, the heterocysts are the site of fixation and their thick walls provide the means whereby the oxygen-sensitive nitrogenase is protected against inactivation by high O_2 concentrations. The non-heterocystous filamentous and unicellular species can fix N_2 only at very low oxygen tensions. The sensitivity of cyanobacteria to O_2 explains why their preferred habitat is slow-moving, eutrophic water with a low level of dissolved oxygen (Gibson & Jordan 1983).

Although world-wide in distribution, cyanobacteria are particularly abundant in the waters and soils of the tropics and subtropics. They are however absent from strongly acid soils, and the optimum pH for N_2 fixation is around neutrality. They are very sensitive to desiccation but their nitrogenase activity recovers rapidly upon rewetting. Although cyanobacteria are photosynthetic, they are capable of growing and fixing N_2 heterotrophically, in the dark, at least for a short time. As Gibson and Jordan (1983) point out, this property is ecologically significant, since it enables them to adapt to poorly lit habitats such as mineral soil. In addition, many species are adapted to growing and fixing N_2 in highly saline environments such as salt marshes.

Cyanobacteria have a recognized role in the nitrogen economy of rice paddies. During the rainy season there is profuse growth in most paddy fields of many nitrogen-fixing genera, and P. K. De and L. N. Mandal in 1956 found that 15–50 kg N ha^{-1} were fixed in six weeks. As primary colonizers in desert regions, they make significant contributions also, for example Y. T. Tchan and N. C. W. Beadle estimated in 1955 that cyanobacteria fixed about 3 kg ha^{-1} yr^{-1} in the arid zone of New South Wales, a substantial amount for such a difficult environment. In the Antarctic, *Nostoc* is common and is the chief component of so-called 'algal peat'. In the littoral zone, W. D. P. Stewart in 1965 found that fixation by cyanobacteria increased from little or none in mid-winter to 8 kg ha^{-1} $month^{-1}$ in spring. Nitrogen fixation in lakes by blue-green algae is well established, reaching a peak in late summer when the algal bloom is commencing, and it is probable that it occurs in the open ocean as well. They have as well been credited with an important role as initiators of plant succession and pedogenesis, as shown for example on new lava flows in Iceland by B. Englund in 1978.

The evidence assembled by D. C. Jordan in 1981 indicates that from 2 to 120 kg ha^{-1} yr^{-1} is fixed by cyanobacteria epiphytic on mosses, from 1 to 122 kg ha^{-1} yr^{-1} is fixed in soils, and nearly 300 kg ha^{-1} yr^{-1} in parts of salt marshes. Although quantitative data on the contribution of these microorganisms to the overall nitrogen economy of the Earth is not available, it is clear that they provide substantial amounts of nitrogen to a diverse group of ecosystems. Their geochemical significance remains to be properly assessed, but it is no doubt considerable.

8.3.2 Non-biological nitrogen fixation

In addition to the fixation of molecular nitrogen by microorganisms, nitrogen may pass from the atmosphere to the biosphere as gaseous ammonia or as nitrogen compounds dissolved in rainwater. A variety of nitrogenous compounds is found in rain. A certain amount of organic nitrogen is always present; it is frequently referred to as 'albuminoid' nitrogen and is associated with particulate matter – dust and organic debris – suspended in the atmosphere. The two major nitrogen components in rain are however, ammonium and nitrate. E. Eriksson in 1952 made a detailed study of the nitrogen content of rain from a large number of meteorological stations throughout the world. Average values for the northern hemisphere were 0.78 mg $litre^{-1}$ NH_4-N and 0.27 mg $litre^{-1}$ NO_3-N. The higher values for NH_4^+ in the northern hemisphere are believed to be due to increased industrial and agricultural activities in that hemisphere. The amount of nitrogen removed from the atmosphere each year in rain depends not only on the concentration of nitrogen compounds in rainwater but also on the annual precipitation. G. E. Hutchinson (1954) estimated that the rate of delivery of nitrogen in rain, averaged over the Earth's surface, is 1.3–3.2 kg ha^{-1} yr^{-1}. For individual land masses, the values are higher, and Eriksson's estimates are 7.5 and 8.1 kg ha^{-1} yr^{-1} for inorganic nitrogen in North America and Europe respectively, and 2.3 kg ha^{-1} yr^{-1} for organic nitrogen.

Early reports of ammonia absorption by strips of filter paper (Ingham 1950a, b) have been substantiated by subsequent work with soils and plants (Malo & Purvis 1964; Hutchinson *et al.* 1972). Free ammonia occurs in the air at concentrations ranging from 10^{-5} to 10^{-7} g m^{-3}, values towards the higher end of the range being common in highly industrialized regions. Inputs from such sources are potentially quite substantial: Malo and Purvis (1964), for example, estimated that agricultural soils in New Jersey absorbed ammonia at rates of 57–255 g N ha^{-1} day^{-1}. If these gains were constant throughout the year, the annual accession of nitrogen would be about 20–80 kg ha^{-1}. It would be unwise to accept this as typical however, and the consensus of opinion is that ammonia absorption is, relative to biological nitrogen fixation, a minor contributor to the nitrogen economy of natural ecosystems. In any event, ammonia is 'cyclical' nitrogen (see next section) so that even if it is accepted as a gain to one ecosystem then it represents a loss to another.

Origin of nitrogen in rainwater

Nitrogen in rainwater could be either cyclical or newly fixed. As G. E. Hutchinson pointed out in 1954, it is important to distinguish between these because newly fixed nitrogen represents an addition to the nitrogen cycle whereas cyclical nitrogen does not. By cyclical nitrogen we mean nitrogen originating from organisms. Ammonium is cyclical nitrogen, that portion which is not clearly anthropogenic being derived from the normal respiratory activities of organisms. The crux of the problem of differentiating between the two kinds of nitrogen lies in determining the source of nitrate. If this is cyclical, then it must be formed from ammonium by oxidation. If on the other hand it is newly fixed nitrogen, the mechanism by which fixation occurs must be established.

Nitrate fixation in thunderstorms

It has long been held that nitrate in rain is fixed by lightning discharges during thunderstorms, an idea first put forward by Justus von Liebig in 1827. This question was thoroughly explored by Hutchinson who concluded, on theoretical grounds, that the major part of the nitrate in rainwater is not fixed electrically. That such fixation can and does occur is undoubtedly true, but its contribution to the nitrogen cycle is minor. This conclusion is confirmed by the knowledge that there is often no correlation between the occurrence of lightning and the presence of nitrate in rain, and indeed sometimes the reverse is true. At Katherine, NT, where the average annual thunder day is the highest in Australia, R. Wetselaar and J. T. Hutton found in 1963 that the nitrate content of rain on days when lightning occurred was 30 per cent lower than on days when no lightning was observed.

If most of the nitrate in rain is not fixed by lightning discharges, it can only have been formed by photochemical oxidation of ammonium derived from the mineralization of organic nitrogen. In other words, it is derived from organisms and is being returned to them; as such it is cyclical nitrogen and does not add any newly fixed nitrogen to the cycle. The photochemical oxidation of ammonium proceeds according to the reactions:

$$NH_4^+ + OH^- + \tfrac{3}{2}O_2 \rightarrow H^+ + NO_2^- + 2H_2O,$$
$$NO_2^- + \tfrac{1}{2}O_2 \rightarrow NO_3^-.$$

Being accompanied by a large decrease in free energy ($\triangle F = -59\,000$ cal mole^{-1}) the first reaction can proceed practically to completion if suitably activated. K. Rao and N. R. Dhar suggested in 1931 that it is catalysed by the presence of certain oxides, notably silica (SiO_2). Their suggestion is supported by the finding that fluctuations in the nitrate concentration in rain are closely paralleled by fluctuations in the silica concentration.

There is much additional evidence which points to the soil and the soil–vegetation system as the ultimate source of nitrate in precipitation. In the vast majority of cases, the NH_4^+/NO_3^- ratio in rainwater is very close to 2/1. This is consistent with the hypothesis that NO_3^- is formed by photochemical oxidation of NH_4^+ and is inconsistent with the fixation of nitrate *de novo* from O_2 and N_2. Furthermore, reports from stations all over the world indicate that the NH_4^+ and NO_3^- content of rain decreases as the amount of precipitation increases. This points to the existence of both NH_4^+ and NO_3^- in the atmosphere before precipitation occurs, and indicates increasing dilution as more and more rain falls. Again, the ammonium level in precipitation and in the air is greatest at seasons when the decomposition rate of soil organic matter is highest, that is in the summer in regions with pronounced seasonal temperature fluctuations. This was clearly shown in a detailed study by C. E. Junge in 1958 of the distribution pattern of NH_4^+ and NO_3^- in precipitation over the United States. Finally, both NO_3^- and NH_4^+ levels in precipitation are high when there are high levels of solid and/or liquid particles in the air, for example in the dry season in monsoon climates. This not only implicates soil as the source of NH_4^+ but supports the contention that photo-oxidation of NH_4^+ takes place in moisture films on the surface of dust or silica particles.

8.3.3 The relative contributions of nitrogen fixing agents: a summing up

G. E. Hutchinson (1954) calculated 7 kg N ha^{-1} as the maximum annual rate of nitrogen fixation for the Earth's surface as a whole. C. C. Delwiche subsequently (in 1965) gave a revised estimate of 0.1 mg cm^{-2}, which is equivalent to 10 kg ha^{-1} yr^{-1}. R. C. Burns and R. W. F. Hardy (1975) estimated that 175 × 10^6 t of nitrogen are fixed annually which, taking the Earth's surface area as 510 × 10 km^2, corresponds to 3.4 kg ha^{-1} yr^{-1}. The true value is unknown but, for the purpose of this discussion, a figure of 5 kg N ha^{-1} yr^{-1} is accepted. Burns and Hardy assign about one-quarter of the total to cropping systems, one-quarter to permanent meadows and grasslands, and one-quarter to forests; the sea accounts for approximately one-fifth and unused land for the remaining 5 per cent.

The major sources of atmospheric nitrogen in cropping systems are herbaceous legumes, free-living and symbiotic cyanobacteria (in rice culture), and associative bacteria (in cereal growing). The relative amounts contributed from these three sources are not known for certain. Hutchinson's (1954) estimate of 0.3 kg ha^{-1} yr^{-1} (sect. 8.3.1) was based on the total surface area of the Earth and may be taken as a minimum value for legumes: the upper end of the range is conjectural but is probably no more than 1 kg ha^{-1} yr^{-1}.

Fixed nitrogen in grassland ecosystems originates mainly from heterotrophic bacteria (free living or associative), though there would be some contribution from herbaceous legumes. In forest ecosystems, non-symbiotic or rhizospheric fixation by heterotrophic bacteria is a major contributor but significant amounts are also fixed by actinorrhizal plants and, especially in the tropics and subtropics, by leguminous trees and shrubs. Only a small fraction of the total amount of N_2 fixed in the biosphere is attributed to unused lands, most of it by the activities of cyanobacteria. These organisms also contribute overwhelmingly to the fixation of nitrogen in oceans, lakes, rivers and estuaries.

Fig. 8.6 is a hypothetical representation of the relative contributions of the various agents of N_2 fixation. The most important, from the geochemical viewpoint, are cyanobacteria.

8.3.4 Nitrogen cycling in the biosphere

The nitrogen incorporated by organisms is eventually returned to the inorganic state by a complex series of reactions involving its stepwise oxidation to NH_4^+, NO_2^- and NO_3^- (Ch. 5). In terrestrial ecosystems, the ammonium ion is normally adsorbed on colloidal

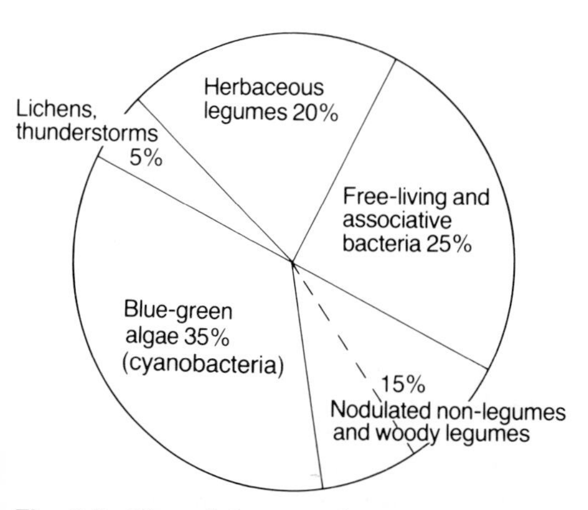

Fig. 8.6 The relative contributions of various nitrogen-fixing agents to the geochemical cycle of nitrogen.

matter near the soil surface, there to be reabsorbed by plants or microbes or further oxidized by the nitrifying bacteria. Although the efficiency of the nitrifiers is not great, they are nevertheless effective biogeochemical agents and have been known to produce several hundred μg g^{-1} NO_3 -N in soils in the space of a few weeks. The nitrate ion is mobile and as a result some nitrate is lost to ground waters by leaching and eventually reaches the sea. Nitrate which is not leached is either assimilated by organisms or else denitrified, that is reduced to gaseous form, and this is the ultimate fate of most of the nitrate that reaches the sea also. Denitrification is the major process by which nitrogen is returned to the atmosphere (Ch. 5), and so constitutes the opposite half of the nitrogen cycle to fixation.

Another way of looking at the geochemical cycle of nitrogen is to consider the equilibrium between it and oxygen, the other major gaseous constituent of the atmosphere:

$$2H_2O + 2N_2 + 5O_2 \leftrightharpoons 4HNO_3; \triangle F = 1780 \text{ cal mole}^{-1}$$

Because the free energy of formation of nitric acid is quite small, one would not expect this reaction to go to completion in either direction, and theoretically, as G. E. Hutchinson (1954) has pointed out, the two major atmospheric gases ought to be in equilibrium with an appreciable quantity of nitric acid. The extreme stability of the N–N bond however, prevents the production of nitrate in detectable amounts at any ordinary temperature and pressure, except by a series of biological processes, namely fixation of N_2

followed by ammonification and subsequent nitrification. Once nitrate has been produced however, it is thermodynamically possible, in accordance with the above equation, for a considerable amount to accumulate in the biosphere. In point of fact it does not, except as isolated deposits in arid regions, for example the Chilean nitrate deposits. It follows that there must be mechanisms operating which are continually returning nitrogen to the atmosphere, the principal one being denitrification.

The whole series of processes whereby the molecular nitrogen of the air enters into other compounds, and is finally delivered again to the atmosphere, is what constitutes the nitrogen cycle. It is illustrated diagrammatically in Fig. 8.7. No attempt is made to indicate in this diagram the relative contributions of the various processes to the overall cycle, except to omit those considered to be insignificant, for example electrical fixation of nitrate in thunderstorms. In many elementary text books, undue and even improper emphasis is sometimes placed on certain facets of the cycle. Thus discussion is usually centred on the classical nitrogen-fixing bacteria *Rhizobium* and *Azotobacter*, and furthermore, the contribution of nitrogen dissolved in rainwater is frequently stressed. In point of fact, with the exception of *Rhizobium*, not one of these agents is particularly important from the geochemical viewpoint.

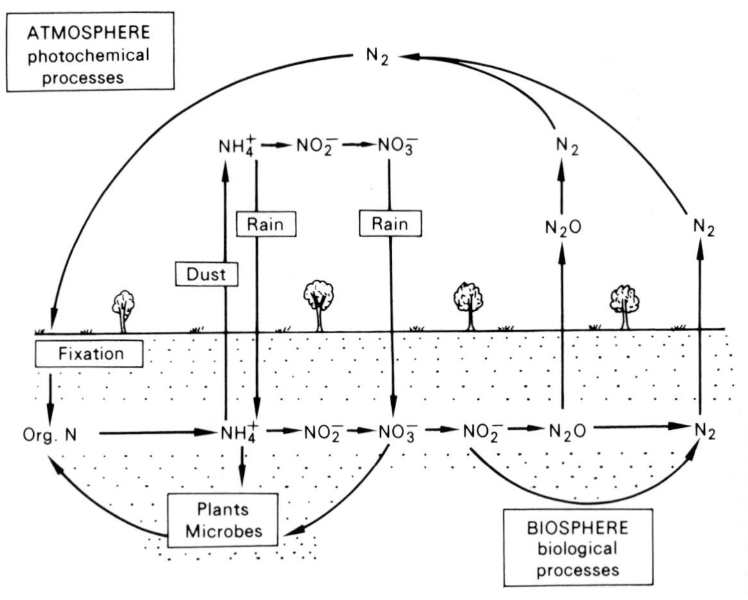

Fig. 8.7 Generalized diagram of the nitrogen cycle.

Rhizobium achieves considerable local significance by adding substantial amounts of newly fixed nitrogen to ecosystems dominated by herbaceous legumes, but whether the woody perennial legumes of the tropical and warm temperate zones contribute more to the world nitrogen budget than actinorrhizal plants remains an open question.

In energy circuit language (Fig. 8.8), nitrogen fixation is seen as the key to the productivity of the biosphere. Ultimately, the rate of capture of solar energy by autotrophic nitrogen-fixing systems, both symbiotic and free-living, determines the amount of nitrogen fixed, but it is this fixed nitrogen which permits accumulation of the organic residues that in turn serve as energy sources for further fixation by heterotrophic nitrogen fixers, and for the mineralization processes which provide nitrogen in a form available to non-fixing green plants. Although the interrelationships among the components of the cycle are complex, the productivity of the latter group is clearly governed by the supply of mineral nitrogen at the work gate through which solar energy flows. Nitrogen fixation thus provides a basis for the complex series of food webs that characterize the biosphere.

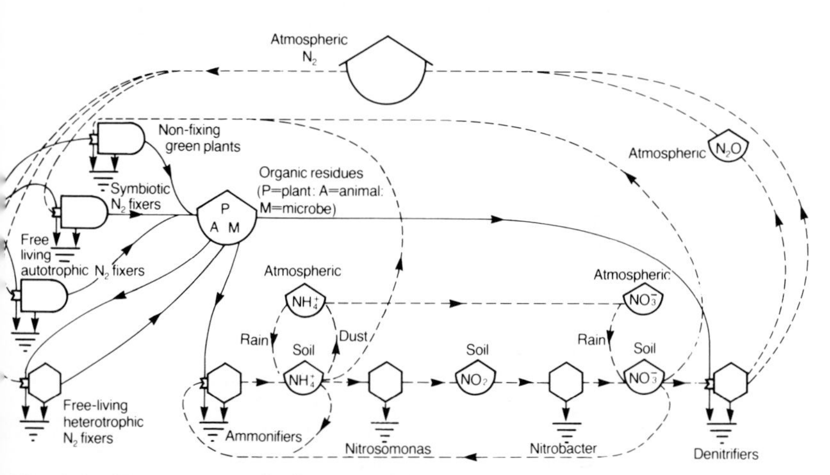

Fig. 8.8 Energy network diagram of the geochemical cycle of nitrogen.

Selected references

Adams, M. A. and Attiwill, P. M. (1982) Nitrogen mineralization and nitrate reduction in forests, *Soil Biol. Biochem.* **14**, 197–202.
Akkermans, A. D. L., Abdulkadir, S. and Trinick, M. J. (1978) Nitrogen-fixing root nodules in Ulmaceae, *Nature, Lond.* **274**, 190.
Alexander, I. J. and Hardy, K. (1981) Surface phosphatase activity of Sitka spruce mycorrhizas from a serpentine site, *Soil Biol. Biochem.* **13**, 301–5.
Alexander, M. (1965) Biodegradation: problems of molecular recalcitrance and microbial fallibility, *Adv. Appl. Microbiol.* **7**, 35–80.
Alexander, M. (1971) *Microbial Ecology*. Wiley, New York.
Alexander, M. (1979) Recalcitrant molecules, fallible micro-organisms, pp. 246–53 in J. M. Lynch and N. J. Poole (eds), *Microbial Ecology: A Conceptual Approach*. Blackwell Scientific Publications, Oxford.
Alexopoulos, C. J. and Mims, C.W. (1979) *Introductory Mycology*. John Wiley, New York.
Allen, M. J. (1982) Influence of vesicular–arbuscular mycorrhiza on water movement through *Bouteloua gracilis* (H.B.K.) Lag ex Steud., New Phytol. **91**, 191–6.
Allen, O. N. and Allen, E. K. (1981) *The Leguminosae*. Univ. Wisconsin Press, Madison.
Anderson, A. J. (1942) Molybdenum deficiency on a South Australian ironstone soil, *J. Aust. Inst. Agric. Sci.* **8**, 73–5.
Anderson, A. J. and Moye, D. V. (1952) Lime and molybdenum in clover development on acid soils, *Aust. J. Agric. Res.* **3**, 95–110.
Anderson, A. J. and Spencer, D. (1950) Sulphur in nitrogen metabolism of legumes and non-legumes *Aust. J. Sci. Res.*, B3, 431–49.
Anderson, J. M. (1977) The organization of soil animal communities, pp. 15–23 in U. Lohm and T. Persson (eds), *Soil Organisms as Components of Ecosystems* (Ecol. Bull. Stockholm 25).
Anderson, J. P. E. and Domsch, K. H. (1975) Measurement of bacterial and fungal contributions to respiration of selected agricultural and forest soils, *Canad. J. Microbiol.* **21**, 314–21.

Anderson, J. P. E. and Domsch, K. H. (1978) A physiological method for the quantitative measurement of microbial biomass in soils, *Soil Biol. Biochem.* **10**, 215–21.
Andrew, C. S. and Norris, D. O. (1961) Comparative responses to calcium of five tropical and four temperate pasture legume species, *Aust. J. Agric. Res.* **12**, 40–55.
Angulo, A. F., Dijk, C. van and Quispel, A. (1976) Symbiotic interactions in non-leguminous root nodules, pp. 475–83 in P. S. Nutman, (ed.), *Symbiotic Nitrogen Fixation in Plants.* Cambridge Univ. Press, Cambridge.
Aprison, M. H. and Burris, R. H. (1952) Time course of fixation of N_2 by excised soybean nodules, *Science* **115**, 264–5.
Ausmus, B. S. and Witkamp, M. (1974) Litter and soil microbial dynamics in a deciduous forest stand. Eastern Deciduous Forest Biome – IBP – 73–10.
Ausmus, B. S., Ferrigini, R. and McBrayer, J. F. (1976) Cited by Witkamp and Ausmus (1976).
Bååth, E. and Söderström, B. (1980) Comparisons of the agar-film and membrane-filter methods for the estimation of hyphal lengths in soil with particular reference to the effect of magnification, *Soil Biol. Biochem.* **12**, 385–7.
Babiuk, L. A. and Paul, E. A. (1970) The use of fluorescein isothiocyanate in the determination of the bacterial biomass of grassland soil, *Can. J. Microbiol.* **16**, 57–62.
Baker, D. and Torrey, J. G. (1979) The isolation and cultivation of actinomycetous root nodule endophytes, pp. 38–56 in J. C. Gordon, C. T. Wheeler and D. A. Perry (eds), *Symbiotic Nitrogen Fixation in the Management of Temperate Forests.* Forest Research Lab., Corvallis, Oregon.
Baker, K. F. and Cook, R. J. (1974) *Biological Control of Plant Pathogens.* W. H. Freeman, San Francisco.
Balandreau, J. and Knowles, R. (1978) The rhizosphere, pp. 243–68 in Y. R. Dommerques and S. V. Krupa (eds), *Interactions between Non-Pathogenic Soil Microorganisms and Plants.* Elsevier, Amsterdam.
Barber, D. A. (1968) Microorganisms and the inorganic nutrition of higher plants, *Ann. Rev. Pl. Physiol.* **19**, 71–88.
Barber, D. A. (1969) The influence of the microflora on the accumulation of ions by plants, pp. 191–200 in I. H. Rorison (ed.), *Ecological Aspects of the Mineral Nutrition of Plants.* Blackwell, Oxford.
Barber, D. A. and Loughman, B. C. (1970) The effect of micro-organisms on the absorption of inorganic nutrients by intact plants. II. Uptake and utilization of phosphate by barley plants grown under sterile and non-sterile conditions, *J. Exp. Bot.* **18**, 170–176.
Barber, D. A. and Martin, J. K. (1976) The release of organic substances by cereal roots into soil, *New Phytol.* **76**, 69–80.
Barley, K. P. (1970) The configuration of the root system in relation to nutrient uptake, *Adv. Agron.* **22**, 159–201.

Bartholomew, W. V. and Clark, F. E. (1950) Nitrogen transformations in soil in relation to the rhizosphere microflora, *Int. Congr. Soil. Sci. Trans.* **4(2)**, 112–3.
Bartlett, E. M. and Lewis, D. H. (1973) Surface phosphatase activity of mycorrhizal roots of beech, *Soil Biol. Biochem.* **5**, 249–57.
Baylis, G. T. S. (1967) Experiments on the ecological significance of phycomycetous mycorrhizas, *New Phytol.* **66**, 231–43.
Beadle, N. C. W. (1954) Soil phosphate and the delimitation of plant communities in eastern Australia, *Ecology* **35**, 370–5.
Beadle, N. C. W. (1966) Soil phosphate and its role in moulding segments of the Australian flora and vegetation, with special reference to xeromorphy and sclerophylly, *Ecology* **47**, 992–1007.
Becking, J. H. (1970) *Frankiaceae* Fam. Nov. (Actinomycetales) with one new combination and six new species of the genus *Frankia* Brunchorst 1886, 174, *Int. J. Syst. Bact.* **20**, 201–20.
Becking, J. H. (1977) Dinitrogen-fixing associations in higher plants other than legumes, pp. 185–275 in R. W. F. Hardy and W. S. Silver (eds), *A Treatise on Dinitrogen Fixation*, Sect. 3 (Biology). Wiley-Interscience, New York.
Becking, J. H., Boer, W. E. and Houwink, A. L. (1964) Electron microscopy of the endophyte of *Alnus glutinosa, Antonie van Leeuwenhoek* **30**, 343–76.
Behara, B: and Wagner, G. H. (1974) Microbial growth rate in glucose amended soil, *Soil Sci. Soc. Amer. Proc.* **38**, 591–7.
Beijerinck, M. W. (1888) Die Bakterien der Papilionaceen-Knollchen, *Bot. Ztg.* **46**, 725–35, 797–804.
Beijerinck, M. W. (1901) Über Oligonitrophile Mikroben, *Zbl. Bakteriol., Parasitenkunde, Infektionskrankh. Hyg.*, Abt. 2 **7**, 561–82.
Benians, G. J. and Barber, D. A. (1974) The uptake of phosphate by barley plants from soil under aseptic and non-sterile conditions, *Soil Biol. Biochem.* **6**, 195–200.
Bergersen, F. J. and Goodchild, D. J. (1973) Cellular location and concentration of leghaemoglobin in soybean root nodules, *Aust. J. Biol. Sci.* **26**, 741–56.
Bergersen, F. J., Kennedy, G. S. and Wittman, W. (1965) Nitrogen fixation in the coralloid roots of *Macrozamia communis* L. Johnson, *Aust. J. Biol. Sci.* **18**, 1135–42.
Bergersen, F. J. and Turner, G. L. (1967). Nitrogen fixation by the bacteroid fraction of breis of soybean root nodules, *Biochim. Biophys. Acta* **141**, 507–15.
Bevege, D. I. (1971) Vesicular–arbuscular mycorrhizas of *Araucaria*: aspects of their ecology, physiology and role in nitrogen fixation, Ph.D. Thesis, University of New England, Armidale, Australia.
Bevege, D. I., Bowen, G. D. and Skinner, M. F. (1975) Comparative carbohydrate physiology of ecto- and endo-mycorrhizas, pp. 149–74 in F. E. Sanders, B. Mosse and P. B. Tinker (eds), *Endomycorrhizas*. Academic Press, London.

Bevege, D. I., Richards, B. N. and Moore, A. W. (1978) Nitrogen fixation associated with conifers. CSIRO Australia, Division of Soils Div. Rep. No. 26, 19 pp.
Biederbeck, V. O. (1978) Soil organic sulfur and fertility, pp. 273–310 in M. Schnitzer and S. U. Khan (eds), *Soil Organic Matter*. Elsevier, Amsterdam.
Bigg, W. L. (1981) Some effects of nitrate, ammonium and mycorrhizal fungi on the growth of Douglas fir and Sitka spruce, Ph.D. Thesis, Aberdeen University. (Cited by Harley and Smith 1983).
Birch, H. F. (1960) Nitrification in soil after different periods of dryness, *Plant Soil* **12**, 81–96.
Birk, E. M. (1979) Disappearance of overstorey and understorey litter in an open eucalypt forest, *Aust. J. Ecol.* **4**, 207–22.
Björkman, E. (1942) Über die Bedingungen der Mykorrhizabildung bei Keifer und Fichte, *Symb. Bot. Upsaliens, VI.* 190 pp.
Björkman, E. (1960) *Monotropa hypopitys* L., an epiparasite on tree roots, *Physiol. Plant* **13**, 308–27.
Blackmer, A. M. and Bremner, J. M. (1978) Inhibitory effect of nitrate on reduction of N_2O to N_2 by soil microorganisms, *Soil Biol. Biochem.* **10**, 187–91.
Bold, H. C., Alexopoulos, C. J. and Delevoryas, T. (1980) *Morphology of Plants and Fungi*. Harper and Row, New York.
Bolin, B., Degens, E. T., Kempe, S. and Ketner, P. (eds) (1981) *The Global Carbon Cycle*. Wiley, New York.
Bollen, W. B. (1953) Mulches and soil conditions. Carbon and nitrogen in farm and forest products, *Agric. Food Chem.* **1**, 379–81.
Bond, G. (1951) The fixation of nitrogen associated with the root nodules of *Myrica gale* L., with special reference to its pH relation and ecological significance, *Ann. Bot.* **15**, 447–59.
Bond, G. (1955) An isotopic study of the fixation of nitrogen associated with nodulated plants of *Alnus, Myrica* and *Hippophaë*, *J. Exp. Bot.* **6**, 303–11.
Bond, G. (1956) Evidence for fixation of nitrogen by root nodules of alder (*Alnus*) under field conditions, *New Phytol.* **55**, 147–53.
Bond, G. (1967) Fixation of nitrogen by higher plants other than legumes, *Ann. Rev. Pl. Physiol.* **18**, 107–26.
Bond, G. and Scott, G. D. (1955) An examination of some symbiotic systems for fixation of nitrogen, *Ann. Bot.* **19**, 57–77.
Bormann, F. H. and Likens, G. E. (1979) *Pattern and Process in a Forested Ecosystem*. Springer-Verlag, New York.
Bowen, G. D. (1980) Misconceptions, concepts and approaches in rhizosphere biology, pp. 283–304, in D. C. Ellwood, J. N. Hedger, M. J. Latham, J. M. Lynch and J. H. Slater (eds), *Contemporary Microbial Ecology*. Academic Press, London.
Bowen, G. D., Bevege, D. I. and Mosse, B. (1975) Phosphate physiology of vesicular–arbuscular mycorrhizas, pp. 241–260 in F. E. Sanders, Mosse and P. B. Tinker (eds), *Endomycorrhizas*. Academic Press, London.

Bowen, G. D. and Rovira, A. D. (1966) Microbial factor in short-term uptake studies with plant roots, *Nature, Lond* **211**, 655–6.
Bowen, G. D. and Rovira, A. D. (1976) Microbial colonization of plant roots, *Ann. Rev. Phytopath.* **14**, 121–44.
Bowen, G. D., Skinner, M. F. and Bevege, D. I. (1974) Zinc uptake by mycorrhizal and uninfected roots of *Pinus radiata* and *Araucaria cunninghamii, Soil Biol. Biochem.* **6**, 141–4.
Bowen, G. D. and Theodorou, C. (1967) Studies on phosphate uptake by mycorrhizas, *Proc. 14th IUFRO Congr., Munich* **24**, 116–38.
Bowen, G. D. and Theodorou, C. (1973) Growth of ectomycorrhizal fungi around seeds and roots, pp. 107–50 in G. C. Marks and T. T. Kozlowski (eds), *Ectomycorrhizae. Their Ecology and Physiology*. Academic Press, New York.
Bowen, G. D. and Theodorou, C. (1979) Interactions between bacteria and ectomycorrhizal fungi, *Soil Biol. Biochem.* **11**, 119–26.
Boyle, J. R., Voigt, G. K. and Sawhney, B. L. (1967). Biotite flakes: alteration by chemical and biological treatment, *Science* **55**, 193–7.
Boyle, J. R. and Voigt, G. K. (1973) Biological weathering of silicate minerals. Implications for tree nutrition and soil genesis, *Plant Soil* **38**, 191–201.
Bray, J. R. (1964) Primary consumption in three forest canopies, *Ecology* **45**, 165–7.
Bremner, J. M. (1965) Nitrogen availability indexes, *Agronomy* **9**, 1324–45.
Bremner, J. M. and Mulvaney, R. L. (1978) Urease activity in soils, pp. 149–196. in R. G. Burns (ed.), *Soil Enzymes*. Academic Press, London.
Brian, P. W. (1957) The ecological significance of antibiotic production, pp. 168–88 in C. C. Spicer and L. E. O. Williams (eds), *Microbial Ecology*. Cambridge Univ. Press, Cambridge.
Briederbeck, V. O. (1978) Soil organic sulfur and fertility, pp. 273–310 in Schnitzer, M. and Khan, S. U. (eds), *Soil Organic Matter*, Elsevier, Amsterdam.
Brock, T. D. (1971) Microbial growth rates in nature, *Bact. Rev.* **35**, 39–58.
Brock, T. D. (1979) *Biology of Microorganisms*. Prentice-Hall, Englewood Cliffs.
Brady, N. C. (1974) *The Nature and Properties of Soils*, 8th edn. Macmillan, New York.
Buller, A. H. R. (1909–34) *Researches on Fungi, I–VI*. Longmans Green, London.
Buol, S. W., Hole, F. D. and McCracken, R. J. (1980) *Soil Genesis and Classification*, 2nd edn., Iowa State University Press, Ames, Iowa.
Burges, A. and Raw, F. (eds) (1967) *Soil Biology*. Academic Press, London and New York.
Burns, R. G. (1977) Soil enzymology, *Sci. Prog. Oxf.* **64**, 275–85.
Burns, R. G. (1983) Extracellular enzyme–substrate interactions in soil, pp. 249–298 in J. H. Slater, R. Whittenbury and

J. W. T. Wimpenny (eds), *Microbes in their Natural Environments* (34th. Symp. Soc. Gen. Microbiol.). Cambridge Univ. Press, Cambridge.

Burns, R. F. and Hardy, R. W. F. (1975) *Nitrogen Fixation in Bacteria and Higher Plants*. Springer-Verlag, Berlin.

Burris, R. H., Arp, D. J., Hageman, R. V., Houchins, J. P., Sweet, W. J. and Tso, Man-yin (1981) Mechanism of nitrogenase action, pp. 56–66 in A. H. Gibson and W. E. Newton (eds), *Current Perspectives in Nitrogen Fixation*. Aust. Acad. Sci., Canberra.

Burris, R. H., Eppling, F. J., Wahlin, H. B. and Wilson, P. W. (1943) Detection of nitrogen fixation with isotopic nitrogen, *J. Biol. Chem.* **148**, 349–57.

Burris, R. H. and Wilson, P. W. (1957) Methods for measurement of nitrogen fixation, pp. 355–66 in S. P. Colowick and N. D. Kaplan (eds), *Methods in Enzymology*. Academic Press, New York.

Butler, E. J. (1939) The occurrence and systematic position of the vesicular–arbuscular type of mycorrhizal fungi, *Trans. Br. Mycol. Soc.* **22**, 274–301.

Butler, J. H. A. and Buckerfield, J. C. (1979) Digestion of lignin by termites, *Soil Biol. Biochem.* **11**, 507–13.

Caldwell, A. G. and Black, C. A. (1958) Inositol hexaphosphate: III. Content in soils, *Soil Sci. Soc. Amer. Proc.* **22**, 296–8.

Callaham, D., Del Tredici, P. and Torrey, J. G. (1978) Isolation and cultivation in vitro of the actinomycete causing root nodulation in *Comptonia, Science* **199**, 899–902.

Callaham, D., Newcombe, W., Torrey, J. G. and Peterson, R. L. (1979) Root hair infection in actinomycete-induced root nodule initiation in *Casuarina, Myria* and *Comptonia, Bot. Gaz. Suppl.* **140**, 51–9.

Callendar, G. S. (1940) Variations of the amount of carbon dioxide in different air currents, *Quart. J. R. Met. Soc.* **66**, 395–400.

Callendar, G. S. (1958) On the amount of carbon dioxide in the atmosphere, *Tellus* **10**, 243–8.

Carnahan, J. E., Mortensen, L. E., Mower, H. F. and Castle, J. E. (1960a) Nitrogen fixation in cell-free extracts of *Clostridium pasteurianum, Biochim. Biophys. Acta* **38**, 188–9.

Carnahan, J. E., Mortensen, L. E., Mower, H. F. and Castle, J. E. (1960b) Nitrogen fixation in cell-free extracts of *Clostridium pasteurianum, Biochim. Biophys. Acta* **44**, 520–35.

Carrodus, B. B. (1966) Absorption of nitrogen by mycorrhizal roots of beech. I. Factors affecting assimilation of nitrogen, *New Phytol.* **65**, 358–71.

Carrodus, B. B. (1967) Absorption of nitrogen by mycorrhizal roots of beech. II. Ammonium and nitrate as sources of nitrogen, *New Phytol.* **66**, 1–4.

Cecconi, S., Ristori, G. G. and Vidrich, V. (1975) The role of octahedral Fe in biological weathering of biotite, pp. 623–8 in

S. W. Bailey (ed.), *Proceedings of the International Clay Conference*, Mexico City, July 1970. Applied Publishing Ltd., Wilmette, Illinois.

Charley, J. L. (1977) Mineral cycling in rangeland ecosystems, pp. 215–56 in R. E. Sosebee (ed.), *Rangeland Plant Physiology*. Range Science Series No. 4, Society for Range Management, Denver, Colorado.

Charley, J. L. and Richards, B. N. (1983) Nutrient allocation in plant communities: mineral cycling in terrestrial ecosystems, Ch. 1, pp. 5–45 in O. L. Lange, P. S. Nobel, C. B. Osmond and H. Ziegler (eds), *Physiological Plant Ecology IV, Encyclopedia of Plant Physiology*. New Series Vol. 12D. Springer-Verlag, Berlin.

Charley, J. L. and West, N. E. (1975) Plant-induced soil chemical patterns in some desert shrub-dominated ecosystems of Utah, *J. Ecol.* **63**, 945–64.

Charley, J. L. and West, N. E. (1977) Micro-patterns of nitrogen mineralization activity in soils of some shrub-dominated semi-desert ecosystems of Utah, *Soil Biol. Biochem.* **9**, 357–65.

Chase, F. E., Corke, C. T. and Robinson, J. B. (1967) Nitrifying bacteria in soil, pp. 593–611 in T. R. Gray and D. Parkinson (eds), *The Ecology of Soil Bacteria*. Liverpool Univ. Press, Liverpool.

Cheshire, M. V. (1979) *Nature and Origin of Carbohydrates in Soils*. Academic Press, London.

Chilvers, G. A. and Gust, L. W. (1982) The development of mycorrhizal populations on pot-grown seedlings of *Eucalyptus st-johnii* R.T. Bak., *New Phytol.* **90**, 677–99.

Chilvers, G. A. and Pryor, L. D. (1965) The structure of eucalypt mycorrhizas, *Aust. J. Bot.* **13**, 245–59.

Clarholm, M. and Rosswall, T. (1980) Biomass and turnover of bacteria in a forest soil and a peat, *Soil Biol. Biochem.* **12**, 49–57.

Clark, F. E. (1965) The concept of competition in microbial ecology, pp. 339–45 in K. F. Baker and W. C. Snyder (eds), *Ecology of Soil-Borne Plant Pathogens*. Univ. California Press, Berkeley.

Clark, F. E. (1967) Bacteria in soil, Ch. 2, pp. 15–49 in A. Burges and F. Raw (eds), *Soil Biology*. Academic Press, London.

Clark, F. E. and Paul, E. A. (1970) The microflora of grassland, *Adv. Agron.* **22**, 375–435.

Clements, F. E. (1936) Nature and structure of the climax, *J. Ecol.* **24**, 252–84.

Clements, F. E. and Shelford, V. E. (1939) *Bioecology*. Wiley, New York.

Coleman, D. C. (1976) A review of root production processes and their influence on soil biota in terrestrial ecosystems, pp. 417–34 in J. M. Anderson and A. Macfadyen (eds), *The Role of Terrestrial and Aquatic Organisms in Decomposition Processes*. Blackwell, Oxford.

Coleman, D. C., Cole, C. V., Anderson, R. V., Blaha, M.,

Campion, M. K., Clarholm, M., Elliott, E. T., Hunt, H. W., Shaefer, B. and Sinclair, J. (1977) An analysis of rhizosphere–saprophage interactions in terrestrial ecosystems, *Ecol. Bull. (Stockholm)* **25**, 299–309.

Conn, H. J. (1918) The microscopic study of bacteria and fungi in soil, *Tech. Bull. N.Y. Agric. Exp. Sta.*, 64.

Connell, J. H. and Slatyer, R. O. (1977) Mechanisms of succession in natural communities and their role in community stability and organization, *Amer. Nat.* **111**, 1119–44.

Cooper, K. M. and Lösel, D. (1978) Lipid physiology of vesicular–arbuscular mycorrhiza. I. Composition of lipids in roots of onion, clover and ryegrass infected with *Glomus mosseae, New Phytol.* **80**, 143–151.

Cooper, K. M. and Tinker, P. B. (1978) Translocation and transfer of nutrients in vesicular–arbuscular mycorrhizas. II. Uptake and translocation of phosphorus, zinc and sulphur, *New Phytol.* **81**, 43–52.

Corke, A. T. K. and Rishbeth, J. (1981) Use of microorganisms to control plant diseases, pp. 717–36 in H. D. Burges (ed.), *Microbial Control of Pests and Plant Diseases 1970–1980*. Academic Press, London.

Cosgrove, D. J. (1977) Microbial transformations in the phosphorus cycle, *Adv. Microb. Ecol.* **1**, 95–134.

Cox, G., Sanders, F. E., Tinker, P. B. and Wild, J. A. (1975) Ultrastructural evidence relating to host–endophyte transfer in a vesicular–arbuscular mycorrhiza, pp. 297–312 in F. E. Sanders, B. Mosse and P. B. Tinker (eds), *Endomycorrhizas*. Academic Press, London.

Cox, G. and Tinker, P. B. (1976) Translocation and transfer of nutrients in vesicular-arbuscular mycorrhizas. I. The arbuscule and phosphorus transfer: a quantitative ultrastructural study. *New Phytol.* **77**, 371–8.

Cox, G. W. and Atkins, M. D. (1979) *Agricultural Ecology. An Analysis of World Food Production Systems.* W. H. Freeman, San Francisco.

Cox, T. L., Harris, W. F., Ausmus, B. F. and Edwards, N. T (1978) The role of roots in biogeochemical cycles in an eastern deciduous forest, *Pedobiologia* **18**, 264–71.

Cress, W. A., Throneberry, G. O. and Lindsey, D. L. (1979) Kinetics of phosphorus absorption by mycorrhizal and non-mycorrhizal tomato roots, *Pl. Physiol.* **64**, 484–7.

Crocker, R. L., (1952) Soil genesis and the pedogenic factors, Quart. *Rev. Biol.* **27**, 139–68.

Crocker, R. L. and Major, J. (1955) Soil development in relation to vegetation and surface age at Glacier Bay, Alaska, *J. Ecol.* **43**, 427–48.

Cromack, K. Jr., Sollins, P., Todd, R. L., Fogel, R., Todd, A. W., Fender, W. M., Crossley, M. E. and Crossley, D. A. Jr. (1977) The role of oxalic acid and bicarbonate in calcium cycling by fungi

and bacteria: some possible implications for soil animals, *Ecol. Bull. (Stockholm)* **25**, 246–52.
Cromack, K., Sollins, P., Graustein, W. C., Speidel, K., Todd, A. W., Spycher, G., Ching, Y-Li and Todd, R. L. (1979) Calcium oxalate accumulation and soil weathering in mats of the hypogenous fungus *Hysterangium crassum, Soil Biol. Biochem.* **11**, 463–8.
Crossley, D. A. Jr. (1977) The roles of terrestrial saprophagous arthropods in forest soils, pp. 49–56 in W. J. Mattson (ed.), *The Role of Arthropods in Forest Ecosystems*. Springer-Verlag, New York.
Crossley, D. A. Jr. and Witkamp, M. (1966) The role of arthropods and microflora in breakdown of white oak litter, *Pedobiologia* **6**, 293–303.
Daft, M. J. and Nicolson, T. H. (1966) Effect of *Endogone* mycorrhiza on plant growth, *New Phytol.* **65**, 343–50.
Daft, M. J. and Nicolson, T. H. (1969) Effect of *Endogone* mycorrhiza on plant growth. II. Influence of inoculum concentration on growth and infection in tomato, *New Phytol.* **68**, 953–63.
Dagley, S. (1975) Microbial degradation of organic compounds in the biosphere, *Amer. Scientist* **63**, 681–9.
Dazzo, F. B. and Hubbell, D. H. (1975) Cross-reactive antigens and lectin as determinants of symbiotic specificity in the *Rhizobium*-clover association, *Appl. Microbiol.* **30**, 1017–33.
De, P. K. and Mandal, L. N. (1956) Fixation of nitrogen by algae in rice soils, *Soil Sci.* **81**, 453–8.
Delwiche, C. E. (1965) The cycling of carbon and nitrogen in the biosphere, pp. 29–58 in C. M. Gilmour and O. N. Allen *Microbiology and Soil Fertility*. Oregon State Univ. Press, Corvallis, Oregon.
Dickinson, C. H., Dawson, D. and Goodfellow, M. (1981) Interactions between bacteria, streptomycetes and fungi from *Picea sitchensis* litter, *Soil Biol. Biochem.* **13**, 65–71.
Dickinson, C. H. and Pugh, G. J. F. (eds) (1974) *Biology of Plant Litter Decomposition Vol. 2.* Academic Press, London and New York.
Dillworth, M. J. (1966) Acetylene reduction by nitrogen-fixing preparations from *Clostridium pasteurianum, Biochim. Biophys Acta* **127**, 285–94.
Doak, K. D. (1928) The mycorrhizal fungus of *Vaccinium, Phytopathology* **18**, 101–8.
Dobbs, C. G. and Hinson, W. H. (1953) A widespread fungistasis in soils, *Nature, Lond.* **172**, 197–9.
Döbereiner, J. (1961) Nitrogen-fixing bacteria of the genus *Beijerinckia* Derx in the rhizosphere of sugar cane, *Plant Soil* **14**, 211–17.
Döbereiner, J. (1974) Nitrogen-fixing bacteria in the rhizosphere, pp. 86–120 in A. Quispel (ed.), *The Biology of Nitrogen Fixation*. Elsevier–North-Holland, Amsterdam.
Döbereiner, J. and Boddey, R. M. (1981) Nitrogen fixation in

association with Gramineae, pp. 305–12 in A. H. Gibson and W. E. Newton (eds), *Current Perspectives in Nitrogen Fixation*. Aust. Acad. Sci., Canberra.

Döbereiner, J. and Day, J. M. (1976) Associative symbioses in tropical grasses: characterization of microorganisms and dinitrogen-fixing sites, pp. 518–38 in W. E. Newton and C. J. Nyman (eds), *Proc. 1st Inter. Symp. Nitrogen Fixation*. Washington State Univ. Press, Pullman, Washington State.

Dommergues, Y. (1963) Evaluation du taux de fixation de l'azote dans un sol dunaire reboïsé en filao (*Casuarina equisetifolia*), *Agrochimica* **7**, 335–40.

Dommerques, Y. R., Belser, L. W. and Schmidt, E. L. (1978) Limiting factors for microbial growth and activity in soil, *Adv. Microb. Ecol.* **2**, 49–104.

Drake, M. Vengris, J. and Colby, W. G. (1951) Cation exchange capacity of plant roots, *Soil Sci.* **72**, 139–47.

Drewes, K. von (1928) Über die Assimilation der Luftstickstoffs durch Blaualgen, *Zbl. Bakteriol., Parasitenkunde, Infektionskrankh. Hyg.*, Abt. 2, **76**, 88–121.

Duddridge, J. and Read, D. J. (1982) An ultrastructural analysis of the development of mycorrhizas in *Monotropa hypopitys L., New Phytol.* **92**, 203–14.

Duddridge, J. A., Malabari, A. and Read, D. J. (1980) Structure and function of mycorrhizal rhizomorphs with special reference to their role in water transport, *Nature, Lond.* **287**, 834–6.

Duff, R. B. and Webley, D. M. (1959) 2-Ketogluconic acid as a natural chelator produced by soil bacteria, *Chem. Inds., Lond.*, 1376–7.

Duff, R. B., Webley, D. M. and Scott, R. O. (1963) The solubilization of minerals and related materials by 2-ketogluconic acid producing bacteria, *Soil Sci.* **95**, 105–14.

Duggin, J. A. (1984) Autotrophic and heterotrophic nitrification in acid forest soils from Hubbard Brook, New Hampshire. Ph.D Dissertation, Yale University.

Edmonds, R. L. (1980) Litter decomposition and nutrient release in Douglas fir, red alder, western hemlock and Pacific fir ecosystems in western Washington, *Can. J. For. Res.* **10**, 327–37.

Edwards, C. A. and Heath, G. W. (1963) The role of soil animals in breakdown of leaf material, pp. 76–84 in J. Doeksen and J. van der Drift (eds), *Soil Organisms*. North-Holland, Amsterdam.

Edwards, C. A., Reichle, D. E. and Crossley, D. A., Jr. (1970) The role of soil invertebrates in turnover of organic matter and nutrients, pp. 147–72 in D. E. Reichle (ed.), *Analysis of Temperate Forest Ecosystems*. Springer-Verlag, New York.

Edwards, N. T. and Harris, W. F. (1977) Carbon cycling in a mixed deciduous forest floor, *Ecology* **58**, 431–7.

Elliott, E. T., Anderson, R. V., Coleman, D. C. and Cole, C. V. (1980) Habitable soil pore space and microbial trophic interactions, *Oikos* **35**,327–35.

Engelmann, M. D. (1964) Energetics, terrestrial field studies and animal productivity, *Adv. Ecol. Res.* **3**, 73–115.
Englund, B. (1978) Algal nitrogen fixation on the lava field of Heimaey, Iceland, *Oecologia* **34**, 45–56.
Epstein, E. (1972) *Mineral Nutrition of Plants: Principles and Perspectives*. Wiley, New York.
Erickson, L. R. (1979) Energetic efficiency of biomass and product formation, *Biotechnol. Bioeng.* **21**, 725–43.
Eriksson, E. (1952) Composition of atmospheric precipitation. I. Nitrogen compounds, *Tellus* **4**, 215–32.
Estermann, E. F. and McLaren, A. D. (1961) Contributions of rhizosphere organisms to the total capacity of plants to utilize organic nutrients, *Plant Soil* **15**, 243–260.
Faye, M., Rancillae, M. and David, A. (1981) Determination of the mycorrhizogenic root formation in *Pinus pinaster* Sol., *New Phytol.* **87**, 557–65.
Fenchel, T. and Blackburn, T. H. (1979) *Bacteria and Mineral Cycling*. Academic Press, London.
Fenton, G. R. (1947) The soil fauna: with special reference to the ecosystem of forest soil, *J. Anim. Ecol.* **16**, 76–93.
Focht, D. D. and Verstraete, W. (1977) Biochemical ecology of nitrification and denitrification, *Adv. Microb. Ecol.* **1**, 135–214.
Fogel, R. (1980) Mycorrhizae and nutrient cycling in natural forest ecosystems, *New Phytol.* **86**, 199–212.
Fogel, R. and Cromack, K. Jr. (1977) Effect of habitat and substrate quality on Douglas fir litter decomposition in western Oregon, *Can. J. Bot.* **55**, 1632–40.
Fogel, R. and Hunt, G. (1979) Fungal and arboreal biomass in a western Oregon Douglas fir ecosystem: distribution patterns and turnover, *Can. J. For. Res.* **9**, 245–56.
Foster, R. C. (1981) The ultrastructure and histochemistry of the rhizosphere, *New Phytol.* **89**, 263–73.
Foster, R. C. (1982) The fine structure of epidermal cell mucilages of roots, *New Phytol.* **91**, 727–40.
Foster, R. C. and Rovira, A. D. (1976) Ultrastructure of wheat rhizosphere, *New Phytol.* **76**, 343–52.
Foster, R. C. and Rovira, A. D. (1978) The ultrastructure of the rhizosphere of *Trifolium subterraneum* L., pp. 282–90 in M. W. Loutit and J. A. R. Miles (eds), *Microbial Ecology*. Springer-Verlag, Berlin.
Foth. H. D. (1978) *Fundamentals of Soil Science*, 6th edn. Wiley, New York.
Frank, B. (1889) Über den gegenwartigen Stand unserer Kenntnis der Assimilation elementaren Stickstoffs durch die Pflanzen, *Ber. Dtsch. Bot. Ces.* **7**, 34–42.
Frankland, J. C. (1974) Importance of phase contrast microscopy for estimation of total fungal biomass by the agar film method, *Soil Biol. Biochem.* **6**, 409–10.

Frankland, J. C. (1975) Estimation of live fungal biomass, *Soil Biol. Biochem.* **7**, 339–40.
Frankland, J. C. (1981) Mechanisms in fungal succession, pp. 403–26 in D. T. Wicklow and G. C. Carroll (eds), *The Fungal Community*. Marcel Dekker, New York.
Freney, J. R., Denmead, O. T. and Simpson, J. R. (1979) Nitrous oxide emission from soils at low moisture contents, *Soil Biol. Biochem.* **11**, 167–73.
Garrett, S. D. (1946) Soil as a medium for multiplication and transfer of disease organisms, *Soil Sci.* **61**, 3–8.
Garrett, S. D. (1951) Ecological groups of soil fungi and survey of substrate relationships, *New Phytol.* **50**, 149–66.
Garrett, S. D. (1956) *Biology of Root-infecting Fungi*, Cambridge Univ. Press, Cambridge.
Garrett, S. D. (1963) *Soil Fungi and Soil Fertility*. Pergamon Press, Oxford.
Gaskins, M. H. and Hubbell, D. H. (1979) Response of nonleguminous plants to root inoculation with free living diazotrophic bacteria, pp. 175–82 in J. L. Harley and R. Scott Russell (eds), *The Soil–Root Interface*. Academic Press, New York.
Gerdemann, J. W. (1955) Relation of a large soil-borne spore to phycomycetous mycorrhizal infections, *Mycologia* **47**, 619–32.
Gerdemann, J. W. (1964) The effect of mycorrhizas on the growth of maize, *Mycologia* **56**, 342–9.
Gerdemann, J. W. and Trappe, J. M. (1974) The Endogonaceae in the Pacific Northwest, *Mycologia, Memoir* 5.
Gibson, A. H. and Jordan, D. C. (1983) Ecophysiology of nitrogen-fixing systems, pp. 301–90 in O. L. Lange, P. S. Nobel, C. B. Osmond and H. Ziegler (eds), *Physiological Plant Ecology III, Encyclopedia of Plant Physiology*, New Series Vol. 12C. Springer-Verlag, Berlin.
Gibson, A. H. and Newton, W. E. (eds) (1981) *Current Perspectives in Nitrogen Fixation*. Aust. Acad. Sci. Canberra.
Goodchild, D. J. and Bergersen, F. J. (1966) Electron microscopy of the infection and subsequent development of soybean nodule cells. *J. Bact.* **92**, 204–13.
Gorham, E., Vitousek, P. M. and Reiners, W. A. (1979) The regulation of chemical budgets over the course of terrestrial ecosystem succession. *Ann. Rev. Ecol. Syst.* **10**, 53–84.
Goudriaan, J. and Atjay, G. L. (1981) The possible effects of increased CO_2 on photosynthesis, pp. 237–49 in B. Bolin, E. T. Degens, S. Kempe and P. Ketner (eds), *The Global Carbon Cycle*. Wiley, New York.
Graham, J. H. and Menge, J. A. (1982) Influence of vesicular–arbuscular mycorrhizae and soil phosphate on take-all disease of wheat, *Phytopathology* **72**, 95–8.
Graustein, W. C., Cromack, K. and Sollins, P. (1977) Calcium oxalate: Occurrence in soils and effect on nutrient and geochemical cycles, *Science* **198**, 1252–4.

Gray, T. R. G. and Williams, S. T. (1971) Microbial productivity in soil, *Symp. Soc. Gen. Microbiol.* **21**, 255–86.
Greaves, M. P. and Darbyshire, J. F. (1972) The ultrastructure of the mucilaginous layer on plant roots, *Soil Biol. Biochem.* **4**, 443–9.
Greaves, M. P. and Webley, D. M. (1965) A study of the breakdown of organic phosphates by micro-organisms from the root region of certain pasture grasses, *J. Appl. Bacteriol.* **28**, 454–65.
Greene, E. M. (1980) Cytokinin production by microorganisms, *Bot. Rev.* **46**, 25–74.
Greenland, D. J. (1958) Nitrate fluctuations in tropical soils, *J. Agric. Sci.* **50**, 82–92.
Greenland, D. J. (1979) The physics and chemistry of the soil–root interface: some comments, pp. 83–98 in J. L. Harley and R. Scott Russell (eds), *The Soil–Root Interface*. Academic Press, London.
Greenland, D. J. and Hayes, M. H. B. (1981) Soil processes, pp. 1–35 in D. J. Greenland, and M. H. B. Hayes (eds), *The Chemistry of Soil Processes*. Wiley, New York.
Greenland, D. J. and Nye, P. H. (1959) Increases in the carbon and nitrogen contents of tropical soils under natural fallows, *J. Soil Sci.* **10**, 284–99.
Grier, C. C. (1978) A *Tsuga heterophylla – Picea sitchensis* ecosystem of coastal Oregon: decomposition and nutrient balances of fallen logs, *Can. J. For. Res.* **8**, 198–206.
Grier, C. C. and Logan, R. S. (1977) Old-growth *Pseudotsuga menziesii* communities of a western Oregon watershed: biomass distribution and production budgets, *Ecol. Monogr.* **47**, 373–400.
Griffin, D. M. (1972) *Ecology of Soil Fungi*. Chapman and Hall, London.
Griffin, D. M. (1960) Fungal colonization of sterile hair in contact with soil, *Trans. Br. Mycol. Soc.* **43**, 583–96.
Griffin, D. M. (1969) Soil water in the ecology of fungi, *A. Rev. Phytopath.* **7**, 289–310.
Grove, T. S., O'Connell, A. M. and Malajczuk, N. (1980) Effects of fire on the growth, nutrient content and rate of nitrogen fixation of the cycad *Macrozamia riedlei, Aust. J. Bot.* **28**, 271–81.
Hacskaylo, E. (1973) Carbohydrate physiology of ectomycorrhizae, pp. 207–30 in G. C. Marks and T. T. Kozlowski (eds), *Ectomycorrhizae, Their Ecology and Physiology*. Academic Press, New York.
Hale, M. G. and Moore, L. D. (1979) Factors affecting root exudation II: 1970–1978. *Adv. Agron.* **31**, 93–124.
Hall, I. R. and Abbott, L. K. (1981) Photographic slide collection illustrating features of the Endogonaceae. Ed. 3, pp. 1–27, plus 400 colour transparencies. Invermay Agricultural Research Centre, and Dept. of Soil Science, University of Western Australia, Nedlands WA.
Halstead, R. L. and McKercher, R. (1975) Biochemistry and cycling of phosphorus, pp. 31–63 in E. A. Paul and A. D. McLaren (eds), *Soil Biochemistry*, Vol. 4. Marcel Dekker, New York.

Handley, W. R. C. (1954) Mull and mor formation in relation to forest soils, *Bull. For. Comm. Lond.* **23**, iv + 115.
Hanlon, R. D. G. and Anderson, J. M. (1979) The effects of collembola grazing on microbial activity in decomposing leaf litter, *Oecologia* **38**, 93–9.
Hardie, K. and Leyton, L. (1981) The influence of vesicular–arbuscular mycorrhiza on growth and water relations of red clover. I. In phosphate deficient soil, *New Phytol.* **89**, 599–608.
Hardy, R. W. F., Holsten, R. D., Jackson, E. K. and Burns, R. C. (1968) The acetylene-ethylene assay for N_2 fixation: laboratory and field evaluation, *Plant Physiol.* **43**, 1185–207.
Harley, J. L. (1964) Incorporation of carbon dioxide into excised beech mycorrhizas in the presence and absence of ammonia, *New Phytol.* **63**, 203–8.
Harley, J. L. (1965) Mycorrhiza, pp. 218–29 in K. F. Baker and W. C. Snyder (eds), *Ecology of Soil-borne Plant Pathogens*. University of California Press.
Harley, J. L. (1969) *The Biology of Mycorrhiza*, 2nd edn. Leonard Hill, London.
Harley, J. L. and Lewis, D. H. (1969) The physiology of ectotrophic mycorrhizas, *Adv. Microb. Physiol.* **3**, 50–81.
Harley, J. L. and McCready, C. C. (1952a) Uptake of phosphate by excised mycorrhiza of the beech. II. Distribution of phosphate between host and fungus, *New Phytol.* **51**, 56–64.
Harley, J. L. and McCready, C. C. (1952b) Uptake of phosphate by excised mycorrhiza of the beech. III. The effect of the fungal sheath on the availability of phosphate to the core, *New Phytol.* **51**, 343–8.
Harley, J. L. and McCready, C. C. (1981) Phosphate accumulation in *Fagus* mycorrhizas, *New Phytol.* **89**, 75–80.
Harley, J. L., McCready, C. C., Brierley, J. K. and Jennings, D. H. (1956) The salt respiration of excised beech mycorrhizas. II. The relationship between oxygen consumption and phosphate absorption, *New Phytol.* **55**, 1–28.
Harley, J. L. and Smith, S. E. (1983) *Mycorrhizal Symbiosis*. Academic Press, London.
Harley, J. L. and Wilson, J. M. (1959) The absorption of potassium by beech mycorrhizas, *New Phytol.* **58**, 281–98.
Harper, J. E. and Webster, J. (1964) An experimental analysis of the coprophilous fungus succession, *Trans. Br. Mycol. Soc.* **47**, 511–30.
Hartigan, R. J. (1981) Soil respiration as an index of forest floor metabolism, Ph.D. Thesis, Univ. New England, Armidale, NSW, Australia.
Hassouna, M. G. and Wareing, P. R. (1964) Possible role of rhizosphere bacteria in the nitrogen nutrition of *Ammophila arenaria, Nature, Lond.* **202**, 467–9.
Hatch, A. B. (1937) The physical basis of mycotrophy in *Pinus, Black Rock Forest Bull.*, **6**, pp. 168.

Havelka, U. D., Boyle, M. G. and Hardy, R. W. F. (1982) Biological nitrogen fixation, pp. 365–422 in F. J. Stevenson (ed.), *Nitrogen in Agricultural Soils*. Amer. Soc. Agronomy, Crop Science Soc. Amer. and Soil Science Soc. Amer., Madison, Wisconsin.
Hayes, M. B. H. and Swift, R. S. (1978) The chemistry of soil organic colloids, pp. 179–320 in D. J. Greenland and M. B. H. Hayes (eds), *The Chemistry of Soil Constituents*. Wiley, New York.
Hellriegel, H. and Wilfarth, H. (1888) Untersuchungen über die Stickstoff-Nahrung der Gramineen und Leguminosen, *Beilageheft Z. Vers. Rübenzükerind*, 1–234.
Henzell, E. F. (1968) Sources of nitrogen for Queensland pastures, *Trop. Grassl.* **2**, 1–17.
Henzell, E. F. and Norris, D. O. (1962) Processes by which nitrogen is added to the soil–plant system, *Commonw. Bur. Pastures and Field Crops. Bull.* **46**, 1–19.
Higuchi, T. (1980) Lignin structure and morphological distribution in plant cell walls, Ch. 1, pp. 1–19 in T. K. Kirk, T. Higuchi and T. Chang (eds), *Lignin Biodegradation: Microbiology, Chemistry, and Potential Applications*, Vol. I. CRC Press, Boca Raton, Florida.
Ho, I. and Trappe, J. M. (1973) Translocation of the ^{14}C from *Festuca* plants to their endomycorrhizal fungi, *Nature, New Biol.* **244**, 30–1.
Ho, I. and Zak, B. (1979) Acid phosphatase activity of six ectomycorrhizal fungi, *Can. J. Bot.* **79**, 1203–5.
Hoover, M. D. and Lunt, H. A. (1952) A key for the classification of forest humus types, *Soil Sci. Soc. Amer. Proc.* **16**, 368–70.
Horn, H. S. (1976) Succession, pp. 187–204 in R. M. May (ed.), *Theoretical Ecology: Principles and Application*. Blackwell Scientific Publ., Oxford.
Hornby, D. (1983) Suppressive soils, *Ann. Rev. Phytopath.* **21**, 65–85.
Hudson, J. (1968) The ecology of fungi on plant remains above the soil, *New Phytol.* **67**, 837–74.
Hutchinson, G. E. (1954) The biogeochemistry of the terrestrial atmosphere, pp. 371–433 in G. P. Kuiper (ed.), *The Earth as a Planet*. Univ. Chicago Press, Chicago.
Hutchinson, G. E. (1958) Concluding remarks, *Cold Spring Harbor Symp. Quant. Biol.* **22**, 415–27.
Hutchinson, G. L., Millington, R. J. and Peters, D. G. (1972) Atmospheric ammonia: absorption by plant leaves, *Science* **175**, 771–2.
Hutchinson, J. (1969) *Evolution and Phylogeny of Flowering Plants*. Academic Press, London.
Ikediugwu, F. E. O. and Webster, J. (1970a) Antagonism between *Coprinus heptemerus* and other coprophilous fungi, *Trans. Br. Mycol. Soc.* **54**, 181–204.
Ikediugwu, F. E. O. and Webster, J. (1970b) Hyphal interference in a range of coprophilous fungi, *Trans. Br. Mycol. Soc.* **54**, 205–10.
Ingham, G. (1950a) The mineral content of air and rain and its importance to agriculture, *J. Agric. Sci.* **40**, *55–6*.

Ingham, G. (1950b) Effect of materials absorbed from the atmosphere in maintaining soil fertility. *Soil Sci.* **70**, 205–12.
Janse, J. M. (1897) Les endophytes radicaux de quelques plants Javanaise, *Annales du Jardin Botanique Buitenzorg* **14**, 53–201.
Jansson, S. L. (1958) Tracer studies on nitrogen transformations in soil with special attention to mineralization–immobilization relationships, *K. Lantbr. Högsk. Annlr.* **24**, 101–361.
Jenkinson, D. S. (1973) Radiocarbon dating of soil organic matter, *Rep. Rothamsted Expl. Sta. 1972, Pt. 1*, 75.
Jenkinson, D. S. (1981) The fate of plant and animal residues in soil, pp. 505–61 in D. J. Greenland and M. H. B. Hayes (eds), *The Chemistry of Soil Processes*. Wiley, New York.
Jenkinson, D. S. and Ladd, J. N. (1981) Microbial biomass in soil – measurement and turnover, pp. 415–71 in E. A. Paul and J. N. Ladd (eds), *Soil Biochemistry Vol. 5*. Marcel Dekker, New York.
Jenkinson, D. S. and Powlson, D. S. (1976) The effects of biocidal treatments on metabolism in soil – V. A method for measuring soil biomass, *Soil Biol. Biochem.* **8**, 209–13.
Jenkinson, D. S. and Rayner, J. H. (1977) The turnover of soil organic matter in some of the Rothamsted classical experiments, *Soil Science* **123**, 298–305.
Jenny, H. (1961) Derivation of state factor equations of soils and ecosystems, *Soil Sci. Soc. Am. Proc.* **25**, 385–8.
Jenny, H., Gessel, S. P. and Bingham, F. T. (1949) Comparative study of decomposition rates of organic matter in temperate and tropical regions, *Soil Sci.* **68**, 419–32.
Jenny, H. and Grossenbacher, K. (1963) Root–soil boundary zones as seen in the electron microscope, *Soil Sci. Soc. Am. Proc.* **27**, 273–7.
Jenny, H. and Overstreet, R. (1939a) Cation interchange between plant roots and soil colloids, *Soil Sci.* **47**, 257–72.
Jenny, H. and Overstreet, R. (1939b) Surface migration of ions and contact exchange, *J. Phys. Chem.* **43**, 1185–96.
Jensen, H. L. (1950) A survey of biological nitrogen fixation in relation to the world supply of nitrogen, *Trans. 4th. Int. Congr. Soil Sci.* **1**, 165–72.
Jensen, H. L. and Swaby, R. J. (1941) Association between nitrogen-fixing and cellulose-decomposing microorganisms, *Nature, Lond.* **147**, 147–8.
Jensen, V. and Holm, E. (1975) Associative growth of nitrogen-fixing bacteria with other microorganisms, pp. 101–19 in W. D. P. Stewart (ed.), *Nitrogen Fixation by Free-living Microorganisms*. Cambridge Univ. Press, Cambridge.
John. D. M. (1973) Accumulation and decay of litter and net production of forest in tropical west Africa, *Oikos* **24**, 430–5.
Johnson, D. W. and Edwards, N. T. (1979) The effects of stem girdling on biogeochemical cycles within a mixed deciduous forest in eastern Tennessee. II. Soil nitrogen mineralization and nitrification rates, *Oecologia* **40**, 259–71.

Jones, D. and Griffiths, E. (1964) The use of thin soil sections for the study of soil microorganisms, *Plant Soil* **20**, 232–40.
Jones, J. M. and Richards, B. N. (1977) Effect of reforestation on the turnover of ^{15}N-labelled nitrate and ammonium in relation to changes in soil microflora, *Soil Biol. Biochem.* **9**, 383–92.
Jones, J. M. and Richards, B. N. (1978) Fungal development and the transformation of ^{15}N-labelled amino- and ammonium-nitrogen in forest soils under several management regimes, *Soil Biol. Biochem.* **10**, 161–8.
Jones, P. C. T. and Mollison, J. E. (1948) A technique for the qualitative estimation of soil micro-organisms, *J. Gen. Microbiol.* **2**, 54–69.
Jordan, D. C. (1981) Nitrogen fixation by selected free-living and associative microorganisms, pp. 317–20 in A. H. Gibson and W. E. Newton (eds), *Current Perspectives in Nitrogen Fixation*. Aust. Acad. Sci., Canberra.
Jordan, C. E., Todd, R. L. and Escalante, G. (1979) Nitrogen conservation in a tropical rain forest, *Oecologia* **39**, 123–8.
Junge, C. E. (1958) The distribution of ammonia and nitrate in rainwater over the United States, *Trans. Am. Geophys. Union* **39**, 241–8.
Jurgensen, M. F. and Davey, C. B. (1971) Non-symbiotic nitrogen fixation in forest and tundra soils, *Plant Soil* **34**, 341–56.
Kaputska, L. A. and Rice, E. L. (1978) Symbiotic and asymbiotic nitrogen fixation in a tall prairie, *Soil Biol. Biochem.* **10**, 553–4.
Katznelson, H., Rouatt, J. V. and Peterson, E. A. (1962) The rhizosphere effect of mycorrhizal and non-mycorrhizal roots of yellow birch seedlings, *Can. J. Bot.* **40**, 377–82.
Kendrick, W. B. and Burges, A. (1962) Biological aspects of the decay of *Pinus sylvestris* leaf litter, *Nova Hedwig* **4**, 313–42.
Kevan, D. K. McE. (1965) The soil fauna – its nature and biology, pp. 33–50 in R. R. Baker and W. C. Snyder (eds), *Ecology of Soil-borne Plant Pathogens*. University of California Press, Berkeley.
Kinden, D. A. and Brown, M. F. (1975) Electron microscopy of vesicular–arbuscular mycorrhizae of yellow poplar. III. Host–endophyte interactions during arbuscular development, *Can. J. Microbiol.* **21**, 1930–9.
Kirk, T. K. (1980) Physiology of lignin metabolism by white-rot fungi, pp. 51–63 in T. K. Kirk, T. Higuchi and H. Chang (eds), *Lignin Biodegradation: Microbiology, Chemistry, and Potential Applications* Vol. II. CRC Press, Boca Raton, Florida.
Kirk, T. K. (1984) Degradation of lignin, pp. 399–437 in D. T. Gibson (ed.), *Microbial Degradation of Organic Compounds*, Marcel Dekker, New York and Basel.
Kirk, T. K., Higuichi, T. and Chang, H. (1980) Lignin biodegradation: summary and perspectives, pp.235–43 in T. K. Kirk, T. Higuchi and H. Chang (eds), *Lignin Biodegradation: Microbiology,*

Chemistry, and Potential Applications Vol. II. CRC Press, Boca Raton, Florida.

Kitchell, J. F., O'Neill, R. V., Webb, D., Gallepp, G., Bartell, S. M., Koonce, J. F. and Ausmus, B. S. (1979) Consumer regulation of nutrient cycling, *Bioscience* **29**, 28–34.

Knowles, R. (1981) The measurement of nitrogen fixation, pp. 327–33 in A. H. Gibson and W. E. Newton (eds), *Current Perspectives in Nitrogen Fixation*. Aust. Acad. Sci., Canberra.

Kreutzer, W. A. (1965) The reinfestation of treated soil, pp. 495–507 in K. F. Baker and W. C. Snyder (eds), *Ecology of Soil-borne Plant Pathogens*. University of California Press, Berkeley.

Kulaj, G. A. (1962) The dissolving of alumino-silicates in the rhizosphere of forest plantations, *Izv. Adad. Nauk. Ser. Biol.* **6**, 915–20.

Lamb, D. (1980) Soil nitrogen mineralization in a secondary rainforest succession, *Oecologia* **47**, 257–63.

Lamb, R. J. (1974) Effect of D-glucose on utilization of single carbon sources by ectomycorrhizal fungi, *Trans. Br. Mycol. Soc.* **63**, 295–306.

Lamb, R. J. and Richards, B. N. (1971) Effect of mycorrhizal fungi on the growth and nutrient status of slash and radiata pine seedlings, *Aust. For.* **35**, 1–7.

Lamb, R. J. and Richards, B. N. (1979) Some mycorrhizal fungi of *Pinus radiata* and *P. elliottii* var. *elliottii* in Australia, *Trans. Br. Mycol. Soc.* **54**, 371–8.

Lebedjantzev, A. N. (1924) Drying of soil as one of the natural factors in maintaining soil fertility, *Soil Sci.* **18**, 419–47.

Lee, K. E. and Wood, T. G. (1971) Physical and chemical effects on soils of some Australian termites, and their pedological significance, *Pedobiologia* **11**, 376–409.

Levi, M. P. and Cowling, E. B. (1969) Role of nitrogen in wood deterioration. VII. Physiological adaptation of wood-destroying and other fungi to substrates deficient in nitrogen. *Phytopathology* **59**, 460–8.

Levy, Y. A. and Krikun, J. (1980) Effect of vesicular–arbuscular mycorrhiza on *Citrus jambhiri* water relations, *New Phytol.* **85**, 25–31.

Lewis, D. H. (1973) Concepts in fungal nutrition and the origin of biotrophy, *Biol. Rev.* **48**, 261–77.

Lewis, D. H. (1975) Comparative aspects of the carbon nutrition of mycorrhizas, pp. 119–48 in F. E. Sanders, B. Mosse and P. B. Tinker (eds), *Endomycorrhizas*. Academic Press, London.

Lewis, D. H. and Harley, J. L. (1965a) Carbohydrate physiology of mycorrhizal roots of beech. I. Identity of endogenous sugars and utilization of exogenous sugars, *New Phytol.* **64**, 224–37.

Lewis, D. H. and Harley, J. L. (1965b) Carbohydrate physiology of mycorrhizal roots of beech. II. Utilization of exogenous sugars by uninfected and mycorrhizal roots, *New Phytol.* **64**, 238–65.

Lewis, D. H. and Harley, J. L. (1965c) Carbohydrate physiology of mycorrhizal roots of beach. III. Movement of sugars between host and fungus, *New Phytol.* **64**, 256–69.
Lindeberg, G. and Lindeberg, M. (1977) Pectinolytic ability of some mycorrhizal and saprophytic Hymenomycetes, *Arch. Microbiol.* **115**, 9–12.
Lipman, J. G. and Coneybeare, A. B. (1936) *New Jersey Agric. Exp. Sta. Bull.* No. 607. (Cited in Hutchinson 1954).
Lochhead, A. G. and Chase, F. E. (1943) Qualitative studies of soil microorganisms. V. Nutritional requirements of the predominant bacterial flora, *Soil Sci.* **55**, 185–95.
Lockwood, J. L. (1964) Soil fungistasis, *A. Rev. Phytopath.* **2**, 341–62.
Lockwood, J. L. (1977) Fungistasis in soils, *Biol. Rev.* **52**, 1–43.
Lockwood, J. L. (1981) Exploitation competition, pp. 319–49 in D. T. Wicklow, and G. C. Carroll (eds), *The Fungal Community.* Marcel Dekker, New York.
Loneragan, J. L. and Dowling, E. J. (1958) The interaction of calcium and hydrogen ions in the nodulation of subterranean clover, *Aust. J. Agric. Res.* **9**, 464–72.
Loneragan, J. F., Meyer, D., Fawcett, F. G. and Anderson, A. J. (1955) Lime pelleted clover seeds for nodulation on acid soils, *J. Aust. Inst. Agric. Sci.* **21**, 264–5.
Louw, H. A. and Webley, D. M. (1959) A study of soil bacteria dissolving certain mineral phosphate fertilizers and related compounds, *J. Appl. Bact.* **22**, 227–33.
Lundeberg, G. (1970) Utilization of various nitrogen sources, in particular bound soil nitrogen, by mycorrhizal fungi, *Stud. For. Suec.* **79**, 1–95.
Macfadyen, A. (1963) The contribution of the microfauna to total soil metabolism, pp. 3–17 in J. Doeksen and J. van der Drift (eds), *Soil Organisms.* North-Holland, Amsterdam.
Macfadyen, A. (1970) Soil metabolism in relation to ecosystem energy flow and to primary and secondary production, pp. 167–72 in J. Phillipson (ed.), *Methods of Study in Soil Ecology.* UNESCO, Paris.
McCauley, B. J. and Thrower, L. B. (1966) Succession of fungi in leaf litter of *Eucalyptus regnans, Trans. Br. Mycol. Soc.* **49**, 509–20.
McLaren, A. D. and Skukins, J. (1971) Trends in the biochemistry of terrestrial soils, pp. 1–15 in A. D. McLaren and J. Skukins (eds), *Soil Biochemistry*, Vol. 2. Marcel Dekker, New York.
McNary, J. E. and Burris, R. H. (1962) Energy requirements for nitrogen fixation by cell-free preparations from *Clostridium pasteurianum, J. Bacteriol.* **84**, 598–9.
Malajczuk, N. and Cromack, K. (1982) Accumulation of calcium oxalate in the mantle of ectomycorrhizal roots of *Pinus radiata* and *Eucalyptus marginata, New Phytol.* **92**, 527–31.
Malajczuk, N., Molina, R. and Trappe, J. M. (1982) Ectomycorrhiza formation in *Eucalyptus*. Pure culture synthesis, host specificity

and mycorrhizal compatability with *Pinus radiata, New Phytol.* **91**, 467–82.
Malo, B.A. and Purvis, E. R. (1964) Soil absorption of atmospheric ammonia, *Soil Sci.* **97**, 242–7.
Malone, C. R. and Reichle, D. E. (1973) Chemical manipulations of soil biota in a fescue meadow, *Soil Biol. Biochem.* **5**, 629–39.
Malquori, A., Ristori, G. and Vidrich, V. (1975) Biological weathering of potassium silicates: I. Biotite, *Agrochimica* **19**, 522–9.
Marks, G. C. and Foster, R. C. (1973) Structure, morphogenesis and ultra-structure of ectomycorrhizae, pp. 1–41 in G. C. Marks and T. T. Kozlowski (eds), *Ectomycorrhizae. Their Ecology and Physiology*. Academic Press, New York.
Marshall, A. J. and Williams, W. D. (eds) (1972) *Textbook of Zoology: Invertebrates*. Macmillan, London.
Martin, J. K. (1975) ^{14}C-labelled material leached from the rhizosphere of plants supplied continuously with $^{14}CO_2$, *Soil Biol. Biochem.* **1**, 395–9.
Martin, J. K. (1977) Effect of soil moisture on the release of organic carbon from wheat roots, *Soil Biol. Biochem.* **9**, 303–4.
Martin, J. K. (1978) The variation with plant age of root carbon available to soil microflora, pp. 299–30 in M. W. Loutit and J. A. R. Miles (eds), *Microbial Ecology*. Springer-Verlag, Berlin.
Martin, M. M. (1979) Biochemical implications of insect mycophagy, *Biol. Rev.* **54**, 1–21.
Marx, D. H. (1969a) The influence of ectotrophic mycorrhizal fungi on the resistance of pine roots to pathogenic infections. I. Antagonism of mycorrhizal fungi to root pathogenic fungi and soil bacteria, *Phytopathology* **59**, 153–163.
Marx, D. H. (1969b) The influence of ectotrophic mycorrhizal fungi on the resistance of pine roots to pathogenic infections. II. Production, identification, and biological activity of antibiotics produced by *Leucopaxillus cerealis* var. *piceina, Phytopathology* **59**, 411–17.
Marx, D. H. (1973) Mycorrhizae and feeder root diseases, pp. 351–82 in G. C. Marks, and T. T. Kozlowski (eds), *Ectomycorrhizae: Their Ecology and Physiology*. Academic Press, New York.
Marx, D. H. (1975) The role of ectomycorrhizae in the protection of pine from root infection by *Phytophthora cinnamomi*, pp. 112–5 in J. W. Bruehl (ed.), *Biology and Control of Soil-borne Plant Pathogens*. American Phytopathology Society.
Marx, D. H. (1979) Synthesis of *Pisolithus* ectomycorrhizae on white oak seedlings in fumigated nursery soil, USDA Forest Service Note, 3E 280.
Marx, D. H. and Bryan, W. C. (1970) Pure culture synthesis of ectomycorrhizae by *Thelephora terrestris* and *Pisolithus tinctorious* on different conifer hosts, *Can. J. Bot.* **48**, 639–43.
Marx, D. H. and Davey, C. B. (1969a) The influence of ectotrophic mycorrhizal fungi on the resistance of pine roots to pathogenic infection by *Phytophthora cinnamomi, Phytopathology* **59**, 549–58.

Marx, D. H. and Davey, C. B. (1969b) The influence of ectotrophic mycorrhizal fungi on the resistance of pine roots to pathogenic infections. IV. Resistance of naturally occurring mycorrhizae to infections by *Phytophthora cinnamomi, Phytopathology* **59**, 559–65.
Marx, D. H., Hatch, A. B. and Mendicino, J. F. (1977) High fertility, decreased sucrose content and susceptibility of loblolly pine roots to ectomycorrhizal infection by *Pisolithus tinctorius, Can. J. Bot.* **55**, 1569–74.
Mason, B. (1966) *Principles of Geochemistry*. Wiley, New York.
Meentemeyer, V. (1978) Macroclimate and lignin control of litter decomposition rates, *Ecology*, **59**, 465–72.
Melin, E. (1954) Growth factor requirements of mycorrhizal fungi of forest trees, *Svensk Bot. Tidskrift.* **48**, 86–94.
Melin, E. (1962) Physiological aspects of mycorrhizae of forest trees, pp. 247–63 in T. T. Kozlowski (ed.), *Tree Growth*. Ronald Press, New York.
Melin, E. (1963) Some effects of forest tree roots on mycorrhizal Basidiomycetes, pp. 125–45 in P. S. Nutman and B. Mosse (eds), *Symbiotic Associations* (13th Symp. Soc. Gen. Microbiol.). Cambridge Univ. Press, Cambridge.
Melin, E. and Das, V. S. R. (1954) The influence of root-metabolites on the growth of tree mycorrhizal fungi, *Pl. Physiol.* **7**, 851–8.
Melin, E. and Nilsson, H. (1950a) Transfer of radioactive phosphorus to pine seedlings by means of mycorrhizal hyphae, *Physiologia Pl.* **3**, 88–92.
Melin, E. and Nilsson, H. (1950b) Transport of labelled nitrogen from an ammonium source to pine seedlings through mycorrhizal mycelium, *Svensk Bot. Tidskr.* **46**, 281–5.
Melin, E. and Nilsson, H. (1953) Transfer of labelled nitrogen from glutamic acid to pine seedlings through the mycelium of *Boletus variegatus* (S. W.) Fr., *Nature, Lond.* **171**, 434.
Melin, E. and Nilsson, H. (1959) ^{45}Ca used as an indicator of transport of cations to pine seedlings by means of mycorrhizal mycelium, *Svensk Bot. Tidskr.* **49**, 119–21.
Melin, E. and Nilsson, H. (1957) Transport of ^{14}C-labelled photosynthate to the fungal associate of pine mycorrhiza, *Svensk Bot. Tidskr.* **51**, 166–86.
Mexal, J. and Reid, C. P. P. (1973) The growth of selected mycorrhizal fungi in response to induced water stress, *Can. J. Bot.* **51**, 1579–88.
Meyer, F. H. (1973) Distribution of mycorrhizae in natural and man-made forests, pp. 79–105 in G. C. Marks and T. T. Kozlowski, (eds), *Ectomycorrhizae. Their Physiology and Ecology*. Academic Press, New York.
Meyer, F. H. (1974) Physiology of mycorrhiza, *Ann. Rev. Plant. Physiol.* **25**, 567–86.
Miller, M. H., Mamaril, C. P. and Blair, G. J. (1970) Ammonium effects on phosphorus absorption through pH changes and phosphorus precipitation at the soil-root interface, *J. Agron.* **62**, 524–7.

Miller, O. K. (1982) Taxonomy of ecto- and ectendomycorrhizal fungi, pp. 91–101 in H. D. Schenck (ed.), *Methods and Principles of Mycorrhizal Research*. American Phytopathological Society, St. Paul, Minnesota.
Minchin, F. R., Witty, J. F., Sheehy, J. E. and Müller, M. (1983) A major error in the acetylene reduction assay: decreases in nodular nitrogenase activity under assay conditions, *J. Exp. Bot.* **34**, 641–9.
Minderman, G. (1968) Addition, decomposition and accumulation of organic matter in forests, *J. Ecol.* **56**, 355–62.
Mishustin, E. and Shil'nikova, V. K. (1971) *Biological Fixation of Atmospheric Nitrogen*. Macmillan, London.
Mitchell, J. E. (1976) The effect of roots on the activity of soil-borne plant pathogens, pp. 104–28 in R. Heitefuss and P. H. Williams (eds), *Physiological Plant Pathology*. Springer-Verlag, Berlin.
Moghimi, A., Lewis, D. G. and Oades, J. M. (1978) Release of phosphate from calcium phosphates by rhizosphere products, *Soil Biol. Biochem.* **10**, 227–81.
Moghimi, A. and Tate, M. E. (1978) Does 2-ketogluconate chelate calcium in the pH range 2.4 to 6.4? *Soil Biol. Biochem.* **10**, 289–92.
Moghimi, A., Tate, M. E. and Oades, J. M. (1978) Characterization of rhizosphere products especially 2-ketogluconic acid, *Soil Biol. Biochem.* **10**, 283–7.
Molina, R. and Trappe, J. M. (1982a) Patterns of ectomycorrhizal host specificity and potential among Pacific Northwest conifers and fungi, *For. Sci.* **28**, 423–58.
Molina, R. and Trappe, J. M. (1982b) Lack of mycorrhizal specificity by ericaceous hosts *Arbutus menziesii* and *Arctostaphylos uva-ursi, New Phytol.* **90**, 495–509.
Montes, R. A. and Christensen, N. L. (1979) Nitrification and succession in the Piedmont of North Carolina, *For. Sci.* **25**, 287–97.
Moore, A. W. (1969) *Azolla*: biology and agronomic significance, *Bot. Rev.* **35**, 17–34.
Moore, D. R. E. and Waid, J. S. (1971) The influence of washings of living roots on nitrification, *Soil Biol. Biochem.* **3**, 69–83.
Morrison, T. M. (1962a) Absorption of phosphorus from soils by mycorrhizal plants, *New Phytol.* **61**, 10–20.
Morrison, T. M. (1962b) Uptake of sulphur by mycorrhizal plants, *New Phytol.* **61**, 21–7.
Morrison, T. M. (1963) Uptake of sulphur by excised beech mycorrhizas, *New Phytol.* **62**, 42–9.
Moser, M. and Haselwandter, K. (1983) Ecophysiology of mycorrhizal symbioses, pp. 391–421 in O. L. Lange, P. S. Nobel and C. B. Osmond and H. Ziegler (eds), *Physiological Plant Ecology III, Encyclopedia of Plant Physiology*, New Series Vol. 12C. Springer-Verlag, Berlin.
Mosse, B. (1953) Fructifications associated with mycorrhizal strawberry roots, *Nature, Lond.* **171**, 974.
Mosse, B. (1956) Fructifications of an *Endogone* species causing endotrophic mycorrhiza in fruit plants, *Ann. Bot.* **20**, 349–62.

Mosse, B. (1957) Growth and chemical composition of mycorrhizal and non-mycorrhizal apples, *Nature, Lond.* **179**, 922–4.
Mosse, B. (1962) The establishment of vesicular–arbuscular mycorrhiza under aseptic conditions, *J. Gen. Microbiol.* **27**, 509–20.
Mosse, B., Powell, C. L. and Hayman, D. S. (1976) Plant growth responses to vesicular–arbuscular mycorrhiza. IX. Interactions between VA mycorrhiza, rock phosphate and symbiotic nitrogen fixation, *New Phytol.* **76**, 331–42.
Newman, E. I. (1978) Root microorganisms: their significance in the ecosystem, *Biol. Rev.* **53**, 511–54.
Nicholas, D. J. D. (1980) Cycling of sulfur through the soil–plant system, pp. 194–202 in J. R. Freney and A. J. Nicolson (eds), *Sulfur in Australia*. Aust. Acad. Sci., Canberra.
Nicholas, D. P. and Parkinson, D. (1967) A comparison of methods for assessing the amount of fungal mycelium in soil samples, *Pedobiologia* **7**, 23–41.
Nicolson, T. H. (1967) Vesicular–arbuscular mycorrhiza – a universal plant symbiosis, *Sci. Progr., Oxf.* **55**, 561–81.
Nicolson, T. H. and Gerdemann, J. W. (1968) Mycorrhizal *Endogone* species, *Mycologia* **60**, 313–75.
Nihlgard, B. (1970) Precipitation, its chemical composition and effect on soil water in a beech and a spruce forest in southern Sweden, *Oikos* **21**, 208–17.
Noble, I. R. and Slatyer, R. O. (1981) Concepts and models of succession in vascular plant communities subject to recurrent fire, pp. 311–35 in A. M. Gill, R. H. Groves and I. R. Noble (eds), *Fire and the Australian Biota*. Aust. Acad. Sci. Canberra.
Norkans, B. (1950) Studies in growth and cellulolytic enzymes of *Tricholoma, Sym. Bot. Upsaliens* **11**, 126 pp.
Norris, D. O. (1956) Legumes and the rhizobium symbiosis, *Emp. J. Exp. Agric.* **24**, 247–70.
Norris, D. O. (1965) Acid production by *Rhizobium*. A unifying concept, *Plant Soil* **22**, 143–66.
Nurmikko, V. (1956) Biochemical factors affecting symbiosis among bacteria, *Experientia* **12**, 245–9.
Nutman, P. A. (1963) Factors influencing the balance of mutual advantage in legume symbiosis, pp. 51–71 in P. S. Nutman and B. Mosse (eds), *Symbiotic Associations* (13th Symp. Soc. Gen. Microbiol). Cambridge Univ. Press, Cambridge.
Nutman, P. S. (ed.) (1976) *Symbiotic Nitrogen Fixation in Plants*. Cambridge Univ. Press, New York.
Nutman, P. S. (1981) Hereditary host factors affecting nodulation and nitrogen fixation, pp. 194–204 in A. H. Gibson and W. E. Newton (eds), *Current Perspectives in Nitrogen Fixation*. Aust. Acad. Sci., Canberra.
Nye, P. H. and Tinker, P. B. (1977) *Solute Movement in the Soil–Plant System*. Blackwell, Oxford.
Nylund, J. E. and Unestam, T. (1982) Structure and physiology of

ectomycorrhizae. I. The process of mycorrhiza formation in Norway spruce *in vitro, New Phytol.* **91**, 63–79.
O'Brien, B. J. and Stout, J. D. (1978) Movement and turnover of soil organic matter as indicated by carbon isotope measurements, *Soil Biol. Biochem.* **10**, 309–17.
O'Connell, A. M., Grove, T. S. and Malajczuk, N. (1979) Nitrogen fixation in the litter layer of eucalypt forests, *Soil Biol. Biochem.* **11**, 681–2.
Odum, E. P. (1969) The strategy of ecosystem development, *Science* **164**, 262–70.
Odum, E. P. (1971) *Fundamentals of Ecology*, 3rd edn. Saunders, Philadelphia.
Odum, H. T. (1967) Work circuits and systems stress, pp. 81–138 in *Primary Productivity and Mineral Cycling in Natural Ecosystems.* Univ. Maine Press, Orono.
Odum, H. T. (1971) *Environment, Power, and Society.* Wiley, New York.
O'Hagan, B. J. (1966) Plant–soil relationships and microbial activity within the *Acacia aneura* alliance in north-western N.S.W. MSc Thesis, University of New England.
O'Hara, G. W., Daniel, R. M. and Steele, K. W. (1983). Effect of oxygen on the synthesis, activity, and breakdown of the rhizobium dentrification system, *J. Gen. Microbiol.* 129, 2405–12.
Olson, J. S. (1963) Energy storage and the balance of producers and decomposers in ecological systems, *Ecology* **44**, 322–31.
Papavizas, G. C. and Davey, C. D. (1961) Extent and nature of the rhizosphere of *Lupinus, Plant Soil* **14**, 215–36.
Park, D. (1976) Carbon and nitrogen levels as factors influencing fungal decomposers, pp. 41–59 in J. M. Anderson and A. Macfadyen (eds), *The Role of Terrestrial and Aquatic Organisms in Decomposition Processes.* Blackwell, Oxford.
Park, J. Y. (1970) Antifungal effect of an ectotrophic mycorrhizal fungus, *Lactarius* sp., associated with basswood seedlings, *Can. J. Microbiol.* **16**, 798–800.
Parker, C. A. (1957) Non-symbiotic nitrogen-fixing bacteria in soil III. Total nitrogen changes in a field soil. *J. Soil. Sci.* **8**, 48–59.
Parker, C. A., Trinick, M. J. and Chatel, D. L. (1977) Rhizobia as soil and rhizosphere inhabitants, pp. 311–52 in R. W. F. Hardy and A. H. Gibson (eds), *A Treatise on Dinitrogen Fixation*, Sect. IV (Agronomy and Ecology). Wiley-Interscience, New York.
Parkinson, D. (1973) Techniques for the study of soil fungi, *Bull. Ecol. Res. Commun.* (*Stockholm*) **17**, 29–36.
Parkinson, D., Domsch, K. H. and Anderson, J. P. E. (1978) Die Entwicklung mikrobieller Biomassen im organischen Horizonteines Fichtenstandortes, *Ecologia Plantarum* **13**, 355–66.
Parkinson, D., Gray, T. R. G. and Williams, S. T. (eds) (1971) *Methods for Studying the Ecology of Soil Micro-organisms*, IBP Handbook No. 19. Blackwell, Oxford.
Parkinson, D., Virrer, S. and Whittaker, J. B. (1979) Effect of

collembolan grazing on fungal colonization of leaf litter, *Soil Biol. Biochem.* **11**, 529–35.
Paul, E. A. and Tu, L. M. (1965) Alteration of microbial activities, mineral nitrogen and free amino acid constituents of soils by physical treatments, *Plant Soil* **22**, 207–19.
Paul, E. A. and Voroney, R. P. (1980) Nutrient and energy flows through soil microbial biomass, pp. 215–37 in D. C. Ellwood, J. N. Hedger, M. J. Latham, J. M. Lynch and J. H. Slater (eds), *Contemporary Microbial Ecology*. Academic Press, London.
Pearson, V. and Read, D. J. (1973) The biology of mycorrhiza in the Ericaceae. I. The isolation of the endophyte and synthesis of mycorrhizas in aseptic culture, *New Phytol.* **72**, 371–9.
Persson, H. (1978) Root dynamics in a Scots pine stand in central Sweden, *Oikos* **30**, 508–19.
Peters, G. A., Toia, R. E., Raveed, D. and Levine, N. J. (1980) The *Azolla–Anabaena azollae* relationship. VI. Morphological aspects of the association, *New Phytol.* **80**, 583–93.
Peyronel, B. (1923) Fructification de l'endophyte à arbuscules et à vesicules des mycorhizes endotrophes, *Bull. Soc. Mycol. France* **39**, 119–26.
Peyronel, B., Fassi, B., Fontana, A. and Trappe, J. M. (1969) Terminology of mycorrhizae, *Mycologia* **61**, 410–11.
Pike, L. H., Tracey, D. M., Sherwood, M. A. and Neilsen, D. (1972) Estimates of biomass and fixed nitrogen of epiphytes from old-growth Douglas fir, pp. 177–87 in J. F. Franklin, L. J. Dempster and R. J. Waring (eds), *Proc. Symp. Res. Coniferous For. Ecosyst.*, Pac. Northwest For. Range Exp. Stn., Portland, Oregon.
Pirt, S. J. (1965) The maintenance energy of bacteria in growing cultures, *Proc. Roy. Soc. Lond., Ser. B.* **163**, 224–31.
Postgate, J. (1974) New advances and future potential in biological nitrogen fixation, *J. Appl. Bacteriol.* **37**, 185–202.
Postgate, J. (1981) Microbiology of the free-living nitrogen-fixing bacteria excluding cyanobacteria, pp. 217–28 in A. H. Gibson and W. E. Newton (eds), *Current Perspectives in Nitrogen Fixation*. Aust. Acad. Sci., Canberra.
Purchase, B. S. (1974a) Evaluation of the claim that grass root exudates inhibit nitrification, *Plant Soil* **41**, 527–39.
Purchase, B. S. (1974b) The influence of phosphate deficiency on nitrification, *Plant Soil* **41**, 541–7.
Quispel, A. and Burggraaf, A. J. P. (1981) *Frankia*, the diazotrophic endophyte from actinorhizas, pp. 229–36 in A. H. Gibson and W. E. Newton (eds), *Current Perspectives in Nitrogen Fixation*. Aust. Acad. Sci., Canberra.
Rao, G. G. and Dhar, N. R. (1931) Bildung von Stickstoffverbindungen im Luft und im Boden unter dern Einfluss des Lichtes. *Z.f. anorg. allg. Chem.*, **199**, 422–6.
Read, D. J. (1974) *Pezizella ericae* sp. nov., the perfect state of a

typical mycorrhizal endophyte of Ericaceae, *Trans. Br. Mycol. Soc.* **63**, 381–3.
Read, D. J. (1978) Biology of mycorrhiza in heathland ecosystems with special reference to the nitrogenous nutrition of Ericaceae, pp. 324–8 in M. W. Loutit and J. A. R. Miles (eds), *Microbiol. Ecology*. Springer-Verlag, Berlin.
Read, D. J. (1982) The biology of mycorrhiza, in *The Ericales* (Proc. 5th North American Conference on Mycorrhizae). (Cited by Harley and Smith 1983).
Reichle, D. E. (1977) The role of soil invertebrates in nutrient cycling, *Ecol. Bull.* (*Stockholm*) **25**, 145–56.
Reichle, D. E., McBrayer, J. F. and Ausmus, B. S. (1975) Ecological energetics of decomposer invertebrates in a deciduous forest and the total respiration budget, pp. 283–92 in J. Vanek (ed.), *Progress in Soil Zoology* (Proc. 5th International Colloquium, Prague, Czechoslovakia, 1973). Junk, The Hague and Academia, Prague.
Reid, C. P. P. (1974) Assimilation, distribution, and root exudation of ^{14}C by ponderosa pine seedlings under induced water stress, *Plant Physiol.* **54**, 44–9.
Reid, C. P. P. and Mexal, J. G. (1977) Water stress effects on root exudation by lodgepole pine, *Soil Biol. Biochem.* **9**, 417–21.
Reid, C. P. P. and Woods, F. W. (1969) Translocation of ^{14}C-labelled compounds in mycorrhiza and its implications in interpreting nutrient cycling, *Ecology* **50**, 179–81.
Reiners, W. A. (1981) Nitrogen cycling in relation to ecosystem succession, in F. E. Clark and T. Rosswall (eds), *Terrestrial Nitrogen Cycles. Ecol. Bull.* (*Stockholm*) **33**, 507–28.
Reiners, W. A. and Lang, G. E. (1979) Vegetational patterns and processes in the balsam fir zone, White Mountains, New Hampshire, *Ecology* **60**, 403–17.
Rennie, D. A., Paul, E. A. and Johns, L. E. (1976) Natural nitrogen-15 abundance of soil and plant samples, *Can. J. Soil Sci.* **56**, 43–50.
Rhodes, L. L. and Gerdemann, J. W. (1978) Influence of phosphorus nutrition on sulphur uptake by vesicular–arbuscular mycorrhizae of onions, *Soil Biol. Biochem.* **10**, 361–4.
Rice, E. L. and Pancholy, S. K. (1972) Inhibition of nitrification by climax vegetation, *Am. J. Bot.* **60**, 691–8.
Rice, E. L. and Pancholy, S. K. (1973) Inhibition of nitrification by climax ecosystems. II. Additional evidence and possible role of tannins, *Am. J. Bot.* **60**, 691–8.
Rice, E. L. and Pancholy, S. K. (1974) Inhibition of nitrification by climax ecosystems. III. Inhibitors other than tannins, *Am. J. Bot.* **61**, 1095–103.
Rich. J. R. and Bird, G. W. (1974) Association of early season vesicular–arbuscular mycorrhizae with increased growth and development of cotton, *Phytopathology* **64**, 1421–5.
Richards, B. N. (1964) Fixation of atmospheric nitrogen in coniferous forests, *Aust. For.* **28**, 68–74.

Richards, B. N. (1965) Mycorrhiza development of loblolly pine seedlings in relation to soil reaction and the supply of nitrate, *Plant Soil* **22**, 187–99.

Richards, B. N. (1973) Nitrogen fixation in the rhizosphere of conifers, *Soil Biol. Biochem.* **5**, 149–52.

Richards, B. N. (1981) Forest floor dynamics, pp. 147–57 in *Productivity in Perpetuity*, Proc. Australian Forest Nutrition Workshop, 10–14 August 1981, Canberra. CSIRO, Melbourne.

Richards, B. N. and Bevege, D. I.(1967) The productivity and nitrogen economy of artificial ecosystems comprising various combinations of perennial legumes and coniferous tree seedlings, *Aust. J. Bot.* **15**, 467–80.

Richards, B. N. and Bevege, D. I. (1969) Critical foliage concentrations of nitrogen and phosphorus as a guide to the underplanting of *Araucaria* to *Pinus, Plant Soil* **31**, 328–36.

Richards, B. N. and Charley, J. L. (1977) Carbon and nitrogen flux through native forest floors, pp. 65–81 in F. H. Hingston (ed.), *Symposium on Nutrient Cycling in Indigenous Forest Ecosystems*, Inst. For. Res. Protect. Como, WA. CSIRO Div. Land Resources Management, Perth, Australia.

Richards, B. N. and Charley, J. L. (1983) Mineral cycling processes and system stability in the eucalypt forest, *For. Ecol. Mangt.* **7**, 31–47.

Richards, B. N., Smith, J. E. N., White, G. J. and Charley, J. L. (1985) Mineralization of soil nitrogen in three forest communities from the New England region of New South Wales, Aust. J. Ecol., **10**, 429–41.

Richards, B. N. and Wilson, G. L. (1963) Nutrient supply and mycorrhiza development in Caribbean pine, *For. Sci.* **9**, 405–12.

Riley, D. and Barber, S. A. (1969) Bicarbonate accumulation and pH changes at soyabean (*Glycine max* (L.) Merr.) root–soil interface, *Soil Sci. Soc. Amer. Proc.* **33**, 905–8.

Riley, D. and Barber, S. A. (1971) Effect of ammonium fertilization on phosphorus uptake as related to root-induced pH changes at the root–soil interface, *Soil Sci. Soc. Amer. Proc.* **35**, 301–6.

Rishbeth, J. (1950) Observations on the biology of *Fomes annosus*, with particular reference to East Anglian pine plantations. I. The outbreaks of disease and ecological status of the fungus, *Ann. Bot.* **14**, 365–83.

Rishbeth, J. (1951) Observations on the biology of *Fomes annosus*, with particular reference to East Anglian pine plantations. III. Natural and experimental infection of pines, and some factors affecting the severity of the disease, *Ann. Bot.* **15**, 221–46.

Rishbeth, J. (1963) Stump protection against *Fomes annosus*. III. Inoculation with *Peniophora gigantea, Ann. Appl. Biol.* **52**, 63–77.

Rivett, M. (1924) The root tubercles in *Arbutus unedo, Ann. Bot.* (*Lond.*) **38**, 661–77.

Robertson, J. G., Lyttleton, P. and Pankhurst, C. E. (1981) Pre-infection and infection processes in the legume–rhizobium symbiosis, pp. 280–91 in A. H. Gibson and W. E. Newton (eds.) *Current*

Perspectives in Nitrogen Fixation. Aust. Acad. Sci., Canberra.
Robertson, N. F. (1954) Studies on the mycorrhiza of *Pinus sylvestris*, *New Phytol*. **53**, 253–83.
Rodin, L. E. and Basilevich, N. I. (1967) *Production and Mineral Cycling in Terrestrial Vegetation*, G. E. Fogg (transl. and ed.). Oliver and Boyd, Edinburgh.
Rose, A. H. (1976) *Chemical Microbiology. An Introduction to Microbial Physiology* Butterworths, London.
Rosswall, T. (1973) *Modern Methods in the Study of Microbial Ecology. Bull. Ecol. Res. Comm.* (*Stockholm*) **17**.
Rovira, A. D. (1956) Plant root excretions in relation to the rhizosphere effect. I. The nature of root exudate from oats and peas, *Plant Soil* **7**, 178–94.
Rovira, A. D. (1965) Plant root exudates and their influence upon soil microorganisms, pp. 170–84 in K. F. Baker and W. C. Snyder (eds), *Ecology of Soil-borne Plant Pathogens*. University of California Press, Berkeley.
Rovira, A. D. (1969a) Diffusion of carbon compounds away from wheat roots, *Aust. J. Biol. Sci.* **22**, 1287–90.
Rovira, A. D. (1969b) Plant root exudates, *Bot. Rev*. **35**, 35–57.
Rovira, A. D. (1973) Zones of exudation along roots and spatial distribution of microorganisms in the rhizosphere, *Pestic. Sci.* **4**, 361–6.
Rovira, A. D. (1979) Biology of the soil–root interface, pp. 145–60 in J. L. Harley and R. Scott Russell (eds), *The Soil–Root Interface*. Academic Press, London.
Rovira, A. D., Bowen, G. D. and Foster, R. C. (1983) The significance of rhizosphere microflora and mycorrhizas in plant nutrition, pp. 61–93 in A. Laüchli and R. L. Bieleski (eds), *Encyclopedia of Plant Physiology*, New Series Vol. 15 A + B, Inorganic Plant Nutrition. Springer-Verlag, Berlin.
Rovira, A. D. and Davey, C. B. (1974) Biology of the rhizosphere, pp. 153–240 in E. W. Carson (ed.), *The Plant Root and its Environment*. Univ. Virginia Press, Charlottesville.
Rovira, A. D., Foster, R. C. and Martin, J. K. (1979) Note on terminology: origin, nature and nomenclature of the organic materials in the rhizosphere, pp. 1–4 in J. L. Harley, and R. Scott Russell (eds), *The Soil–Root Interface*. Academic Press, London.
Rovira, A. D., Newman, E. I., Bowen, H. J. and Campbell, R. (1974) Quantitative assessement of the rhizoplane microflora by direct microscopy, *Soil Biol. Biochem*. **6**, 211–6.
Rowell, D. L. (1981) Oxidation and reduction, pp. 401–61 in D. J. Greenland and M. H. B. Hayes (eds), *The Chemistry of Soil Processes*. Wiley, New York.
Runge, M. (1974) Die Stickstoff-Mineralisation im Boden eines Sauerhumus-Buchenwaldes. II. Die Nitratproduktion, *Oecol. Plant*. **9**, 219–30.
Runge, M. (1983) Physiology and ecology of nitrogen nutrition, pp. 163–200 in O. L. Lange, P. S. Nobel, C. B. Osmond and

H. Ziegler (eds), *Encyclopedia of Plant Physiology Vol. 12C, Physiological Plant Ecology III: Responses to the Chemical and Biological Environment*. Springer-Verlag, Berlin.

Russell, E. W. (1973) *Soil Conditions and Plant Growth*, 10th edn. Longman, London and New York.

Safir, G. R., Boyer, J. S. and Gerdemann, J. W. (1971) Mycorrhizal enhancement of water transport in soybean, *Science* **172**, 581–3.

Safir, G. R., Boyer, J. S. and Gerdemann, J. W. (1972) Nutrient status and mycorrhizal enhancement of water transport in soybean, *Pl. Physiol.* **49**, 700–3.

Sanders, F. E. and Tinker, P. B. (1973) Phosphate flow into mycorrhizal roots, *Pestic. Sci.* **4**, 385–95.

Satchell, J. E. (1974) Litter interface of animate/inanimate matter, Introduction, pp. xiii–xliv in C. H. Dickinson and G. J. F. Pugh (eds), *Biology of Plant Litter Decomposition* Vol. 1. Academic Press, New York.

Schmidt, E. L. (1978) Nitrifying microorganisms and their methodology, pp. 288–91 in D. Schlessinger (ed.), *Microbiology 1978*. Amer. Soc. Microbiology, Washington, DC.

Schöllhorn, R. and Burris, R. H. (1966) Study of intermediates in nitrogen fixation, *Fed. Proc.* **25**, 710.

Shanks, R. E. and Olsen, J. S. (1961) First-year breakdown of leaf litter in Southern Appalachian forests, *Science* **134**, 194–5.

Shields, J. A., Paul, E. A. and Lowe, W. E. (1973) Turnover of microbial tissue in soil under field conditions, *Soil Biol. Biochem.* **5**, 753–64.

Silver, W. S., Centrifanto, Y. M. and Nicholas, D. J. D. (1963) Nitrogen fixation by the leaf-nodule endophyte of *Psychotria bacteriophila, Nature, Lond.* **199**, 396–7.

Silverman, M. P. and Munoz, E. F. (1970) Fungal attack on rock: solubilization and altered infrared spectra, *Science* **169**, 985–7.

Silvester, W. B. (1978) Nitrogen fixation and mineralization in kauri excluding legumes, pp. 141–90 in R. W. F. Hardy and A. H. Gibson (eds), *A Treatise on Dinitrogen Fixation*, Sect. IV (Agronomy and Ecology). Wiley-Interscience, New York.

Silvester, W. B. (1978) Nitrogen fixation and mineralization in kauri (*Agathis australis*) forest in New Zealand, pp. 138–43 in M. W. Loutit and J. A. R. Miles (eds), *Microbial Ecology*. Springer-Verlag, Berlin.

Simpson, J. R. (1954) Mineral nitrogen fluctuations in soils under improved pasture in southern New South Wales, *Aust. J. Agric. Res.* **13**, 1059–72.

Skinner, M. F. and Bowen, G. D. (1974) The penetration of soil by mycelial strands of ectomycorrhizal fungi, *Soil Biol. Biochem.* **6**, 57–81.

Skujins, J. and McLaren, A. D. (1968) Persistence of enzymatic activities in stored and geologically preserved soils, *Enzymologia* **34**, 213–25.

Slankis, V. (1973) Hormonal relationships in mycorrhizal development,

pp. 231–98 in G. C. Marks and T. T. Kozlowski (eds), *Ectomycorrhizae*. Academic Press, New York.
Smith, D. Muscatine, L. and Lewis, D. (1969) Carbohydrate movement from autotrophs to heterotrophs in parasitic and mutualistic symbiosis, *Biol. Rev.* **44**, 17–90.
Smith, F. A. (1972) A comparison of the uptake of nitrate, chloride, and phosphate by excised beech mycorrhizas, *New Phytol.* **71**, 875–82.
Smith, J. E. N. (1974) Mineralization of soil nitrogen in the New England National Park, M. Nat. Res. Thesis, Univ. New England, Armidale, NSW, Australia.
Smith, S. E., Nicholas, D. J. D. and Smith, F. A. (1979) The effect of early mycorrhizal infection on nodulation and nitrogen fixation in *Trifolium subterraneum*, L., *Aust. J. Pl. Physiol.* **6**, 305–11.
Smith, W. H. (1970) Technique for collection of root exudates from mature trees, *Plant Soil* **32**, 328–41.
Smith, W. H. (1972) Influence of artificial defoliation on exudates of sugar maple, *Soil Biol. Biochem.* **4**, 11–13.
Smith, W. H. (1976) Character and significance of forest tree root exudates, *Ecology* **57**, 324–31.
Sollins, P., Cromack, K. Jr., Fogel, R. and Li, C. Y. (1981) Role of low-molecular-weight organic acids in the inorganic nutrition of fungi and higher plants, pp. 607–19 in D. T. Wicklow, and G. C. Carroll, (eds), *The Fungal Community*. Marcel-Dekker, New York.
Springett, B. P. (1978) On the ecological role of insects in Australian eucalypt forests. *Aust. J. Ecol.* **3**, 129–39.
Stanier, R. Y. (1953) Adaptation, evolutionary and physiological: or Darwinism among the microorganisms, pp. 1–14 in R. Davies and E. F. Gale (eds), *Adaptation in Micro-organisms*, (Soc. Gen. Microbiol. 3rd Symp.) Cambridge Univ. Press, Cambridge.
Stanier, R. Y., Adelberg, E. A. and Ingraham, J. L. (1977) *General Microbiology*. Macmillan, London.
Stanier, R. Y. and Cohen-Bazire, G. (1977) Phototrophic prokaryotes: the cyanobacteria, *Ann. Rev. Microbiol.* **31**, 225–74.
Starc, A. (1942) Mikrobiologische untersuchungen einiger podsoliger Böden Kroatiens, *Arch, Mikrobiol.* **12**, 329–52.
Stevenson, G. (1959) Fixation of nitrogen by non-nodulated seed plants, *Ann. Bot.* (NS) **23**, 622–35.
Stewart, W. D. P. (1965) Nitrogen turnover in marine and brackish habitats, *Ann. Bot.* (NS) **29**, 229–39.
Stewart, W. D. P. (1966) *Nitrogen Fixation in Plants*. Athlone Press, London.
Stone, E. L. (1950) Some effects of mycorrhizae on the phosphorus nutrition of Monterey pine seedlings, *Soil Sci. Soc. Amer. Proc.* **14**, 340–5.
Stribley, D. P. and Read, D. J. (1974) The biology of mycorrhizae in Ericaceae. III. Movement of carbon-14 from host to fungus, *New Phytol.* **73**, 731–41.
Stribley, D. P. and Read, D. J. (1975) Some nutritional aspects of the

biology of ericaceous mycorrhizas, pp. 195–207 in F. E. Sanders, B. Mosse and P. B. Tinker (eds), *Endomycorrhizas*. Academic Press, London.

Swaby, R. J. and Fedel, R. (1973) Microbial production of sulphate and sulphide in some Australian soils, *Soil Biol. Biochem.* **5**, 773–81.

Swaby, R. J. and Fedel, R. (1977) Sulphur oxidation and reduction in some tropical Australian soils, *Soil Biol. Biochem.* **9**, 327–31.

Swift, M. J. (1973) Estimation of mycelial growth during decomposition of plant litter, pp. 323–8 in T. Rosswall (ed.), *Modern Methods in the Study of Microbial Ecology. Bull. Ecol. Res. Comm. (Stockholm)* **17**.

Swift, M. J. (1977a) The ecology of wood decomposition, *Sci. Prog. Oxf.* **64**, 175–99.

Swift, M. J. (1977b) The roles of fungi and animals in the immobilisation and release of nutrient elements from decomposing branch-wood, *Ecol. Bull. (Stockholm)* **25**, 193–202.

Swift, M. J., Heal, O. W. and Anderson, J. M. (1979) *Decomposition in Terrestrial Ecosystems*, Studies in Ecology Vol. 5. Blackwell, Oxford.

Switzer, G. L. and Nelson, L. E. (1972) Nutrient accumulation and cycling in loblolly pine (*Pinus taeda* L.) plantation ecosystems: the first twenty years, *Soil Sci. Soc. Amer. Proc.* **36**, 143–7.

Tan, K. H. and Nopamornbodi, O. (1979) Election microbeam scanning of element distribution zones in soil rhizosphere and plant tissue, *Soil Sci.* **127**, 235–41.

Tchan, Y. T. and Beadle, N. C. W. (1955) Nitrogen economy in semi-arid plant communities, *Proc. Linn, Soc. N.S.W.* **80**, 97–104.

Teng, J. and Whistler, R. L. (1973) Cellulose and chitin, pp. 249–69 in L. P. Miller (ed.), *Phytochemistry* Vol. I. van Nostrand Reinhold Co., New York.

Theodorou, C. (1971) The phytase activity of the mycorrhizal fungus *Rhizopogon roseolus, Soil Biol. Biochem.* **3**, 89–90.

Theodorou, C. and Bowen, G. D. (1969) Influence of pH and nitrate on mycorrhizal associations of *Pinus radiata* D. Don., *Aust. J. Bot.* **17**, 59–67.

Thompson, E. J. and Black, C. A. (1970) Changes in organic phosphorus in soil in the presence and absence of plants. II. Soil in a simulated rhizosphere, *Plant Soil* **32**, 161–8.

Till, A. R. (1980) Sulfur cycling in soil-plant-animal systems, pp. 204–216 in J. R. Freney and A. J. Nicolson (eds), *Sulfur in Australia*. Aust. Acad. Sci., Canberra.

Tjepkema, J. D., Omerod, W. and Torrey, J. G. (1980) Vesicle formation and acetylene reduction activity in *Frankia* sp. CPI 1 cultured in defined nutrient media, *Nature, Lond.* **287**, 633–5.

Tommerup, I. C. and Abbott, L. K. (1981) Prolonged survival and viability of V–A mycorrhizal hyphae after root death, *Soil Biol. Biochem.* **13**, 431–3.

Towle, G. A. and Whistler, R. L. (1973) Hemicelluloses and gums,

pp. 198–248 in L. P. Miller (ed.), *Phytochemistry* Vol. I. van Nostrand Reinhold Co., New York.
Tribe, H. T. (1957) Ecology of microorganisms in soils as observed during their development upon buried cellulose film, pp. 287–98 in C. C. Spicer and R. E. O. Williams (eds), *Microbial Ecology* (Soc. Gen. Microbiol. 7th Symp.). Cambridge Univ. Press, Cambridge.
Trinick, M. J. (1973) Symbiosis between *Rhizobium* and the non-legume, *Trema aspera, Nature, Lond*. **244**, 459–60.
Trinick, M. J. (1979) Structure of nitrogen-fixing nodules formed by *Rhizobium* on roots of *Parasponia andersonii* Planch., *Can. J. Microbiol*. **25**, 565–8.
Turner, J. C. and Singer, M. J. (1976) Nutrient distribution and cycling in a sub-alpine coniferous ecosystem, *J. Appl. Ecol*. **13**, 295–301.
van Berkum, P. and Bohlool, B. B. (1980) Evaluation of nitrogen fixation by bacteria in association with roots of tropical grasses, *Microbiol. Rev*. **44**, 491–517.
van Dijk, C. (1978) Spore formation and endophyte diversity in root nodules of *Alnus glutinosa* (L) Vill., *New Phytol*. **81**, 601–15.
van Schreven, D. A. (1964) A comparison between the effect of fresh and dried organic materials added to soil on carbon and nitrogen mineralization, *Plant Soil* **20**, 149–65.
van Veen, J. A. and Paul, E. A. (1979) Conversion of biovolume measurements of soil organisms grown under various moisture tensions, to biomass and their nutrient content, *Appl. Environ. Microbial*. **37**, 686–92.
Vancura, V. (1964) Root exudates of plants. I., *Plant Soil* **21**, 231–8.
Vancura, V., Prikryl, Z., Kalachova, L. and Wurst, M. (1977) Some quantitative aspects of root exudation, *Ecol. Bull*. (*Stockholm*) **25**, 281–6.
Verma, D. P. S., Legocki, R. P. and Auger, S. (1981) Expression of nodule-specific host genes in soybean, pp. 205–8 in Gibson, A. H. and Newton, W. E. (eds), *Current Perspectives in Nitrogen Fixation*. Aust. Acad. Sci., Canberra.
Verstraete, W. (1981) Nitrification, pp. 303–314 in Clark, F. E. and Rosswall, T. (eds), *Terrestrial Nitrogen Cycles. Processes, Ecosystem Strategies and Management Impacts. Ecol. Bull.* (*Stockholm*) **33**.
Vitolins, I. and Swaby, R. J. (1967) Activity of sulphur-oxidising microorganisms in some Australian soils, *Aust. J. Soil Res*. **7**, 171–93.
Vitousek, P. M. (1977) The regulation of element concentrations in mountain streams in the northeastern United States, *Ecol. Monogr*. **47**, 65–87.
Vitousek, P. M., Gosz, J. R., Grier, C. C., Mellilo, J. M., Reiners, W. J. and Todd, R. L. (1979) Nitrate losses from disturbed ecosystems, *Science* **204**, 469–74.
Vitousek, P. M. and Reiners, W. A. (1975) Ecosystem succession and nutrient retention: a hypothesis, *BioScience* **25**, 376–81.
Vogt, K. A., Grier, C. C., Meier, C. E. and Edmonds, R. L. (1982)

Mycorrhizal role in net primary production and nutrient cycling in *Abies amabalis* (Dougl.) Forbes ecosystems in western Washington, *Ecology* **63**, 370–80.
Voigt, G. K. (1965) Biological mobilization of potassium from primary minerals, pp. 33–46 in C. T. Youngberg, (ed), *Forest–Soil Relationships in North America*. Oregon State Univ. Press, Corvallis, Oregon.
Waid, J. S. (1974) Decomposition of roots, Ch. 6, pp. 175–211 in C. H. Dickinson and G. J. F. Pugh (eds), *Biology of Plant Litter Decomposition*, Vol. I. Academic Press, London.
Wainwright, M. (1978) Microbial sulphur oxidation in soil, *Sci. Prog. Oxf.* **65**, 459–75.
Wainwright, M. (1981) Mineral transformations by fungi in culture and in soils, *Z. Pflanzenernaehr. Bodenkd.* **144**, 41–63.
Wallace, R. H. and Lochhead, A. G. (1949) Qualitative studies of soil microorganisms. VIII. Influence of various crop plants on the nutritional groups of soil bacteria, *Soil Sci.* **67**, 63–9.
Wallwork, J. A. (1970) *Ecology of Soil Animals*. McGraw-Hill, London.
Wallwork, J. A. (1976) *The Distribution and Diversity of Soil Fauna*. Academic Press, London.
Warcup, J. H. (1965) Growth and reproduction of soil microorganisms in relation to substrate, pp. 52–67 in K. F. Baker and W. C. Snyder (eds), *Ecology of Soil-borne Plant Pathogens*. University of California Press, Berkeley.
Warcup, J. H. (1981) The mycorrhizal relationship of Australian orchids, *New Phytol.* **87**, 371–87.
Warembourg, F. R. and Billes, G. (1979) Estimating carbon transfers in the plant rhizosphere, pp. 183–96 in J. L. Harley and R. Scott Russell (eds), *The Soil–Root Interface*. Academic Press, London.
Warembourg, F. R. and Morrall, R. A. A. (1978) Energy flow in the plant–microorganism system, pp. 205–242 in Y. R. Dommerques and S. V. Krupa (eds), *Interactions between Non-Pathogenic Soil Organisms and Plants*. Elsevier, Amsterdam.
Watanabe, I. (1981) Biological nitrogen fixation associated with wetland rice, pp. 313–6 in A. H. Gibson and W. E. Newton (eds), *Current Perspectives in Nitrogen Fixation*. Aust. Acad. Sci., Canberra.
Watt, A. S. (1947) Pattern and process in the plant community, *J. Ecol.* **35**, 1–22.
Webley, D. M. and Duff, R. B. (1965) The incidence, in soils and other habitats, of microorganisms producing 2-ketogluconic acid, *Plant Soil* **22**, 307–13.
Webster, J. (1970) Coprophilous fungi, *Trans. Br. Mycol. Soc.* **54**, 161–80.
Webster, J. (1980) *Introduction to Fungi*. Cambridge Univ. Press, Cambridge.
Weigert, R. G., Coleman, D. C. and Odum, E. P. (1970) Energetics of the litter-soil subsystem, pp. 93–8 in J. Phillipson (ed.), *Methods of Study in Soil Ecology*. UNESCO, Paris.

Wetselaar, R. and Hutton, J. T. (1963) The ionic composition of rainwater at Katherine, NT, and its part in the cycling of plant nutrients, *Aust. J. Agric, Res.* **14**, 319–29.

White, P. S. (1979) Pattern, process and natural disturbance in vegetation, *Bot. Rev.* **45**, 299–99.

White, R. E. (1979) *Introduction to the Principles and Practice of Soil Science.* Blackwell, Oxford.

Whittaker, R. H. (1959) On the broad classification of organisms, *Quart. Rev. Biol.* **34**, 210–26.

Whittaker, R. H. (1975) *Communities and Ecosystems*, 2nd edn. Macmillan, London and New York.

Whittaker, R. H., Levin, S. A. and Root, R. B. (1973) Niche, habitat and ecotope, *Amer. Naturalist* **107**, 321–38.

Whittaker, R. H. and Margulis, L. (1978) Protist classification and the kingdoms of organisms, *BioSystems* **10**, 3–18.

Whittaker, R. H. and Woodwell, G. M. (1969) Structure, production and diversity of the oak–pine forest at Brookhaven, New York, *J. Ecol.* **57**, 155–74.

Wicklow, D. J. (1981) Interference competition and the organization of fungal communities, pp. 351–75 in D. T. Wicklow and G. C. Carroll (eds), *The Fungal Community*. Marcel Dekker, New York.

Williams, S. T. and Gray, T. R. G. (1974) Decomposition of litter on the soil surface, pp. 611–32 in C. H. Dickinson and G. J. F. Pugh (ed), *Biology of Litter Plant Decomposition*, Vol. 2. Academic Press, London.

Wilson, P. W. and Burris, R. H. (1947) The mechanism of biological nitrogen fixation, *Bact. Rev.* **11**, 41–73.

Witkamp, M. and Ausmus, B. S. (1976) Processes in decomposition and nutrient transfer in forest systems, pp. 375–96 in J. M. Anderson and A. Macfadyen (eds) *The Role of Terrestrial and Aquatic Organisms* in *Decomposition Processes*. Blackwell, Oxford.

Witkamp, M. and Olson, J. S. (1963) Breakdown of confined and non-confined oak litter, *Oikos* **14**, 138–47.

Woodwell, G. M. and Botkin, D. B. (1970) Metabolism of terrestrial ecosystems by gas exchange techniques: the Brookhaven approach, pp. 73–85 in D. E. Reichle (ed), *Analysis of Temperate Forest Ecosystems.* Springer-Verlag, New York.

Zak, B. (1964) Role of mycorrhiza in root disease, *Ann. Rev. Phytopath.* **2**, 377–92.

Zak, B. (1976) Pure culture synthesis of Pacific madrone ectendomycorrhizae, *Mycologia* **68**, 362–9.

Zak, B. and Bryan, W. C. (1963) Isolation of fungal symbionts from pine mycorrhizae, *For. Sci.* **9**, 270–8.

Zak, B. and Marx, D. H. (1964) Isolation of mycorrhizal fungi from roots of individual slash pines, *For. Sci.* **10**, 214–21.

Zinke, P. J. (1962) The pattern of influence of individual forest trees on soil properties. *Ecology* **43**, 130–3.

Index

Page numbers in italics refer to tables or illustrations